APPLIED NONLINEAR DYNAMICS AND STOCHASTIC SYSTEMS NEAR THE MILLENIUM

APPLIED NONLINEAR DYNAMICS AND STOCHASTIC SYSTEMS NEAR THE MILLENIUM

San Diego, CA July 1997

EDITORS
James B. Kadtke
*Institute for Pure and Applied Physical Science,
University of California, San Diego*

Adi Bulsara
*Naval Command, Control,
and Ocean Surveillance Center
RDT&E Division, San Diego*

AIP CONFERENCE
PROCEEDINGS 411

American Institute of Physics   Woodbury, New York

Authorization to photocopy items for internal or personal use, beyond the free copying permitted under the 1978 U.S. Copyright Law (see statement below), is granted by the American Institute of Physics for users registered with the Copyright Clearance Center (CCC) Transactional Reporting Service, provided that the base fee of $10.00 per copy is paid directly to CCC, 222 Rosewood Drive, Danvers, MA 01923. For those organizations that have been granted a photocopy license by CCC, a separate system of payment has been arranged. The fee code for users of the Transactional Reporting Service is: 1-56396-736-7/ 97 /$10.00.

© 1997 American Institute of Physics

Individual readers of this volume and nonprofit libraries, acting for them, are permitted to make fair use of the material in it, such as copying an article for use in teaching or research. Permission is granted to quote from this volume in scientific work with the customary acknowledgment of the source. To reprint a figure, table, or other excerpt requires the consent of one of the original authors and notification to AIP. Republication or systematic or multiple reproduction of any material in this volume is permitted only under license from AIP. Address inquiries to Office of Rights and Permissions, 500 Sunnyside Boulevard, Woodbury, NY 11797-2999; phone: 516-576-2268; fax: 516-576-2499; e-mail: rights@aip.org.

L.C. Catalog Card No. 97-77035
ISBN 1-56396-736-7
ISSN 0094-243X
DOE CONF- 970728

Printed in the United States of America

Contents

Preface ix
Chairman and Speaker List xi

FLUIDS/TURBULENCE

Convection, Stability, and Low Dimensional Dynamics 3
    C. R. Doering
Coupled Map Lattices Simulating Fully Developed Turbulent Flows 11
    A. Hilgers and C. Beck
Dynamics of Heterogeneous Catalysis: Understanding Spatial Coupling 18
    C. D. Lund, C. M. Surko, B. M. Maple, and S. Y. Yamamoto
Optimal Control of Wave Propagation Governed by Nonlinear Partial Differential Equations 24
    V. Khasilev
Chaos Due to the Interaction of Capillary and Gravity Waves on the Surface of Deep Water 30
    A. D. Grishchenko

TIME SERIES ANALYSIS AND SYNCHRONIZATION

Synchronization Transitions in Coupled Chaotic Oscillators 39
    J. Kurths, A. S. Pikovsky, and M. G. Rosenblum
Cycles from Short Time Series 47
    S. Allie and A. Mees
Diagnosing Intermittency 51
    S. Hammel, J. Barnett, and N. Platt
Using Delay Differential Equations as Dynamical Classifiers 57
    M. N. Kremliovsky and J. B. Kadtke
Weak and Strong Synchronization of Chaos 63
    K. Pyragas
Extraction of Dynamics from Non-Stationary Time Series Data 69
    L. Cao
The Identification of Time-Delay Systems 75
    M. J. Bünner, Th. Meyer, A. Kittel, and J. Parisi
Synchronizing Hyperchaotic Circuits 81
    A. Tamaševičius, A. Čenys, A. Namajūnas, G. Mykolaitis, and E. Lindberg
Solving Inverse Problems of Identification Type by Optimal Control Methods 87
    S. Lenhart, V. Protopopescu, and J. Yong

DYNAMICAL CONTROL AND SPATIO-TEMPORAL PHENOMENA

Karhunen-Loeve Mode Control in Chaotic Reaction-Diffusion System 97
    I. Triandaf and I. B. Schwartz
Stability of Bound States of Pulses in the Ginzburg-Landau Equations 103
    V. V. Afanasjev and B. A. Malomed
Controlling Symmetric Vortex Configurations 109
    Á. Péntek, J. B. Kadtke, and Z. Toroczkai
Quantifiers for Spatio-Temporal Bifurcations in Coupled Map Lattices 117
    N. Chatterjee and N. Gupte
**Reconstruction of a Set of Differential Equations Modeling
an Experimental Homoclinic Chaos in the Belousov-Zhabotinskii** 125
    C. Letellier, J. Maquet, L. Le Sceller, G. Gouesbet, F. Argoul, and A. Arnéode
Multistability and Invariants in Delay-Differential Equations 131
    B. Mensour and A. Longtin

NONLINEAR STOCHASTIC PHENOMENA

Adiabatically Rocked Quantum Ratchets 139
    P. Reimann, M. Grifoni, and P. Hänggi
Quantum Steps in Hysteresis Loops 145
    M. Thorwart, P. Reimann, and P. Jung
Synchronization in Ensembles of Stochastic Resonators 151
    A. Neiman, F. Moss, L. Schimansky-Geier, and W. Ebeling
Nonlinear Dynamics of Large Fluctuations 157
    M. I. Dykman, V. N. Smelyanskiy, C. J. Lambert, D. G. Luchinsky,
    and P. V. E. McClintock
**Ratchet Effects for Simple and Compound Objects Driven
by Correlated Noise** 163
    T. E. Dialynas, K. Lindenberg, and G. Tsironis
Optimal Control of Large Fluctuations by Nonadiabatic Fields 169
    M. I. Dykman, H. Rabitz, V. N. Smelyanskiy, and B. E. Vugmeister
Stochastic Resonance at Molecular Level: the Poisson Wave Model 175
    S. M. Bezrukov and I. Vodyanoy
Theoretical Foundations for Nervous Networks 179
    B. Hasslacher and M. W. Tilden

NONLINEAR MODELS OF SOCIETAL SYSTEMS

**Complex Issues of Military Capability: Measurement, Assessment,
Simulation** 187
    L. D. Miller, M. F. Sulcoski, and B. A. Farmer
Degree of Correlation Inside a Financial Market 197
    R. N. Mantegna
Structuring the IW Diagnosis Problem 203
    M. J. Coombs, A. Taha, and D. Birx

Tests for Robustness of Dynamical Models of Arms Races 213
    G. Mayer-Kress

DEVICES/EXPERIMENTS

Stochastic Resonance and Resonant Trapping in Schmidt Trigger............ 221
    L. Gammaitoni, F. Marchesoni, and S. Santucci
New Regime in the Stochastic Resonance Dynamics of SQUIDs.............. 227
    A. D. Hibbs and B. R. Whitecotton
Quasicontinuous Control Using Time-Delay Coordinates.................... 237
    A. Schenk zu Schweinsberg and U. Dressler
Coupled Brownian Rectifiers 243
    R. Häußler, R. Bartussek, and P. Hänggi
Generation and Processing of Chaotic Signals in Range Measurement
Systems.. 249
    A. Bauer
Spatio-Temporal Simulation in Subthreshold CMOS 255
    J. Neeley and J. G. Harris
Nonlinear Quantum Mechanical Ground State SQUID Magnetometer
Dynamics 261
    R. R. Whiteman, J. Diggins, V. Schöllmann, T. D. Clark, R. J. Prance,
    H. Prance, J. F. Ralph, and M. Everitt
Nonlinear Signal Transformation in the Regime of SR 267
    I. A. Khovanov and V. S. Anishchenko

NEUROSCIENCE AND BIOMEDICAL ENGINEERING

Noise in Any Frequency Range Can Enhance Information
Transmission in a Sensory Neuron 275
    J. E. Levin
The Neuromodulatory Properties of "Noisy Neuronal Oscillators" 281
    H. A. Braun, M. T. Huber, M. Dewald, and K. Voigt
Traveling Waves in a Circular Array of Integrate-and-Fire Neurons......... 287
    P. C. Bressloff and S. Coombes
Subthreshold Coincidence Detection 293
    A. Longtin and J. M. Laniel
Similarity Regime in the Brain Activity 299
    E. Novikov, A. Novikov, D. Shannahoff-Khalsa, B. Schwartz,
    and J. Wright
Transition to Subthreshold Activity With the Use of Phase Shifting
in a Model Thalamic Network 303
    E. Thomas and T. Grisar
Mass Separation by Ratchets 309
    B. Lindner, L. Schimansky-Geier, P. Reimann, and P. Hänggi
Stability and Chaos in an Inertial Two Neuron System 315
    D. W. Wheeler and W. C. Schieve

Random and Chaotic Oscillations in a Model of Childhood Epidemics Caused by Seasonal Variations of the Contact Rate 321
    P. S. Landa and A. A. Zaikin

SIGNAL AND IMAGE PROCESSING

Chaotic Encoding of Information Without Synchronization 329
    V. B. Ryabov, P. V. Usik, and D. M. Vavriv
Enlightenment in Shadows 335
    I. Gilmour and L. A. Smith
Noise Induced Effects and Stochastic Resonance in an El Nino Model 341
    L. Stone and P. I. Saparin
Processing the Vertebral CT-images and Quantification of their Structure by Measures of Complexity. 347
    P. I. Saparin, W. Gowin, J. Kurths, and D. Felsenberg
Some Applications of Nonlinear Diffusion to Processing of Dynamic Evolution Images 353
    A. N. Goltsov and S. A. Nikishov
Low Dimensional Model of Heart Rhythm Dynamics as a Tool for Diagnosing the Anaerobic Threshold. 359
    O. L. Anosov, O. Ya. Butkovskii, J. B. Kadtke, Yu. A. Kravtsov, and V. Protopopescu
Multiple-Scales Analysis of Non-Stationary Biological Signals 366
    C.-K. Peng

Author Index 373

PREFACE

The past decade has seen an explosion of new ideas in the field of nonlinear dynamics. Increasingly, it has become clear from such current applications in signal processing, lasers, molecular motors, and the control of biomedical anomalies, that nonlinear dynamics pervades systems that we deal with in everyday life. In addition, since (stochastic) noise is also ubiquitous in nearly every environment, the role of background fluctuations in the performance of nonlinear dynamical systems cannot be ignored.

Despite the large amount of research and development in nonlinear dynamics to date, however, there has been very little transition of these fundamental results to practical, "real-world" applications. In some areas (e.g. nanosystems, molecular motors) there is clearly a need for some additional research before practical applications can be realized; in others such as stochastic resonance and chaos control, the time is ripe for the incorporation of some proven and well-established ideas into actual devices.

Thus, we conceived the central objective of this meeting: a discussion of nonlinear phenomena in random environments, with a view to their implementation in real devices and applications. In addition, we attempted to give greater exposure to younger, promising researchers who seldom are able to speak at international meetings, as well as soliciting several review talks by established researchers to provide education and a solid background. It was decided that the program should emphasize applications of nonlinear dynamical systems theory in fields as diverse as Neuroscience and Biomedical Engineering, Fluids, Chaos Control, Nonlinear Signal/Image Processing, Stochastic Resonance, Devices, and Nonlinear Dynamics in Socio-Economic Systems. In particular, we tried to choose presentations which described methods being actively implemented to solve real problems.

San Diego presented a wonderful site for such a meeting. Besides the physically beautiful environment conducive to intense discussion, it is home to some of the best research and development institutions in the US. These include the Naval Command Control and Ocean Surveillance Center (a leader in nonlinear signal processing, filtering, and communications), the University of California at San Diego (having a strong history in nonlinear research), the Salk Institute, Scripps Institute of Oceanography (specializing in geo-physical fluid dynamics and geo-magnetic remote sensing), and numerous (private) R&D companies. Most of these organizations actively participated in the meeting. The resulting meeting was intended to bridge the gaps between other established nonlinear science meetings (e.g. Dynamics Days, the Experimental Chaos Conferences, and the recent series of Stochastic Resonance meetings), and has ultimately lead to this collection of proceedings papers that hopefully will benefit numerous researchers in the general area of applied nonlinear dynamics.

In the wake of this conference, the chairmen wish to offer thanks to the numerous individuals who helped make the meeting a huge success. A large amount of effort and scientific expertise was provided by those who served on the technical committees, including Kurt Wiesenfeld, Luca Gammaitoni, Mark Sulcoski, Lou Pecora, Vadim Anischenko, Juergen Kurths, Jim Collins, Dave Broomhead, Katja Lindenberg, and Andre Longtin. In addition, we particularly thank those that supplied financial (and moral) support, such as Frank Gordon (NCCOSC), Mike Shlesinger (ONR), Dave Brown (California Coordinating Committee for Nonlinear Science), Mark Sulcoski (NGIC, ARO), and the Dean's office of UCSD. Finally, an enormous and capable effort was provided by the administrative support people, such as Donna Bott, Linda Higgins, and Amy Diaz (IPAPS); Tina Pagan, JoAnn Pagan, and Anita Sorgenfrey (UCSD Conference Services); and Claudia Lainscsek, Mike Kremliovsky, and Aron Pentek. In particular, Mr. Pentek is responsible for the heroic effort put forth in electronically compiling all of the manuscripts included in this volume, and also for maintaining the conference Web page.

In conclusion, we hope that this volume provides a reasonably useful and interesting summary of the technical content of the conference. We sincerely hope that it will provide some small impetus to further the development of practical applications derived from the exciting ideas which pervade this field.

A. Bulsara
Naval Command, Control and Ocean Surveillance Center
RDT&E Division, San Diego

J. Kadtke
University of California, San Diego

San Diego, California 12 September, 1997

CHAIRMEN

K. Lindenberg
V. Anischenko
L. Pecora
D. Brown
S. Vohra
M. Dykman
M. Sulcoski
M. Spano
P. Krylstedt
J. Collins
D. Broomhead
K. Wiesenfeld

SPEAKER LIST

C. Doering
A. Hilgers
K. Fine
C. Lund
Y. Khasilev
G. Grischenko
L. Gammaitoni
J. Kurths
A. Mees
S. Hammel
M. Kremliovsky
K. Pyragas
L. Cao
M. Buenner
U. Parlitz
A. Tamasevicius
V. Protopopescu
S. Strogatz
I. Triandaf
B. Malomed
A. Pentek
N. Chatterjee
C. Letellier
B. Mensour
C. van den Broeck
M. Stemmler
P. Reimann

P. Jung
A. Neiman
P. V.E. McClintock
L. Reichl
G. Tsironis
V. Smelyanskiy
S. Bezrukov
M. Tilden
L. Miller
A. Woodcock
R. Mantegna
S. Jackson
M. Coombs
G. Mayer-Kress
T. Brown
A. Hibbs
M. Inchiosa
A. Schenck
R. Bartussek
M. Locher
A. Bauer
J. Neeley
R. Whiteman
I. Khovanov
J. Garcia-Ojalvo
F. Moss
J. Levin
H. Braun
P. Bressloff
A. Longtin
P. Gailey
D. Astumian
M. Shlesinger
D. Vavriv
I. Gilmore
L. Stone
P. Saparin
A. Goltsov
J. Huke
O. Butkovskii
R. Zoller
C. Peng
T. Tel
P. Rapp

E. Novikov
E. Thomas
L. Schimansky-Geier
D. Wheeler
A. Zaikin

FLUIDS/TURBULENCE

Convection, Stability, and Low Dimensional Dynamics

Charles R. Doering[1]

Department of Mathematics, University of Michigan
Ann Arbor, MI 48109-1109

Abstract. Recent developments concerning the connection between notions of hydrodynamic stability—usually associated with stationary laminar flows—and dynamics, most notably turbulent fluid flows, are reviewed. Based on a technical device originally introduced by Hopf in 1941, a rigorous mathematical relationship between criteria for nonlinear energy stability and bounds on global transport by steady, unsteady, or even turbulent flows, has been established. The optimal "marginal stability" criteria for the best bound leads to a novel variational problem, and the differential operator associated with the stability condition generates an adapted basis in which turbulent flow fields may naturally be decomposed. The application and implications of Galerkin truncations in these bases to produce low dimensional dynamical systems models is discussed in the context of thermal convection in a saturated porous layer.

INTRODUCTION

Hydrodynamic stability is usually considered a characteristic of steady flow whereas turbulence is the antithesis of steady flow. The goals of this presentation are, first, to describe a physical and mathematical relationship between stability and turbulence and, second, to propose that this connection may lead to some new insights regarding low dimensional models of turbulent dynamics. Such stability considerations in the context of turbulence have been applied to many fundamental problems in fluid dynamics including convective heat flux, momentum flux across a shear layer, and mass flux in pipe and channel flows. In each case the basic observation is that global transport is controlled by the thickness of a laminar boundary layer near a surface driving the flow by heating or shearing, and in each case the basic physical assumption is that the layer thickness adjusts itself so that, considered as a system unto itself,

---

[1] A report on results from an ongoing project and current and past collaborations with P. Constantin (Chicago), J.M. Hyman (Los Alamos), M. Orwoll (Clarkson) and D. Wick (Clarkson/Los Alamos). Research supported in part by NSF and DOE.

the layer is marginally stable. This idea has been around for a long time. The issue of stability for convective thermal boundary layers was raised by Malkus [1] in the 1950's. In the early 1960's, Howard [2] used a marginal stability argument to derive a quantitative scaling relationship, including prefactor estimate, for global heat transport in a fluid confined between parallel plates. There is now a mathematically rigorous way to implement these ideas starting from a mathematical idea introduced by Hopf in the 1940's [3].

Hopf's motivation was to derive *a priori* bounds on the time averaged rate of energy dissipation for solutions of the Navier-Stokes equations in the presence of rigid boundaries, a necessary technical step in the construction of weak solutions to the nonlinear partial differential equations. Hopf derived finite estimates for the energy dissipation rate in some some general cases which were, however, so unrealistically large as to be physically irrelevant. But as established in the 1990's [4–6], Hopf's method can be applied to moderately simple geometries where sharper calculations lead to a rigorous derivation of Kolmogorov-type scaling estimates for turbulent transport. A careful formulation of Hopf's techniques leads naturally to novel variational problems for the optimal estimates [7,8], and these recent technical developments have been the focus of a number papers: the method has been generalized for use with time-dependent boundary conditions [9] and more complex geometries [10,11], and there have been improvements in the anaysis and formulation of the optimization problem [12,13]. For the problem of turbulent transport in relatively symmetric geometries [14,15], the variational problem makes contact with the large body of work, initiated by Howard [16,17] and Busse [18,19] in the 1960's and 1970's, on bounds for statistically stationary turbulence. Much recent experimental work has focused on convection in a fluid layer [20–22], but in this talk I will outline these ideas in the context of the problem of thermal convection of an incompressible Newtonian fluid in a saturated porous layer [23–26]. The next section describes the set-up and presents the model, with a brief review of the system's stability and dynamics given in the following section. The subsequent section contains a brief discussion of heuristic and rigorous connections between stability and turbulence, and in the final section implications for low order dynamical systems models are discussed.

CONVECTION IN A POROUS LAYER

Consider a fluid-saturated layer of porous material occupying a region with horizontal surfaces parallel to the $x-y$ plane and thickness h in the vertical (z) direction. The layer is heated from below so that the bottom of the slab is held at temperature $T_0 + \delta T$ while the top is cooled uniformly to temperature T_0. The horizontal planes at $z = 0$ and $z = h$ are impenetrable to the fluid and, for the purposes of this presentation, we consider periodic boundary conditions in the (finite) horizontal directions. We use Darcy's law to model the relationship between the incompressible velocity vector field $\mathbf{u}(\mathbf{x},t) = (u,v,w)$ and the local temperature in the slab, $T(\mathbf{x},t)$, presumed to evolve according to the convection-diffusion equation. Then the dynamical equations are

$$\frac{\partial T}{\partial t} + \mathbf{u}\cdot\nabla T = \kappa\Delta T, \quad \frac{\nu}{K}\mathbf{u} = -\nabla p + g\alpha\mathbf{k}T, \quad \nabla\cdot\mathbf{u} = 0, \qquad (1)$$

where κ is the thermal diffusivity of the fluid-material mixture, ν is the fluid viscosity, K is the Darcy permeability coefficient, $\mathbf{k}$ is the vertical unit vector, g is the acceleration of gravity, α is the thermal expansion coefficient, and $p(\mathbf{x},t)$ is the pressure as determined by incompressibility condition. Neglect of inertial terms in the velocity equation corresponds to the infinite Darcy-Prandtl number limit of the model [23,24].

The physical heat current vector is proportional to

$$\mathbf{J} = \mathbf{u}T - \kappa\nabla T. \qquad (2)$$

A quantity of particular interest is the space-time averaged vertical heat flux,

$$\langle \mathbf{k}\cdot\mathbf{J}\rangle = \langle J_3\rangle = \left\langle \mathbf{k}\cdot\mathbf{u}T - \kappa\frac{\partial T}{\partial z}\right\rangle = \langle wT\rangle + \kappa\frac{\delta T}{h}, \qquad (3)$$

as a function of the applied temperature gradient and other system parameters.

These equations are naturally nondimensionalized in terms of the vertical length scale h, the thermal diffusion time scale $\frac{h^2}{\kappa}$, and the overall temperature drop δT to become

$$\frac{\partial T}{\partial t} + \mathbf{u}\cdot\nabla T = \Delta T, \quad \mathbf{u} = -\nabla p + Ra\mathbf{k}T, \quad \nabla\cdot\mathbf{u} = 0, \qquad (4)$$

with boundary conditions

$$T = 1, w = 0 \text{ when } z = 0, \text{ and } T = 0, w = 0 \text{ when } z = 1. \qquad (5)$$

The Rayleigh number $Ra = \frac{g\alpha\delta T K h}{\nu\kappa}$ is the nondimensional measure of the applied temperature drop. The nondimensional heat flux is the Nusselt number Nu, the ratio of the total heat flux to the purely conductive heat flux:

$$Nu \equiv \left\langle J_3^{nondimensional}\right\rangle = \frac{\left\langle J_3^{dimensional}\right\rangle}{\kappa\frac{\delta T}{h}} = 1 + \langle wT\rangle. \qquad (6)$$

Note that both the velocity vector field and the pressure field are linear (albeit nonlocal) functionals of the instantaneous temperature distribution. A simple computation based on the equations of motion allows us to express the Nusselt number in terms of the mean squared temperature gradient:

$$Nu = \left\langle |\nabla T|^2\right\rangle. \qquad (7)$$

STABILITY AND CONVECTION

The pure conduction state is the solution with no flow and a linear temperature profile: $\mathbf{u}_{cond} = 0$, $T_{cond} = 1 - z$. For low values of Ra it is both linearly and nonlinearly stable, so the onset of convection in this model is a forward pitchfork bifurcation.

This is seen by writing an arbitrary solution of the initial value problem as $T(\mathbf{x}, t) = T_{cond}(z) + \theta(\mathbf{x}, t)$ where the "fluctuation" temperature field θ satisfies homogeneous boundary conditions on the top and bottom of the slab. The linearized stability analysis of the pure conduction profile $T_{cond}(z)$ consists of looking for solutions to the linearized evolution operator (dropping $\mathbf{u} \cdot \nabla \theta$) with a time dependence $\sim e^{-\lambda t}$:

$$\lambda \theta = -\Delta \theta - w, \quad \Delta w = Ra(\Delta - \partial_z^2)\theta. \tag{8}$$

This is a linear constant coefficient system and it is straightforward to see that all the eigenvalues are real with $\lambda > 0$ for $Ra < Ra_c^{lin} = 4\pi^2 \lesssim 40$, in which case there are no unstable modes. Linearized stability gives a sufficient criteria for *instability*, so strictly speaking this shows that the pure conduction solution looses stability at $Ra_c^{lin} = 4\pi^2$.

Nonlinear stability is established via the energy method [27,28]. The fully nonlinear equation of motion for a fluctuation away from the pure conduction solution is

$$\frac{\partial \theta}{\partial t} + \mathbf{u} \cdot \nabla \theta = \Delta \theta + w, \quad \Delta w = Ra(\Delta - \partial_z^2)\theta, \tag{9}$$

and multiplying the θ-equation by θ and integrating over the slab making use of the boundary homogeneous conditions leads to

$$\frac{d}{dt}\frac{1}{2}\int \theta^2 d\mathbf{x} = -\int (|\nabla \theta|^2 - w\theta) d\mathbf{x}. \tag{10}$$

As long the right hand side is negative, arbitrarily large deviations from the pure conduction state will vanish as $t \to \infty$. This will be the case if

$$0 < \lambda \equiv \min \int (|\nabla \theta|^2 - w\theta) d\mathbf{x}, \tag{11}$$

where the minimization is over all test functions $\theta(\mathbf{x})$ satisfying homogeneous boundary conditions, normalization $\int \theta^2 d\mathbf{x} = 1$, and where $w(\mathbf{x})$ is constrained to be the linear functional of θ defined by the solution of the Poisson. After Fourier transforming in the horizontal plane, the Euler-Lagrange equations for this variational problem of nonlinear energy stability are found to be

$$\lambda \hat{\theta}(z) = (-\partial_z^2 + k^2)\hat{\theta} - \frac{1}{2}\hat{w} - \frac{1}{2}k^2\hat{v}, \quad (-\partial_z^2 + k^2)\hat{v} = Ra\,\hat{\theta}, \tag{12}$$

where k is the magnitude of the horizontal wavenumber, the "eigenvalue" λ is the Lagrange multiplier enforcing the normalization, and $\hat{v}(z)$, the Lagrange multiplier enforcing the constraint $(-\partial_z^2 + k^2)\hat{w} = Ra\ k^2\hat{\theta}$, satisfies homogeneous Dirichlet conditions at $z = 0$ and $z = 1$. Solutions of these differential equation corresponds precisely to solutions of the linearized stability equations, so the critical Rayleigh number Ra_c^{nonlin} of nonlinear stability—ensuring absolute stability for $Ra < Ra_c^{nonlin}$—corresponds exactly to the critical Rayleigh number Ra_c^{lin}. In terms of the heat transport, these stability analyses imply

$$Nu = 1 \text{ for } Ra \leq Ra_c^{nonlin} = Ra_c^{lin} = 4\pi^2 \equiv Ra_c. \tag{13}$$

Stationary roles appear immediately beyond the onset of convective motion, and they may be studied by means of amplitude equations for $Ra \gtrsim Ra_c$ not too far above transition. Subsequent bifurcations, patterns, and dynamics depend on details of the container shape and sidewall conditions but eventually, for sufficiently high Ra, one expects to enter what may be referred to as a turbulent regime. Convective turbulence may be identified by the loss of coherent flow structures correlated over long ranges (i.e., the disappearance of rolls). This is accompanied by the formation of thermal boundary layers across which most of the (horizontally averaged) temperature drop occurs, and an isothermal (on average) core where the heat is transported by rising or falling "plumes" or "blobs" of hot or cold fluid that have broken away from the hot or cold boundary layers. Scaling with is observed in convective turbulence, and the boundary layer thicknesses δ varies inversely as a power of Ra.

A chief aim of theories of convection is to predict the relationship between the heat current and the applied temperature drop, i.e., the function $Nu(Ra)$. The $Nu-Ra$ relationship in the turbulent regime is determined by the $\delta-Ra$ scaling because the heat transport through each horizontal layer must be the same in a statistically stationary situation. Near the boundaries where $w \to 0$, the vertical heat current $J_3 \approx -\frac{\partial T}{\partial z} \approx \frac{1}{2\delta}$ because of the temperature drop of $\frac{1}{2}$ across the layer of thickness δ. Hence $Nu = \frac{1}{2\delta}$.

The observed [25] high Rayleigh number scaling is $\delta \sim Ra^{-1}$, i.e.,

$$Nu \sim Ra \text{ for } Ra >> Ra_c \tag{14}$$

We refer to this particular scaling, $Nu \sim Ra$, as *Kolmogorov* scaling for this problem because in this situation the physical heat flux is independent of the microscopic heat conduction coefficient κ. Indeed, $Nu \sim Ra$ implies

$$\left\langle J_3^{dimensional} \right\rangle \sim \kappa \frac{\delta T}{h} Ra = \frac{g\alpha(\delta T)^2 K}{\nu}. \tag{15}$$

This is in direct analogy to the fundamental Kolmogorov scaling hypothesis for incompressible Navier-Stokes turbulence that the rate of energy dissipation is independent of the molecular viscosity at high Reynolds numbers.

STABILITY, TURBULENCE AND BOUNDS

Howard's marginal stability argument [2] applied to convection in a porous layer starts with the observation that most of the horizontally averaged temperature drop occurs across thermal boundary layers in which the fluid is effectively at rest due to the impenetrable boundaries. How thick are the boundary layers? If they are so thin as to be stable when considered as convection layers, then they may grow. If they are so thick that they are ustable as convection layers, then they will break up and shed a plume. Thus one expects the average thickness to self-adjust so that the boundary layer is just marginally stable. This means that the Rayleigh number based on half the total temperature drop across a layer of (dimensional) thickness δ is just about equal to the critical Rayleigh number:

$$O(40) = \frac{g\alpha(\frac{1}{2}\delta T)K\delta}{\nu\kappa} = \frac{1}{2}Ra\frac{\delta}{h} \qquad (16)$$

Hence the nondimensionalized boundary layer thickness is expected to scale like $\delta \sim Ra^{-1}$ so that

$$Nu \approx c\, Ra \qquad (17)$$

with a coefficient $c = O(.01)$.

This idea is takes on a rigorous realization in the following theorem [14,26]: Let $\tau(z)$ be a "background" temperature profile that satisfies the physical boundary conditions, $\tau(0) = 1$ and $\tau(1) = 0$. Then the bulk heat transfer is bounded from above by the heat transfer in $\tau(z)$, i.e., $Nu < \int_0^1 \tau'(z)^2 dz$, when the profile is marginally energy stable at parameter value $2Ra$.

In the theorem, energy stability is defined by the natural generalization of the criteria we used for the nonlinear stability analysis of the pure conduction solution. For an arbitrary profile $\tau(z)$ at parameter value $2Ra$, energy stability means

$$0 < \lambda \equiv min\, \int (|\nabla\theta|^2 + 2\tau'(z)w\theta)d\mathbf{x}, \qquad (18)$$

where the minimization is performed with the same constraints as before. For the proof of this theorem see the references [14,26].

Stable profiles may be constructed by considering function $\tau(z)$ which is constant, so $\tau' = 0$, throughout most of the slab. This is incompatible with the boundary conditions, but allowing τ' to vary strongly near the boundaries where the "test" θ's and w's vanish can be made to work. A piecewise linear stable temperature profile produces the rigorous upper bound [26]

$$Nu < \frac{1}{24}Ra \approx .04\, Ra, \qquad (19)$$

just an $O(4)$ factor above the observed [24,25] behavior. Optimization, i.e., solving the variational problem for the optimal profile $\tau_{opt}(z)$, improves the upper bound considerably [24,29,15,26].

LOW DIMENSIONAL MODELS

The self-adjoint linear operator associated with the nonlinear stability condition for $\tau_{opt}(z)$ produces a natural basis for the function space of temperature fluctuations $\theta(\mathbf{x}, t) = T(\mathbf{x}, t) - \tau_{opt}(z)$. Indeed, we may write

$$\theta(\mathbf{x}, t) = \sum_{k_x, k_y} e^{ik_x x + ik_y y} \sum_j \Theta_j^{(k_x, k_y)}(z) a_j^{(k_x, k_y)}(t), \qquad (20)$$

or more compactly,

$$\theta(\mathbf{x}, t) = \sum_n \Theta_n(x, y, z) a_n(t), \qquad (21)$$

where the functions $\Theta_n(x, y, z)$ are the eigenfunctions of the linear (albeit nonlocal) operator in the quadratic form in Eq.(18), ordered in ascending order of the magnitude of the eigenvalues. The spectrum of the linear operator of nonlinear stability for the optimal "marginally stable" profile is nonnegative definite, i.e., each $\lambda_n \geq 0$. These eigenvalues contribute to the linear term in the evolution equations for the modal amplitudes, the $a_n(t)$'s, either neutrally (for the marginal modes with $\lambda_n = 0$) or dissipatively (for the modes with $\lambda_n > 0$).

Once the optimal "marginally stable" background, $\tau_{opt}(z)$, profile has been computed, it is natural to look for corrections to the upper bound it produces. In terms of the modal amplitudes of the optimal eigenfunctions, the exact heat transfer is

$$Nu = \int_0^1 \tau'(z)^2 - \sum_{n=0}^\infty \lambda_n \left\langle \left| a_n^2 \right| \right\rangle, \qquad (22)$$

so knowledge of the exact modal dynamics leads to the exact heat flux. Of course the full set of coupled ode's for the modal amplitudes are of the same complexity as the original nonlinear pde's, but the ordering of the eigenvalues suggests that low dimensional truncations, a.k.a. Galerkin truncations, in this basis may yield useful approximations. Sensible truncations will include the weakly damped modes, i.e., those corresponding to small values of λ_n, with the expectation that the strongly damped modes with large λ_n will be small in magnitude and slaved to the active modes. Certainly the marginally stable ($\lambda_n = 0$) modes should be included in any such dynamical systems model.

Now we can use a result from the variational problem for $\tau_{opt}(z)$ to get some insight into meaningful low dimensional dynamical systems models. It has been observed both theoretically [18,24,14] and numerically [29] that for increasing Ra, the marginal subspace (the span of the eigenvectors with $\lambda_n = 0$) becomes increasingly high dimensional. That is, the eigenvalue 0 becomes increasingly degenerate and the number of independent modes with $\lambda_n = 0$ increases without bound as $Ra \to \infty$. This means that there can be "no free lunch" for low order truncations: if we seek quantitative corrections to the upper bound as in Eq.(22), then such models must necessarily be of increasing complexity, i.e., of higher and higer dimension, as the turbulence intensifies.

REFERENCES

1. Malkus, W.V.R., *Proc. R. Soc. London Ser. A* **225**, 185 (1954).
2. Howard, L.N., "Convection at high Rayleigh number", in *Applied Mechanics*, Proc. 11th Cong. Appl. Mech., ed. H. Görtler, Berlin: Springer, 1966, pp. 1109-1115.
3. Hopf, E., *Mathematische Annalen* **117**, 764 (1941).
4. Doering, C.R., and P. Constantin, *Phys. Rev. Lett.* **69**, 1648 (1992); erratum: *Phys. Rev. Lett.* **69**, 3000(E) (1992).
5. Doering, C.R., and P. Constantin, *Phys. Rev. E* **49**, 4087 (1994).
6. Constantin, P., and C.R. Doering, *Phys. Rev. E* **51**, 3192 (1995).
7. Constantin, P., and C. R. Doering, *Physica D* **82**, 221 (1995).
8. Constantin, P., and C.R. Doering, *Nonlinearity* **9**, 1049 (1996).
9. Marchioro, C., *Physica D* **74**, 395 (1994).
10. Kerswell, R., *J. Fluid Mech.* **321**, 335 (1996).
11. Wang, X., *Physica D* **99**, 555 (1997).
12. Gebhardt, T., S. Grossmann, M. Holthaus, and M. Löhden, *Phys. Rev. E* **51**, 360 (1995).
13. Nicodemus, R., S. Grossmann and M. Holthaus, *Physica D* **101**, 178 (1997).
14. Doering, C.R., and P. Constantin, *Phys. Rev. E* **53**, 5957 (1996).
15. Kerswell, R., *Physica D* **100**, 355 (1997).
16. Howard, L.N., *J. Fluid Mech.* **17**, 405 (1963).
17. Howard, L.N., *Ann. Rev. Fluid Mech.* **4**, 473 (1972).
18. Busse, F.H., *J. Fluid Mech.* **37**, 457 (1969).
19. Busse, F.H., *Adv. Appl. Mech.* **18**, 77 (1978).
20. Heslot, F., B. Castaing, and A. Libchaber, *Phys. Rev. A* **36**, 5870 (1987).
21. Belmonte, A., A. Tilgner, and A. Libchaber, *Phys. Rev. E* **50**, 269 (1994).
22. Cioni, S., S. Ciliberto, and J. Sommeria, *J. Fluid Mech.* **335**, 111 (1997).
23. Busse, F.H., and D.D. Joseph, *J. Fluid Mech.* **54**, 521 (1972).
24. Gupta, V. P., and D.D. Joseph, *J. Fluid Mech.* **57**, 491 (1973).
25. Graham, M. D., and P. H. Steen, *J. Fluid Mech.* **272**, 491 (1994).
26. Constantin, P., and C. R. Doering, *in preparation* (1997).
27. Joseph, D.D., *Stability of Fluid Motions*, Berlin: Springer-Verlag, 1976.
28. Straughan, B., *The Energy Method, Stability and Nonlinear Convection*, Berlin: Springer-Verlag, 1992.
29. Doering, C.R., and J. M. Hyman, *Phys. Rev. E* **55**, 7775 (1997).
30. Holmes, P., J. L. Lumley, and G. Berkooz, *Turbulence, Coherent Structures, Dynamical Systems and Symmetry*, Cambridge: Cambridge University Press, 1996.

Coupled map lattices simulating fully developed turbulent flows

Angela Hilgers and Christian Beck

School of Mathematical Sciences
Queen Mary and Westfield College
University of London
Mile End Road, London E1 4NS
England

Abstract. Recently a coupled map lattice has been introduced that simulates the time evolution of velocity differences in fully developed turbulent flows. It is based on an extension of the Langevin theory to more complicated stochastic processes generated by chaotic dynamical systems. These act on the various levels of a self-similar cascade. It has been shown that with this model probability distributions of velocity differences measured in high-Reynolds number flows can be reproduced correctly. We introduce an extension of the model to spatially extended systems and present an application to the dynamics of currency exchange rates.

INTRODUCTION

Coupled map lattices are spatially extended nonlinear dynamical systems. The first examples of coupled map lattices were introduced by Kaneko [1] and Kapral [2]. Meanwhile several physical applications have been studied [3]. A new application to hydrodynamical systems was described in [4]. The probability distributions of velocity differences measured in fully developed turbulent flows are typically observed to deviate from Gaussians in a characteristically asymmetric way. So far there has been neither a theory nor a model that explains this asymmetry in a quantitatively correct way. In [4] a coupled map lattice has been introduced as a cascade model for fully developed turbulent flows. The model is based on dynamical systems of Langevin type, where the fluctuating force is not Gaussian white noise but a nonlinear dynamical system. The dynamics of the noise is generated by iterates of maps conjugated to the Bernoulli shift. It has been shown that despite its simplicity the model leads to probability distributions which are in perfect agreement with those of velocity differences measured in turbulent flows. The model also turned out to be suitable to describe probability densities of price changes in economic systems.

In this talk we will introduce an extension of this model to spatially extended cascades. In the first section the model will be introduced. In the second section results will be presented for the probability densities of velocity differences in turbulent fluids. Finally, an application to the simulation of probability densities of currency exchange rates will be shown.

DESCRIPTION OF THE MODEL

The difference of the radial component of the velocity field, observed at two points with distance r in a liquid

$$u(r,t) = v_r(\vec{x}+\vec{r},t) - v_r(\vec{x},t) \qquad (1)$$

is a measure for the velocity of an eddy with diameter r. It has been shown experimentally, that probability densities of velocity differences deviate from a Gaussian distribution [5,6]. They are slightly asymmetric; all odd moments are negative but the average $\langle u(r) \rangle$ is zero. For small distances r, the distributions show stretched exponential tails caused by intermittent behaviour of the turbulent fluid.

The model we use is based on a generalization of the Langevin theory to more complex stochastic processes generated by chaotic dynamical systems [7]. The dynamics of velocity differences is described by

$$\dot{u} = -\gamma u + F(t) \qquad (2)$$

with damping $\gamma > 0$. The first term on the right hand side describes the relaxation of velocity differences to the laminar state $u = 0$, the second one time-dependent turbulent forces in the liquid which replace Gaussian white noise.

We choose the chaotic kick force

$$F(t) = \sum_{n=0}^{\infty} x_{n-1} \delta(t - n\tau) \qquad (3)$$

where the evolution of the kick amplitude is given by a mapping $T(x)$. The time constant τ indicates how much the chaotic process deviates from a Gaussian process.

After integration this leads to the map of Kaplan–York type

$$\begin{aligned} x_{n+1} &= T(x_n) \\ u_{n+1} &= \lambda u_n + x_n, \qquad \lambda = e^{-\gamma\tau} \end{aligned} \qquad (4)$$

where u_n is the stroboscopic velocity difference.

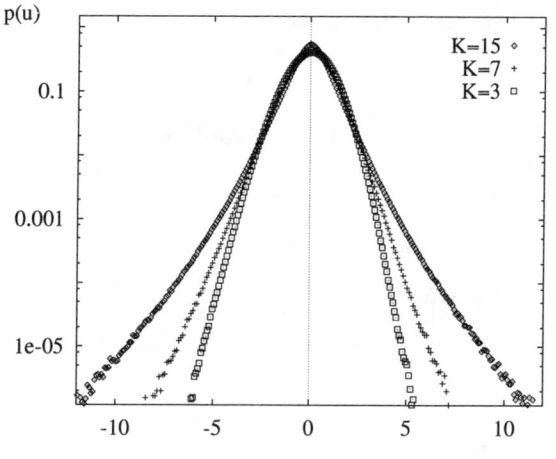

FIGURE 1. Histogram of velocity differences for $\gamma\tau = 0.03, g = 0.02, K = 3, 7, 15$

So far this is just a low-dimensional map. In order to achieve a more realistic description of a turbulent liquid we will extend it by introducing various levels of a self-similar cascade and a diffusive spatial coupling.

Similar to the β-model introduced by Frisch et al. [8] and Benzi et al. [9], we presume that fully developed turbulence is characterised by the existence of a self-similar set of eddies on various scales $r_k = 2^{-k}$. Energy is transferred from 'mother eddies' at level k to the smaller 'daughter eddies' at level $k+1$. For fully developed turbulence it is assumed that this cascade possesses a self-similar hierarchical structure at least in the inertial range. The basic assumption of our model is that the momentum loss at level k serves as a chaotic driving force at level $k+1$ and that each daughter eddy gets a random fraction $\xi \in [0, 1]$ of this momentum.

The spatial coupling is carried out by adding a diffusive term at each level which takes into account the influence of the nearest neighbours. Its strength can be varied by changing a coupling parameter g.

Let $u_n^k(i)$ denote the velocity difference at level k at time step n and position i. The model then has the following form in d dimensions:

$$x_{n+1}(i) = T(x_n(i))$$
$$u^{(1)}_{n+1}(i) = \lambda_1 u_n^{(1)}(i) + x_n(i)$$
$$u^{(k)}_{n+1}(i) = \lambda_k u_n^{(k)}(i) + \frac{g}{2d} \sum_{\substack{nearest \\ neighbours\ \sigma}} u_n^{(k)}(i+\sigma) + \xi(1-\lambda_{k-1})u_n^{(k-1)}(i)$$
$$k = 2, \ldots, K. \qquad (5)$$

Since viscous effects increase on smaller scales, we have chosen λ_k as

$$\lambda_k = e^{-\gamma\tau k}. \qquad (6)$$

With $r_k = 2^{-k}$ it follows that (6) is equivalent to

$$\lambda(r) = r^{\frac{\gamma\tau}{\ln 2}}. \tag{7}$$

The exponent is of the same order of magnitude as the inverse Reynolds number Re^{-1} since

$$\gamma\tau \sim \frac{\nu\tau}{L^2} \sim \frac{\nu}{vL} = Re^{-1}.$$

Here ν denotes the kinematic viscosity, while v and L are a typical velocity and a typical length scale (average stream velocity and Taylor scale). The free parameters of the model are $\gamma\tau$, the coupling parameter g and the maximum number of energy levels K. The smaller the distances r investigated are, the larger K has to be chosen.

APPLICATION TO TURBULENT FLUIDS AND FOREIGN EXCHANGE RATES

A system without spatial coupling has been used in [4] to simulate the distribution of velocity differences in high Reynolds-number flows. The Ulam map $T(x) = 1 - 2x^2$ has been chosen as the chaotic dynamics at the top of the cascade. It has been shown in [4] that the simulated probability distributions are in very good agreement with those measured in jet [5] and tunnel experiments [6]. The simulated probability distributions show the characteristic asymmetry and the transition from almost Gaussian distributions for large distances r to distributions with stretched exponential tails for small r. This simple model already reproduces important aspects of high Reynolds number flows correctly.

In the following we present results produced on a two-dimensional 30×30-lattice with periodic boundary conditions. Figure 1 shows probability distributions for three different levels of the lattice. Just as in the one-dimensional case, the distributions are slightly asymmetric. Whereas for $K = 3$ the shape does not differ much from a Gaussian, there are exponential tails for $K = 7$ and strongly leptocurtic distributions for $K = 15$. In agreement with the experimental observations, the average of the distribution vanishes and all odd moments are negative.

The model may be used to visualise the flow pattern. Figure 2 shows a typical two-dimensional field of velocity differences for the same parameters as in figure 1 and $K = 15$. The diffusive coupling leads to similar behaviour of the velocities in small subregions of the grid. This effect is more pronounced for higher levels of the lattice, i.e. smaller distances. The spatial correlation length has been determined numerically to be $r_c = 1.6$ for $K = 5$ and $r_c = 2.0$ for $K = 15$.

The model without spatial coupling has also been used to describe price changes in financial markets. Recently there has been a lot of interest in the

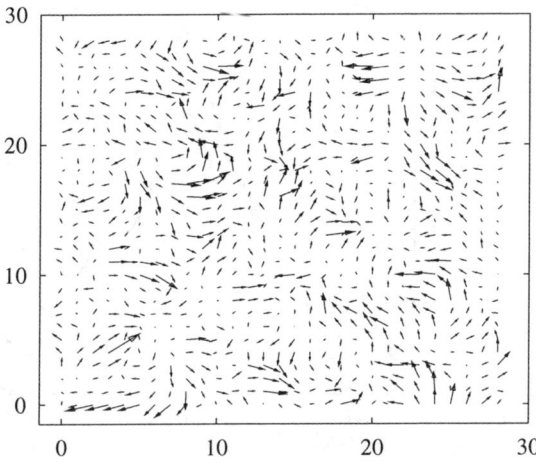

FIGURE 2. Field of velocity differences for $\gamma\tau = 0.03, g = 0.02, K = 15$

dynamics of economic indices [10] and similarities between the dynamics of turbulent flows and of stock exchange indices or foreign currency exchange rates [11,12]. It has been shown that the shape of the probability densities depends on the time difference between the observation of price differences. For long intervals these distributions are close to a Gaussian, whereas for short sampling times the tails become more pronounced. It is therefore natural to describe financial systems with our model based on a coupled map lattice.

Let $u_n^{(k)}$ denote the price change at time n, observed at times separated by a time difference t_k. The higher k, the smaller is the sampling time. Price changes at a time $n+1$ are influenced by previous price changes at time n. The strength of this influence is parametrised by a damping constant λ. We further assume that price changes on the level k are influenced by those on the larger time scale $k-1$. This influence of larger scales is weighted by a random number $\xi \in [0,1]$, representing the random element brought in by traders deciding to take information from events happening on larger time scales into account or not. As before we have chosen a damping which decreases exponentially with the index k. The results do not depend significantly on this choice.

Figure 3 shows probability distributions for the changes of the exchange rate US-Dollar–German Mark obtained by Ghashghaie et al. [11]. The sampling time is 640s and 40960s respectively. The figure also shows histograms obtained from our simulation for the parameter values $\gamma = 0.022$, $K = 20$ and $K = 4$. The simulation results agree well with the data. Not only the centre of the distribution, but also the stretched tails are fitted quite well for long and short sampling times. For the simulations of the currency exchange rates we have again chosen the dynamics $T(x) = 1 - 2x^2$ at the top of the cascade. It turned out that changes of other economic indices, like the Standard & Poor's

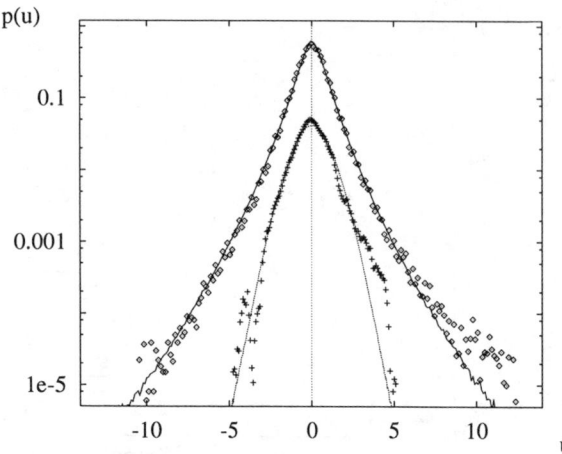

FIGURE 3. Probability density of price changes u for the exchange rate US-Dollar–German Mark, $t = 640s$ (rhombs), $40960s$ (triangles, the data have been rescaled by a factor 0.1), and simulation results for $\gamma\tau = 0.022$, $K = 19$ (solid line) and $K = 4$ (dashed line).

index investigated by Mantegna et. al. [10], lead to symmetric probability densities. In this case a good agreement between the simulation and the data can be achieved by using either odd mappings or equally distributed random numbers at the top of the cascade.

In summary, cascade models of type (5) seem to be a very promising approach for the study of the dynamics of turbulent flows as well as of economic systems. They are based on a plausible dynamics and allow for a simple computer simulation.

REFERENCES

1. Kaneko K., *Progr. Theor. Phys.* **72**, 480 (1984).
2. Kapral R., *Phys. Rev.* **31A**, 3868 (1985).
3. Kaneko K., *Theory and Applications of Coupled Map Lattices*, New York: John Wiley and Sons (1993).
4. Beck C., *Phys. Rev.* **49E**, 3641 (1994).
5. Castaing B., Gagne Y. and Hopfinger E. J., *Physica* **46D**, 177 (1990).
6. Anselmet F., Gagne Y., Hopfinger E. J. and Antonia R. A., *J. Fluid Mech.* **140**, 63 (1984).
7. Beck C., *Physica* **233 A**, 419 (1996).
8. Frisch U., Sulem P. L. and Nelkin M., *J. Fluid Mech.* **87**, 719 (1978).
9. Benzi R., Paladin G., Parisi G. and Vulpiani A., *J. Phys.* **A17**, 352 (1984).
10. Mantegna R. N. and Stanley H. E., *Nature* **376**, 46-49 (1995).

11. Ghashghaie S., Breymann W., Peinke J., Talkner P. and Y. Dodge, *Nature* **381**, 767 (1996).
12. Hilgers A. and Beck C., to appear in *Int. J. Bif. Chaos* (1997).

Dynamics in Heterogeneous Catalysis: Understanding spatial coupling

C. D. Lund, C. M. Surko, M. B. Maple, and S. Y. Yamamoto

Department of Physics
University of California, San Diego
La Jolla, CA 92093-0319

Abstract. Infrared imaging is used to study spatiotemporal behavior associated with the catalytic oxidation of CO on platinum at atmospheric pressures, in the regime where the rate of reaction is oscillatory. In this regime, gas-phase diffusion of the reactants has been established as the dominant spatial coupling mechanism. By changing the gas flow rate, we are able to vary this coupling, and through the addition of small amounts of CO locally to the catalyst, we are able to trigger the reaction.

In a variety of binary oxidation reactions catalyzed on metal surfaces, the rate of reaction oscillates in time. One of the most extensively studied reactions is the oxidation of carbon monoxide (CO) on platinum. The reaction rate oscillations are observed in two regimes. In the "surface science" regime, experiments are carried out under UHV conditions on single-crystal platinum surfaces, and the observed oscillations have been found to be due to a reconstruction of the platinum surface [1]. Photoemission electron microscopy (PEEM) has revealed a variety of spatiotemporal phenomena in this regime including target patterns, spirals, and solitons [2]. In a regime of practical interest, oxidation takes place at atmospheric pressure on catalysts which are typically polycrystalline wires, foils, or films [3]. Here, the *in situ* tools which are used under UHV conditions are no longer available. Within this regime, there are a number of open questions regarding the nature of the oscillations and associated spatiotemporal dynamics. In addition, this system is, in principle, sufficiently simple to allow one to study with precision a number of dynamical phenomena, such as the propagation of reacting pulses and fronts.

Here, we report a study in which reaction-rate oscillations are observed on a platinum thin-film catalyst at atmospheric pressure. Since the oxidation reaction is exothermic, fluctuations in the temperature of the catalyst can be used as a measure of the rate at which the reaction is proceeding. Thus, the oscillations can be studied non-invasively using infrared imaging of the platinum surface. We have previously found that spatial coupling in this system is due to gas-phase diffusion of the reactants [4]. In this paper, we discuss the results of two experiments designed to vary the spatial coupling in the system. In the first experiment, the composition of the gas stream is changed in order to vary the degree of spatial coupling. In the second, perturbations in the local CO concentration are used to directly affect the local reaction conditions and induce oscillations.

The experiment is conducted in a continuous flow reactor which consists of a long quartz tube mounted in a furnace. The oxygen and CO reactants, along with helium and/or argon buffer gases, are filtered through a glass bead trap to remove transition-metal carbonyl impurities. Downstream from the bead trap, the entire environment is constructed of quartz to minimize the potential for impurities in the gas stream. In the main reactor, near the catalyst, the gas flow is believed to be fully developed with a parabolic flow profile and laminar with a Reynold's number of ~ 10. The mean velocity of the gas stream in the main reactor is ~ 1 cm/s. The apparatus is described in more detail elsewhere [5].

The catalyst is in the form of a platinum thin-film deposited onto a 0.5 mm thick, washer-shaped quartz substrate with a diameter of 3.0 cm. The catalyst is deposited by e-beam evaporation in a UHV chamber with a base pressure of 10^{-9} torr. In the reactor, the catalyst is mounted perpendicular to the gas stream with the platinum side facing the oncoming gas. The catalyst is supported by three K-type thermocouples evenly spaced around its periphery, which both support the catalyst in the reactor and provide a temperature reference for calibrating the infrared imager. Although a variety of catalyst geometries can be studied, typical experiments are conducted on a narrow annulus. This geometry provides a quasi-one-dimensional system with periodic boundary conditions.

Images are acquired using an Amber model AE4256 infrared camera mounted outside the reactor that looks through a sapphire window at the end of the main reactor chamber. Imaging of the platinum is accomplished by measuring emission from the quartz substrate. Quartz has a much higher emissivity (0.93) than platinum (0.03), which results in a much higher temperature sensitivity than if the platinum was imaged directly. Since the substrate is thin, temperature fluctuations in the platinum propagate to the back side of the substrate before significant lateral diffusion can occur.

To analyze the data, a sequence of around 500 images are taken at a typical rate of 2 Hz. A reference image is formed by averaging all the data images together. This reference is then subtracted from each of the data images to yield a precise measure of the temperature fluctuations. After this, the data are displayed in the form of spacetime plots with the temperature of the catalyst shown in greyscale as a function of the angular position and time.

Since the spatial coupling is governed by diffusion through the gas stream,

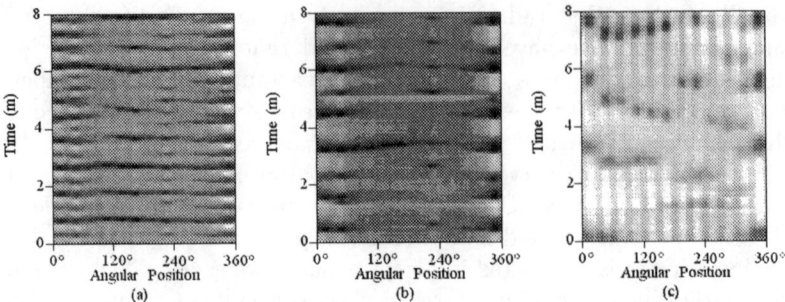

FIGURE 1. The effects of varying the buffer gas species and flow rate. In (a) the helium flow rate is 400 cc/min. In (b) the buffer is argon flowing at 400 cc/min, and in (c) argon flows at 3000 cc/min.

we attempted to control the degree of coupling present in the system by varying the composition of the gas stream. For this experiment, a catalyst in the form of twelve platinum patches arranged in a ring was used. Due to the physical separation between the patches, this will, by itself, tend to limit spatial coupling in the system. In the first test, oscillations were initiated at a catalyst temperature of 150 °C with $P_{CO}/P_{O_2} = 0.02$ at a total flow rate of 800 cc/min. The gas stream included, in addition to the reactants, a helium buffer flowing at 400 cc/min. The resulting spacetime plot is shown in Fig. 1(a). In this figure, the patches show up as the lighter colored stripes on a dark background. The oscillations here are not as well synchronized as oscillations observed on a continuous platinum catalyst, but there is still good coupling between different parts of the ring.

In the second test, the helium buffer was replaced with an argon buffer at the same flow rate. Since argon has a much lower diffusivity than helium, this is expected to decrease the degree of coupling in the system. The resulting spacetime plot is shown in Fig. 1(b). As expected, the coupling is indeed reduced in comparison to the helium buffer. There are still times, however, where all parts of the catalyst undergo an oscillation nearly simultaneously.

In the third test, the flow rate of the argon was increased to 3000 cc/min. Since coupling in this system is due to gas phase diffusion of the reactants, by increasing the total flow rate of the gas stream, fluctuations in reactant concentration should be swept downstream from the catalyst before they can propagate significantly laterally, resulting in a greatly decreased spatial coupling. The results of this test are shown in Fig. 1(c). As expected, the degree of synchronization was much less than with the slower buffer gas streams. Coupling now only occurred between a few patches at any given time with no simultaneous oscillations on all parts of the catalyst.

In this experiment, we have shown that we can effectively control the degree of synchronization of the oscillations. In order to actually trigger oscillations, local perturbations in CO concentration were used. In the regime of atmospheric pressure, the reaction oscillates between two different rates of reaction. On the high reaction-rate branch, the surface is characterized by a high oxygen coverage and a low CO coverage. On the low reaction-rate branch, the CO coverage is high, while the oxygen coverage is low. It is known that at atmospheric pressure, the surface coverage of CO is linked to the amount of CO in the surrounding gas. When a catalyst is in the high reaction-rate branch, adding CO to the gas stream will increase the coverage of CO until a transition to the lower branch occurs. As the perturbation is swept downstream and the newly adsorbed CO reacts away, the catalyst returns to the steady-state, high reaction-rate regime.

The perturbations are in the form of small puffs of pure CO applied near the surface of one part of the catalyst. To apply the perturbations, a small stainless steel hypodermic tube is mounted just upstream from the catalyst surface. This tube is connected to a gas-tight syringe located outside the reactor. A 1 to 200 μL puff of CO can be applied to the catalyst surface with high accuracy and repeatability. In order to apply the perturbation, a valve upstream from the syringe is opened to flood the puff tube with CO after which the valve is closed. This flooding of the tube results in transient effects in the catalyst (i.e., from exposure to excess CO), which die away within a few minutes. During the time the transients are dying away, there is some diffusion of CO through the open end of the tube, but due to the length of the relevant portion of the tube ($\sim$30 cm), we expect the effect of this diffusion to be small. After these transients have died away, the contents of the syringe are rapidly ejected (i.e. in less than 0.1 s), thereby applying the CO puff to a local region of the catalyst.

In the first perturbation test, the catalyst was exposed to a gas stream with $P_{CO}/P_{O_2} = 0.02$ at a temperature of 280 °C. Under these conditions, the catalyst is in a regime in which the rate of reaction is high, but without oscillatory behavior. When 200 μL of CO were applied to the catalyst surface, part of the catalyst underwent a transition to the low reaction rate branch, and after a few seconds returned to the higher reaction rate branch. The test was repeated, with the same gas stream composition, at temperatures of 270 °C and 260 °C with qualitatively the same results. As the temperature of the system is decreased, the amount of CO required to cause a transition in the catalyst decreases. This was evident as an angular broadening of the width of the transition. In all three tests, the duration of the transition was $\sim$5 s. This test demonstrates that the perturbation is truly localized in both space and time.

In the next test, the catalyst was exposed to a gas stream with $P_{CO}/P_{O_2} = 0.03$ at a temperature of 230 °C. Under these conditions, the rate of reaction oscillates with a period of about 12 minutes. Most of the oscillatory cycle ($\sim$95%) is spent in the state where the rate of reaction is high. A 10 μL puff of CO was applied at the time and place indicated by the arrows in Fig. 2(a),

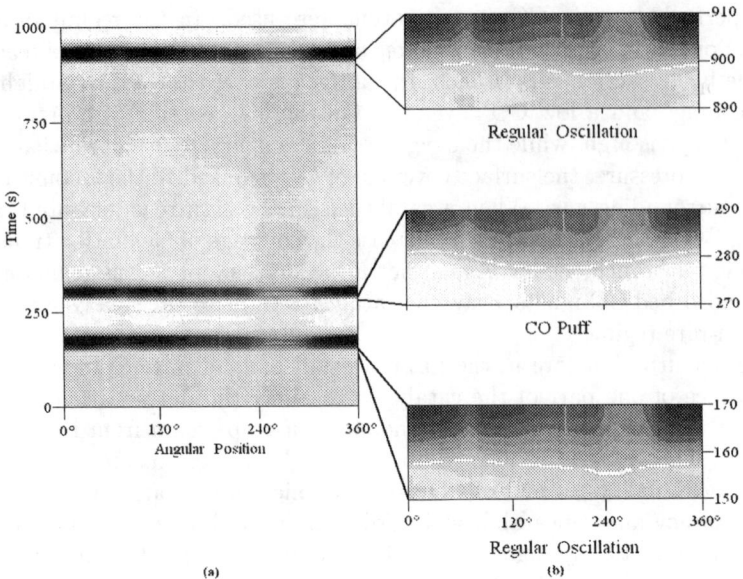

FIGURE 2. In (a), a localized CO perturbation induces a global oscillation. The leading edge of the oscillations (white line) shown in (b) confirms that the oscillation is not coincidental.

just after a natural oscillation of the catalyst. Unlike the perturbations applied to a catalyst under steady-state conditions, the response to this perturbation was what appeared to be a normal oscillation by the entire catalyst. After some time had passed, the catalyst then underwent another natural oscillation. To show that the oscillation which accompanied the CO puff was an induced event, Fig. 2(b) shows the downward transition of each of the three oscillations on an expanded time scale. In each of the natural oscillations, the downward transition has a somewhat irregular shape with the transition beginning near $\theta = 220°$. The induced oscillation, on the other hand, has a smoother shape and begins near the point where the perturbation was applied, at $\theta = 180°$. This indicates that this middle oscillation is indeed induced, rather than being a coincidental event.

Other than the shape of the initial downward transition, all other aspects of the induced oscillation appear the same as the natural oscillations. The upward transitions at the end of the oscillation show qualitatively little difference from one another. The only other difference is in the period of oscillation. The time the induced oscillation spent in the lower state is $\sim$31 s. While this is somewhat less than the $\sim$37 s time of the natural oscillations, it is still significantly longer than the effects of the perturbation in the steady state ($\sim$5 s).

Another question is whether the phase of the oscillations, as observed by the timing between successive oscillations, can be affected by the CO perturbations. Several runs were conducted in which a perturbation induces an oscillation between two natural oscillations of the catalyst. In every case, the CO puff is applied at a different time, but the time between the two natural oscillations was 720 ± 30 s. Thus, while the CO perturbation induces global oscillations, it does not affect the phase of the oscillations. This suggests that the mechanism involved in inducing oscillations with CO perturbations is different than that responsible for the natural oscillations.

We have found that by varying the contents of the buffer gas stream, we have an effective method of controlling the degree of coupling within the system. Using CO perturbations, oscillations can be induced in the system. The reaction fronts induced by these perturbations are observed to travel at ~ 1 cm/s. This is consistent with previously investigated front-propagation phenomena in this system. For example, we have previously observed a rotating pulse which travels at a linear velocity of 0.8 cm/s [6]. We have observed this propagation speed on several catalysts and under varying conditions. This suggests that there is one underlying mechanism controlling the propagation of all these fronts. We are currently studying the relationship of these results to appropriate theoretical models [7,8].

We would like to thank R. K. Herz and A. La Porta for helpful discussions and E. A. Jerzewski for technical assistance. We thank A. Shapiro for assistance in fabricating the platinum catalysts using equipment made available to us by Professor F. Hellman. This work is supported by the Office of Naval Research under grant #N00014-95-1-0179.

REFERENCES

1. Imbihl, R., Cox, M. P., and Ertl, G., *J. Chem. Phys.* **84**, 3519–3534 (1986).
2. Jakubith, S., Rotermund, H. H., Engel, W., von Oertzen, A., and Ertl, G., *J. Chem. Phys.*, **65**, 3013–3016 (1990).
3. Lobban, L. and Luss, D., *J. Chem. Phys.* **93**, 6530–6533 (1989).
4. Yamamoto, S. Y., Surko, C. M., and Maple, M. B., *J. Chem. Phys.* **103**, 8209–8515 (1995).
5. Yamamoto, S. Y., Surko, C. M., Maple, M. B., and Pina, R.K., *J. Chem. Phys.* **102**, 8614–8625 (1995).
6. Yamamoto, S. Y., Surko, C. M., Maple, M. B., and Pina, R.K., *Phys. Rev. Lett.* **74**, 4071–4074 (1995).
7. Levine, H. and Zou, X., *J. Chem. Phys.* **95**, 3815–3825 (1991).
8. Liauw, M. A., Ning, J., and Luss, D., *J. Chem. Phys.* **104**, 5657–5662 (1996).

Optimal Control of Wave Propagation Governed by Nonlinear Partial Differential Equations

Vladimir Khasilev

Courant Institute of Mathematical Sciences, New York University
New York, NY 10012 E-mail: khasilev@cims.nyu.edu

Abstract. The theory of optimal control of distributed media is applied for optimization of soliton propagation in nonlinear media. The goal of optimization was to eliminate nonsoliton radiation which arises during soliton amplification (attenuation) and disturbs the soliton. An analytical solution of the optimal control problem describing ideal soliton amplification was obtained. Methods of optimal control may be applied to other physical systems, such as an oscillator with stochastic resonance.

Introduction

The dynamics of physical systems are described, as a rule, in terms of the initial (boundary) value problem for ordinary or partial differential equations. The solution of the equations usually depends on one or several parameters, and the values of the parameters for optimal implementation of the system can be determined by a simple method. However, if the solution also depends on one or several functions (variable coefficients), then the optimization of physical system on the basis of initial value problem requires considerable effort. The theory of optimal control enables us to solve the problem of optimization with comparatively small computational efforts. This paper reports the results of applying optimal control to equations describing the dynamics of nonlinear waves in fiber-optical waveguides, hydrodynamics, and stochastic resonance.

OPTIMAL CONTROL IN NONLINEAR DISTRIBUTED MEDIA

The use of soliton pulses in optical fiber communication systems may lead to ultrahigh transmission rates [1]. Methods for handling fiber loss use all-optical amplifiers to strengthen the soliton every several tens of kilometers. When solitons propagate in the optical fiber with inserted optical amplifiers, they lose their ideal soliton shape and nonsoliton radiation appears. We obtained a solution to the problem which describes the amplification of the solitons in an optimal manner without formation of nonsoliton radiation.

We consider the Nonlinear Schrödinger Equation (NSE) with variable coefficients:

$$iU_z + \alpha(z)U_{tt} + \beta(z)|U|^2 U = i\gamma(z)U$$

where the subscripts denote partial differentiation with respect to the spatial coordinate z and time t, $\gamma(z)$ is gain (loss). The energy of pulse can be expressed in the following form: $E(z) = \vartheta \int_{-\infty}^{\infty} |U(z,t)|^2 \, dt$, where $U(z,t)$ is a solution of the above NSE equation, and ϑ is constant. This expression includes both soliton E_s and nonsoliton E_r components, $E(z) = E_s(z) + E_r(z)$. Using NSE, we can obtain for $E(z)$ the following equation $\partial E/\partial z = 2\gamma(z)E$, which has the solution: $E(z) = E(z_1)\exp(2\int_{z_1}^{z} \gamma(z) \, dz)$

On the other hand, we can determine an energy of ideal soliton. Let it have the shape: $|U_{opt}(z,t)| = A(z)\cosh^{-1}[A(z)t]$. Then its energy is $E_s(z) = \vartheta \int_{-\infty}^{\infty} |U_{opt}(z,t)|^2 \, dt = 2\vartheta A(z)$

Goal of optimization

In the ideal situation, the nonsoliton component is to be vanished, $E_r(z) = 0$, or, $E_r(z_2) = 0$. Thus, in this ideal case we have $A(z) = A(z_1)\exp(2\int_{z_1}^{z} \gamma(z) \, dz)$. However, the real solution of the NSE may deviate from the goal. We want this deviation to be minimal. Thus, we can express the goal of optimization in the following form: $\int_{-\infty}^{\infty}(|U(t,z_2)|^2 - |U_{opt}(t,z_2)|^2)^2 dt = $ min, where $U(t,z)$ is the real solution of the NSE, $U_{opt}(t,z)$ is the ideal solution, and z_2 is the output point of the fiber-optical amplifier. The above expression is considered as the cost functional to be minimized. As will be shown, the ideal solution is possible in the special case $\alpha(z) = 1 = \beta(z)$.

Optimal control problem

We consider the variational problem, to minimize the functional:

$$J = \int_{-\infty}^{\infty} \int_{z_1}^{z_2} (|U(z,t)|^2 - |U_{opt}(z,t)|^2)^2 dt dz$$

subject to the following control NSE equation with variable coefficients:

$$iU_z + \alpha(z)U_{tt} + \beta(z)|U|^2 U = i\gamma(z)U$$

where $\alpha(z), \beta(z), \gamma(z)$ are the distributed dispersion, nonlinearity and loss (gain) parameters,

$$\gamma(z) = \begin{cases} -\gamma_1 = const & z_0 < z < z_1 \\ \gamma_2(z) & z_1 < z < z_2 \end{cases}$$

initial condition: $|U(z_1,t)| = A(z_1)\cosh^{-1}(A(z_1)t)$, with gain function $\gamma_2(z)$ as control.

Method of solution

The Lagrange multiplier rule [4,5,7] and the new modifications of the Inverse Scattering Transform (IST) method with variable spectral parameter [8,9,13] enabled us to obtain integrable equations with time- and spatial-dependent coefficients describing the dynamics of spatially inhomogeneous systems [9,3]. The remarkable fact is that the solution of the optimal control problem can be obtained analytically. These solutions describe the propagation of the solitons in dispersive nonlinear media with inserted amplification regions. It was found that J possesses a minimum and, for some special gain functions $\gamma_2(z)$, J vanishes, which implies soliton amplification by the optimal manner without formation of nonsoliton radiation. The analytical solution for the soliton in the fiber with the variable gain coefficient $\gamma(z)$ can be expressed as follows: $\alpha(z) = \beta(z) = 1$; $\gamma(z) = -1/(2(z+z_3))$;

$$U(z,t) = \frac{a}{z+z_3}\text{sech}\left(\frac{a(t+t_o)}{\sqrt{2}(z+z_3)}\right)\exp\left\{\frac{i[(t+t_o)^2 - 2a^2]}{4(z+z_3)}\right\}$$

where a, t_o, and z_3 are the arbitrary constants. The advantage of this method of ideal soliton amplification is that it does not require optical fiber with varying dispersion or nonlinear coefficient.

The vector NSE describing parallel process in nonlinear media is also considered [2]. Soliton solutions in planar waveguides display interesting signal dynamics. Interaction of these solitons can be used as multiplexers and demulitplexers in a number of potential soliton communication applications.

OPTIMAL CONTROL OF BENNEY EQUATIONS

The interaction between long and short waves was described by Benney [10]. Let L denote the amplitude of the long wave, and S the envelope of a

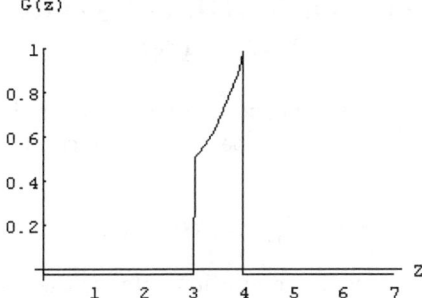

FIGURE 1. Soliton propagation in fiber-optical amplifier with optimized gain.

short wave. The equations describing the dynamics of these waves have the following form

$$L_t + c_l L_x = \alpha(SS^*)_x$$
$$S_t + c_g S_x = i\beta S_{xx} + i\gamma S^2 S^* - i\delta LS$$

where c_l is the phase velocity of the long wave; c_g, the group velocity of the short wave; $\alpha, \beta, \gamma, \delta$ interaction coefficients. This equation possesses soliton solutions and can describe tsunami wave propagation. We considered optimal control problem for this Equation. The goal of optimization was to vanish the soliton wave. Interaction between the short and the long waves with transfer of energy requires large amplitude of the long wave to control the soliton amplitude. A parametric interaction may lead to control of the soliton at the small amplitude of the long wave. The long wave can be considered as a perturbation of the media. It is known that periodic perturbation leads to change of the dispersion coefficient $\beta(x)$. Slow variations of $\beta(x)$ can reduce the soliton amplitude and enlarge its width, providing all energy is trapped into the soliton. After that, sharp changes of $\beta(x)$ lead to separation of the soliton into series of waves of small amplitude.

Recently considerable attention was given to the system of coupled equations describing multi component flow of liquids [11]

$$\eta_t^k + (\eta^k v^k)_x = \gamma(x,t) v^k$$
$$v_t^k + v^k v_x^k + h_x = 0, \ h = \sum_{l=1}^{N} \eta_l$$

where v^k and η^k are velocity and height of k-th layer of flow. This equation was investigated extensively at $\gamma(x,t) = 0$. We considered the more general case $\gamma \neq 0$, that corresponds to a possibility of gain of amplitude of waves. The goal of optimization was trapping the energy of amplified waves into a single localized state. The results obtained are of value in problems of ocean surveillance and storm prediction.

OPTIMAL CONTROL OF STOCHASTIC RESONANCE

Periodically modulated stochastic systems have received considerable attention recently [12]. We may write the dynamics of such systems in the "particle-in-potential" form

$$\dot{x} = -\frac{\partial U(x)}{\partial x} + S(t) + N(t) \qquad (1)$$

where the dot denotes time-differentiation, $U(x)$ is the potential function (often taken to be even and bistable), $S(t)$ is a deterministic signal (usually taken to be time-periodic), $N(t)$ is zero-mean Gaussian exponentially correlated noise.

Let $x = \pm x_0$ be the minima of the potential function, initial condition is $x(0) = +x_0$, and signal detection corresponds to the switching into another state $x(t) = -x_0$. An average of $x(t)$ over period T can be estimated as follows

$$X(t) = T^{-1} \int_{t-T/2}^{t+T/2} x(\tau) \, d\tau$$

Thus, the probability of signal detection will be maximum if the deviation of $X(t)$ from $-x_0$ is minimum,

$$J(t) = (X(t) + x_0)^2 = \min \qquad (2)$$

So, the optimal control problem (2), subject to the control equation (1), at the initial condition $x(t) = x_0$, and $U(x)$ and $S(t)$ as control can be considered. The solution of the problem gives the optimum of the shape of potential function and the deterministic signals. We suppose that at those optimum shapes the sensitivity of the detector can be increased.

The new results have interdisciplinary character and include the soliton theory, theory of optimal control of distributed media, and numerous applications to the mathematical models of distributed systems, such as parallel and distributed computing.

We are indebted to Dr. Adi Bulsara for the kind hospitality, the valuable discussions and support at the ANDM'97. This research is supported in part by the NCCOSC travel grant. We thank Dr. Will Klump for critical reading the manuscript.

REFERENCES

1. L. F. Mollenauer, R. H. Stolen, and J. P. Gordon, "Experimental observation of picosecond pulse narrowing and solitons in optical fibers," *Phys. Rev. Lett.*, **45**, pp. 1095-1098, 1980.
2. V. Y. Khasilev, "Envelope solitons with different carrier frequencies in a dispersive nonlinear medium", *JETP Lett.* **56**, pp. 194-197, 1992."

3. V. Y. Khasilev Optimal control of all-optical communication soliton systems " *SPIE Proceedings, All-Optical Communication Systems: Architecture, Control, and Network Issues* II, **2919**, pp.177-188, 1996.
4. Bryson, Arthur E. and Ho, Yu-Chi, Applied optimal control, Hernisphere Publishing Corporation, 1988.
5. Sussman H., Jurdjevic V., Controllability of nonlinear systems.- *J. Differential Equations*, **12**, 95-116, 1972.
6. Wendell H. Fleming, Future Directions in Control Theory. A Mathematical Perspective. Siam. Philadelphia, 1988.
   Stochastic Optimal Control. Springer Verlag,1975.
7. Butkovskiy A.G. Distributed Control Systems.-New-York, Elsevier. 1969
8. S.P. Burtsev, A.V. Mikhailov, V.E. Zakharov, *Sov. TMP*, **70**, 227,1987.
9. S. Burtsev, D.J.Kaup, B.A.Malomed, Optimum reshaping of optical soliton by a nonlinear amplifier. *J. Opt. Soc. Am. B*, **3**, 888-893, 1996.
10. D.J. Benney, A General theory for Interaction Between Short and Long Waves. *Stud. Appl. Math.* **56**, 81-94, 1997.
11. D.J. Benney, Some Properties of Long Nonlinear Waves. *Stud. Appl. Math.*, **LII**, 45-50, 1973.
12. A. R. Bulsara and A. Zador, Threshold Detection of Wideband Signals: A Noise-Induced Maximum in the Mutual Information, *Physical Review E*, **54**, R2185-88, 1996.
13. V. Khasilev, Inverse scattering transform method and solitons in Nonlinear wave propagation. In Proceedings of the International Symposium on Methods and Application of Analysis, Hong-Kong, December 16-19, 1994, p. 24.

Chaos Due to the Interaction of Capillary and Gravity Waves on the Surface of Deep Water

A.D. Grishchenko

Kharkov State University
4 Svobody Sq., 310077 Kharkov, Ukraine
E-mail agrish@glue.umd.edu

Abstract. The interaction of capillary and gravity waves excited by wind on the surface of deep water is studied. It is shown that the essential difference of the spatial and time scales of these waves does not prevent their mutual interaction which may exert a strong influence on the surface pattern. In particular, this interaction is responsible for the appearance of irregular ripples on the surface of large-scale waves. It is shown that the formation of such ripples is the manifestation of the interaction of capillary and gravity waves. The theory of this phenomenon is developed, and analytical approaches are proposed for the prediction of the chaotic states arising.

INTRODUCTION

It is well known that when the wind velocity V_W is greater than the following critical value $V_{cr} = \sqrt[4]{4g\alpha/\rho} \approx 23$ cm/s , capillary and gravity waves are excited simultaneously provided that the spatial spectrum components with wave numbers equal to

$$k_{1,2} = \frac{V_W^2 \mp \sqrt{g\alpha/\rho}}{2\alpha/\rho} \tag{1}$$

exist in the wind spectrum. Here α is the coefficient of the surface tension of water, g is the free fall acceleration, ρ is the water density. Even if the wind velocity comparatively small the spatial scales of these waves are essentially different, and capillary waves are observed as ripples on the surface of large-scale gravity waves. Although the excitation of these waves has been intensively studied [1–3], the interaction between them is not usually taken into account. In this work, the interaction of capillary and gravity waves is examined in the framework of a two-scale model.

MATHEMATICAL MODEL

Assuming that both capillary and gravity waves interact in a resonance manner with the corresponding wind spectral components, which satisfy spatial and time resonant conditions, we write down equations for the amplitudes and phases of the interacting waves [1],

$$\frac{dA_1}{dt} = -\alpha_1 A_1 - B_1 \cos\varphi_1, \tag{2a}$$

$$\frac{d\varphi_1}{dt} = -\omega_1 - 2W_1 A_2^2 - W_2 A_1^2 - \frac{B_1}{A_1}\sin\varphi_1, \tag{2b}$$

$$\frac{dA_2}{dt} = -\alpha_2 A_2 - B_2 \cos\varphi_2, \tag{2c}$$

$$\frac{d\varphi_2}{dt} = \Delta - 4V A_1 \cos\varphi_1 - 2W_1 A_1^2 - W'_2 A_2^2 + \frac{B_2}{A_2}\sin\varphi_2, \tag{2d}$$

where $A_{1,2}$, $\varphi_{1,2}$ are the amplitudes and phases of gravity and capillary waves, respectively, $B_{1,2}$ are the amplitudes of wind components which are responsible for the excitation of these waves, $\alpha_{1,2}$ are the coefficients of dissipation of gravity and capillary waves, respectively, ω_1 is the natural frequency of gravity waves, $\Delta = \omega' - \omega_2$ is the difference between the frequency ω' of the resonant wind component and that of the capillary wave (ω_2), W_1, W_2, W'_2, V are coefficients which, in the case of the collinear propagation of these waves, are given by the following expressions [1],

$$W_1(k_1, k_2) = -\frac{3\tilde{\alpha}}{32\pi^2}\frac{(k_1 k_2)^{5/2}}{\sqrt{g\tilde{\alpha}}} + \frac{k_1 k_2}{64\pi^2}\left(6k_1 + 2k_2 - 4\sqrt{\frac{gk_1}{\tilde{\alpha}k_2^3}}(k_1 - k_2)\right),$$

$$W_2(k_1) = -\frac{3\tilde{\alpha}}{32\pi^2}\frac{k_1^5}{g} + \frac{k_1^3}{8\pi^2},$$

$$W'_2(k_1) = \frac{k_2^3}{32\pi^2},$$

$$V(k_1, k_2) = \frac{k_1 k_2}{4\sqrt{2}\pi} g,$$

where $\tilde{\alpha} = \alpha/\rho$. It should be noted that the natural frequencies of the gravity and capillary waves read $\omega_1 = \sqrt{gk_1}$, $\omega_2 = \sqrt{\tilde{\alpha}k_2^3}$, respectively.

To simplify system (2), the following fact may be taken into account. The characteristic time interval of the variation of the amplitude of the gravity wave is much greater than that for the capillary wave, inasmuch as $\alpha_1 \ll \alpha_2$ [3]. This fact allows as to neglect the variation of the amplitude of the gravity waves on the time interval $\propto 1/\alpha_1$. Then, according to the Eq. (2a), we have $A_1 = -(B_1/\alpha_1)\cos\varphi_1 = const$. The solution of the second equation can be written in the form $\varphi_1 = -\omega_1 t - W_2 A_1^2 t$ if the second and the last terms are not taken into account. This procedure is justified because nonlinear terms are small in comparison with the other. Due to these assumptions, Eqs. (2a-d) can be transformed to the following non autonomous system of the second order,

$$\frac{dA_2}{dt} = -\alpha_2 A_2 - B_2 \cos \varphi_2, \tag{3a}$$

$$\frac{d\varphi_2}{dt} = \Delta - 4V A_1 \cos\left(\omega_1 - W_2 A_1^2\right) t - 2W_1 A_1^2 - W_2' A_2^2 + \frac{B_1}{A_1} \sin \varphi_2. \tag{3b}$$

This set of equations is amenable to analytical study in some limits which may be of practical interest.

CONDITIONS FOR CHAOS

System (3) admits stationary states of two types corresponding to periodic and chaotic oscillations. As far as we know the chaos onset due to the interaction of capillary and gravity waves has not been studied until now. To determine analytical conditions for the chaos to arise, we use the obvious fact that the strange attractors in the phase space of this system can be formed only due the transverse intersection of stable and unstable manifolds of some saddle orbits. Thus, the appearance of a saddle orbit in the phase space of system (3) is a necessary condition for the chaos to arise. In order to find this condition, let us neglect for a while the time-dependent term in system (3), considering this term as a fast varying and relatively small one. In other words, it means that we apply an averaging procedure to system (3) to remove this term. It results in an autonomous system which possesses a singular saddle point in its phase space if the following conditions are met,

$$\Delta > \sqrt{3}\alpha_2, \tag{4a}$$

$$B_2 > 16\pi/3^{3/4} \left(\alpha_2/k_2\right)^{3/2}. \tag{4b}$$

Now if we again take into account the periodical perturbation, this saddle will transform into a saddle orbit, which may give birth of a strange attractor. Hence, Eqs. (4a,b) can be considered as necessary conditions for the chaos onset.

Next, we note that Eqs. (3a,b) are similar to those describing an oscillator with frequency-modulated forcing. According to [4] we can apply Melnikov's technique to this system considering the time dependent term and dissipation as small perturbations. Finally, we come to the following condition for the chaos onset with respect to the amplitude A_1,

$$A_1 > A_{cr} \equiv \alpha_2 \frac{W_2' \sinh(\pi\sigma)}{8V\pi\omega_1 \sinh(\sigma\theta^{\pm})}, \qquad (5)$$

where

$$\sigma = 2\omega_1/W_2'\mu,$$
$$\mu^2 = -2fu_s - b^2/4,$$
$$b = -4\left(\delta - u_s^2\right),$$
$$f = 4B_2/W_2',$$
$$q = \pm\sqrt{-2fu_s},$$

and u_s is the smallest real root of the cubic equation $W_2'u^3 + \Delta'u - B_2 = 0$, where $\Delta' = 2W_1A_1^2 - \Delta$, and $\theta^{\pm} = \pm\arccos(-b/q)$.

Thus, if the amplitude of the gravity wave is greater than critical value A_{cr}, then the capillary wave may go into chaotic motion. This results in a small scale chaotic modulation of the gravity wave. We should emphasize that the value of A_{cr} is proportional to the coefficient of dissipation α_2 of the capillary wave. Furthermore, as it follows from the analysis of the Eq. (5), A_{cr} reaches its minimum when the frequency ω_1 is approximately equal to the coefficient of dissipation α_2. Hence, the wind spectral components with the frequencies

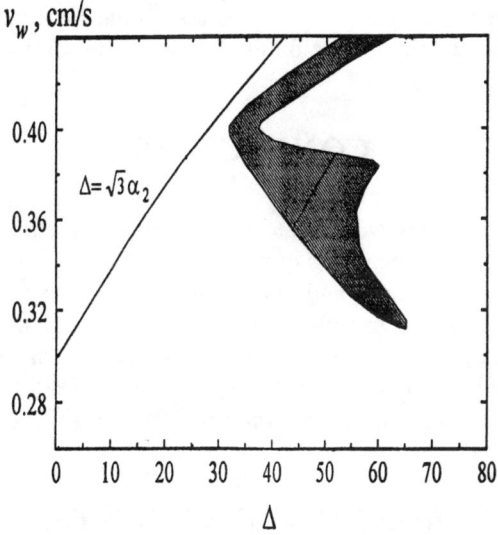

FIGURE 1. State diagram of system (3) at $A_1 = 100$. Condition (4a) is satisfied on the right side of the curve given. Region of chaos found from the result of computer simulation is shaded.

which are close to the α_2-value have the greatest influence on the dynamic of the system. The expression for α_2 can be written as follows, $\alpha_2 = 2\nu k_2^2$, where ν is the coefficient of the kinematic viscosity of water. Now, taking into account the definition of the frequency ω_1 and Eq. (1), we can rewrite the condition $\alpha_2 \approx \omega_1$ for the chaotic states arising in the following form

$$V_W \approx g^{1/5}\alpha^{2/5}\nu^{-1/5}\rho^{-2/5}. \qquad (6)$$

At this value of the wind velocity the chaotization of the waves occurs with the smallest value of the amplitude of the gravity wave. For the typical values of the parameters involved here, one can find from Eq. (6) that $V_W = 56 cm/s$.

To illustrate the conditions for the chaos onset, a state diagram on the parameter plane (V_W, Δ) is shown in Fig.1. The critical curve $\Delta = \sqrt{3}\alpha_2$ is also plotted on the plane. According to the condition (4a) chaotic states may arise on the right hand side with respect to this curve. One can see that chaotic states arise even when the wind velocity is slightly greater than the critical value $V_{cr} = 23$ cm/s. These states exist in a wide region of the variation of parameters, and they produce evidence that the appearance of chaotic states is not critical phenomenon with respect to the frequency and amplitude of the resonant spectral components of wind. It should be noted that the region of chaos found from the analysis of system (3) correlates well with that found from the solution of the initial system (2). Hence, the chaos onset due to the interaction of the capillary and gravity waves should be considered as typical phenomena leading to the formation of chaotic ripples on the sea surface.

CONCLUSION

We have shown in this work that the interaction of capillary and gravity waves may lead to the appearance of periodic or chaotic modulation of gravity waves even if the both time and spatial scales of these types of waves are essentially different in magnitude. Under natural situations this modulation is known as the appearance of ripples on the surface of large-scale waves. We have proposed a model for describing the formation of such ripples and developed analytical approaches for the prediction of their chaotization.

ACKNOWLEDGMENTS

The author would like to thank Prof. V. Kontorovich and Prof. D. Vavriv for helpful discussions. This work was supported in part by EC under contract ERBIC15CT960816.

REFERENCES

1. Zakharov, V.Ye.,*Prikl. Matem. i Tekh. Fizika* **2**, 86 (1968),(In Russian).
2. Kadomtsev, B.B., and Kontorovich, V.M., *Izv Vuzov- Radiofizika* **17**, 511 (1974), (In Russian).
3. Phillips, *The Dynamics of the Upper Ocean*, Cambrige University Press, 1977.
4. Vavriv, D.M., Ryabov, V.B., and Sharapov, S.A., *Radiotech. i Electr.* **38**, 464 (1993), (In Russian).
5. Zakharov, V.Ye., L'vov, V.S., Falkovich, G., and Kolmogorov, A.N., *Spectra of Turbulence (Wave Turbulence)*, New York: Springer, 1992.

TIME SERIES ANALYSIS AND SYNCHRONIZATON

Synchronization Transitions in Coupled Chaotic Oscillators

Jürgen Kurths, Arkady S. Pikovsky,
and Michael G. Rosenblum

*Department of Physics, University of Potsdam,
PF 60 15 53, 14415 Potsdam, Germany
Home page: http://www.agnld.uni-potsdam.de*

Abstract. We discuss the notion of phase for chaotic systems and describe synchronization transitions in a system of two coupled self–sustained chaotic oscillators. We demonstrate that with the increase of coupling strength the system first undergoes the transition to phase synchronization. With a further increase of coupling, a new synchronous regime is observed, where the states of both oscillators are nearly identical, but one system is delayed in time with respect to the other. We describe this regime as a state with correlated amplitudes and a constant phase shift. These transitions are traced in the Lyapunov spectrum.

INTRODUCTION

Synchronization effects appearing due to a comparatively strong interaction of chaotic oscillators have been intensively studied; one differentiates between complete (full) [1–4]; and generalized synchronization [5–7]. We have recently described the new phenomenon of phase synchronization in weakly coupled chaotic systems [8]. We define it as the occurrence of a certain relation between the phases of interacting systems while the amplitudes can remain chaotic and are, in general, uncorrelated. The effect has been studied for the case of external driving [9], self–synchronization in a large population of globally coupled nonidentical oscillators [10] and synchronization in a lattice [11]. This effect appears to be robust enough to be observed in physical experiment [12,13] and even in living nature [14]. It can be used for coherent summation of outputs of chaotic generators and as a theoretical basis for new techniques in time series analysis.

In the present work we describe the transition route from non-synchronous via phase-locked state to lag synchronization [15]. We show that both transitions occur via intermittency.

PHASE OF A CHAOTIC SYSTEM

Roughly speaking, the phase of an autonomous self-sustained oscillatory system is related to the symmetry with respect to time shifts. Therefore, the phase disturbances do not grow or decay, what corresponds to the zero Lyapunov exponent. If the oscillations are periodic, the phase rotates nearly uniformly, while in the chaotic case the dynamics of the phase is effected by chaotic changes of the amplitude. So one can expect a Brownian (random–walk–like) behavior of the phase. In the simplest case, when a projection of a strange attractor on some plane (x, y) can be found with the phase point always rotating around some point that can be taken as the origin, we can define the phase as $\phi = \arctan(y/x)$. For comparison of different approaches to the phase definition see [9].

We start with a rather artificial example where in a chaotic system one can rigorously define the phase satisfying $\dot\phi = \omega_0$ and the possibility of phase synchronization is obvious. We consider the oscillator described by

$$\begin{aligned}
\dot x &= \frac{e}{2}x - \omega y - z\frac{x^2}{x^2+y^2}, \\
\dot y &= \frac{e}{2}y + \omega x - z\frac{xy}{x^2+y^2}, \\
\dot z &= f + z(x-c),
\end{aligned} \quad (1)$$

where $e, f, c,$ and ω are parameters. This system can be considered as a modification of the Rössler system with the same parameters, and its attractor (fig. 1a) is similar to the Rössler attractor. By substitution $x = A\cos\phi$ and $y = A\sin\phi$, the Eqs. (1) can be re-written as

$$\begin{aligned}
\dot A &= \frac{e}{2}A - z\cos\phi, \\
\dot\phi &= \omega, \\
\dot z &= f + z(A\cos\phi - c).
\end{aligned} \quad (2)$$

The second equation in (2) coincides with the equation governing the dynamics of the phase of periodic oscillators; the phase of the chaotic system (1) introduced in this way obviously corresponds to the zero Lyapunov exponent. Suppose now that two oscillators of this kind are coupled via the phase variables, i.e. the second equation in (2) has the form $\dot\phi_{1,2} = \omega_{1,2} + \varepsilon\sin(\phi_{2,1} - \phi_{1,2})$, where $\omega_{1,2} = \omega_0 \pm \Delta$. Then, of course, the phases $\phi_{1,2}$ are locked if the coupling strength satisfies $\varepsilon \geq \Delta$. If the frequency mismatch is small, $\Delta \to 0$, the locking happens for vanishing coupling. Hence, similar to the synchronization of periodic oscillators and contrary to other types of synchronization of chaotic systems, phase synchronization can appear *without a threshold*.

To demonstrate the chaotic character and independence of amplitudes $A_{1,2}$ in the synchronous state, we calculate them at time moments $2\pi/\omega_0 \cdot n$, i.e. for

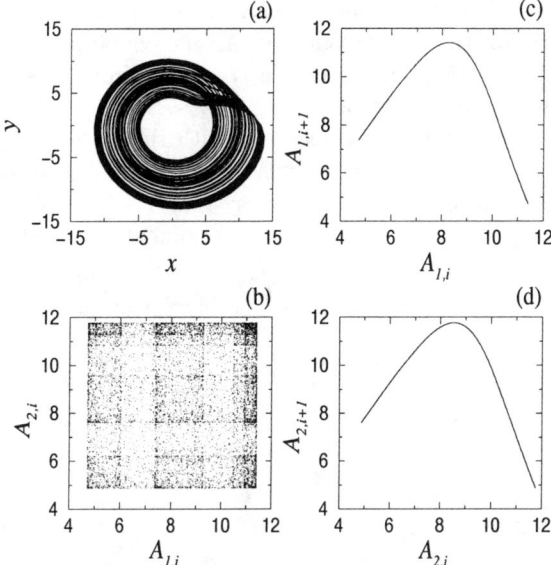

FIGURE 1. (a): Projection of the attractor of the system (1) for the parameter values $f = 0.4$, $e = 0.15$, $c = 8.5$, and $\omega = 1$. The amplitudes $A_{1,i}$ and $A_{2,i}$ of both synchronized systems (2) for $\omega_0 = 1$, $\Delta = 0.02$, $\varepsilon = 0.05$ are independent (b), although each amplitude remains chaotic, as can be seen from the next amplitudes plots (c) and (d).

constant values of $\phi_{1,2}$, construct the next amplitude plots (partial Poincaré maps) and plot one amplitude vs the other one (Figure 1).

This example is rather exceptional, because generally one cannot expect that one variable in a chaotic system separates from the other ones, like in Eq. (2). In the general case, we use phase definitions applicable to a large class of systems, e.g. the definition based on the Hilbert transform (see [9]).

SMALL COUPLING: TRANSITION TO PHASE SYNCHRONIZATION

Next we consider a more realistic example of two coupled non–identical Rössler oscillators [16]:

$$\begin{aligned}\dot{x}_{1,2} &= -\omega_{1,2} y_{1,2} - z_{1,2} + \varepsilon(x_{2,1} - x_{1,2}), \\ \dot{y}_{1,2} &= \omega_{1,2} x_{1,2} + e y_{1,2}, \\ \dot{z}_{1,2} &= f + z_{1,2}(x_{1,2} - c),\end{aligned} \quad (3)$$

where e, f, and c are the standard parameters of the Rössler model. The additional parameters $\omega_{1,2} = 1 \pm \Delta$ and ε govern the frequency mismatch and the strength of coupling, respectively. First we describe the appearance of phase synchronization in system (3). As the coupling is increased (while the mismatch Δ is fixed), we observe a transition from a regime, where the phases rotate with different velocities $\phi_1 - \phi_2 \sim \Delta \cdot t$, to a synchronous state, where the phase difference does not grow with time, $|\phi_1 - \phi_2| < $ const, and $\Delta\Omega = <\dot\phi_1 - \dot\phi_2> = 0$ (Figure 2a). We emphasize that in contrast to the other types of synchronization of chaotic systems [1–3,17,18], here the instant vectors (x_1, y_1, z_1) and (x_2, y_2, z_2) do not coincide. Moreover, the correlations between the amplitudes $A_{1,2} = \sqrt{x_{1,2}^2 + y_{1,2}^2}$ are pretty small, although the phases are completely locked and in this respect the motions are highly coherent.

An interesting feature is the appearance of intermittency at the onset of

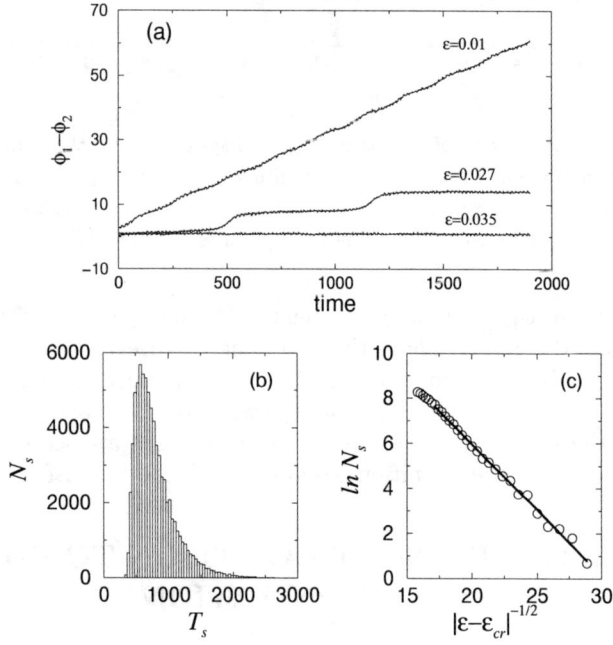

FIGURE 2. (a): Phase difference of two coupled Rössler systems (eq. (3)) vs. time for non–synchronous, ($\varepsilon = 0.01$), nearly synchronous, or intermittent, ($\varepsilon = 0.027$) and synchronous ($\varepsilon = 0.035$) states. In the last case the amplitudes $A_{1,2}$ remain chaotic, their cross-correlation is less than 0.2. The frequency mismatch is $\Delta = 0.015$. Parameter values: $e = 0.15$, $f = 0.2$, $c = 10$. (b): The distribution of the number of phase slips N_s with the interval between slips T_s for $\varepsilon = 0.027$; it demonstrates that the slips occur irregularly. (c) The number of phase slips per constant time N_s on the coupling strength in the vicinity of the transition point. The slips are exponentially rare.

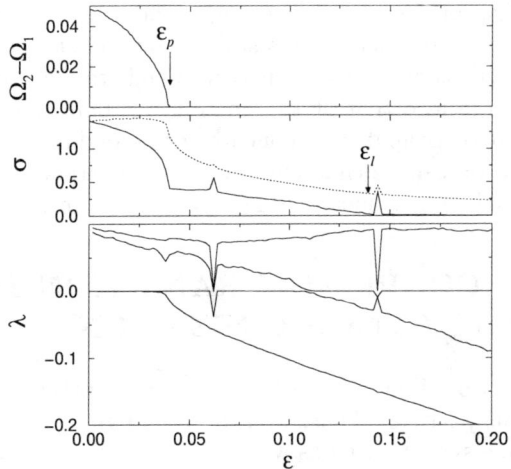

FIGURE 3. The frequency difference $\Omega_1 - \Omega_2$, the minimum of the similarity function σ and the four largest Lyapunov exponents λ of two coupled Rössler oscillators vs. the coupling ε. Three different regions are clearly seen in the σ vs ε plot correspondent to non–synchronous state, phase and lag synchronization respectively. The transitions between these regimes are reflected in the spectrum of Lyapunov exponents: at the first transition one of the zero exponents becomes negative, while the second transition corresponds to the zero crossing of one of the positive exponents. The dashed line shows dependence of $S(0)$ on the coupling; from this plot one can see that comparison of states of interacting systems without time shift does not reveal the transition to LS. Two "outbursts" on σ vs ε plot at $\varepsilon \approx 0.06$ and $\varepsilon \approx 0.145$ correspond to period 3 windows.

synchronization. Indeed, as one can see from Figure 2a, at the border of the region of complete phase locking, the phases are almost locked. It means that from time to time phase slips occur, where during a rather small interval of time the phase difference changes by 2π. The time intervals between these slips are irregular, as one can see from their distribution (Figure 2b). The slips are exponentially rare, and the dependence of the number of phase slips per constant time N_s on the coupling strength obeys a relation $N_s \sim \exp(-|\varepsilon - \varepsilon_c|^{-1/2})$ [19] (Figure 2c).

In order to describe this phase synchronization transition in the framework of transitions in chaotic systems, we have studied the Lyapunov exponents. In Figure 3 we present the 4 largest Lyapunov exponents for the system (3) in dependence on the coupling strength ε. In the uncoupled case each oscillator has one positive, one zero, and one negative Lyapunov exponent, the zero ones corresponding to the phases. For $\varepsilon < 0.04$ the phases are not locked, and two Lyapunov exponents are nearly zero, i.e. we cannot distinguish them from zero within numerical resolution. We see from Figure 3 that the transition to phase synchronization happens, when one of these zero Lyapunov exponents becomes negative, corresponding to a stable relation between the phases (one Lyapunov exponent is exactly zero, it corresponds to a simultaneous shift of both phases). Note also that at the phase synchronization transition there remain two positive Lyapunov exponents corresponding to the amplitudes. Thus this transition can be characterized as a transition inside chaos (to be more precise, inside hyperchaos) and not as a "chaos–order" transition.

LARGE COUPLING: TRANSITION TO LAG SYNCHRONIZATION

As one can see from Figure 3, for large enough coupling only one Lyapunov exponent remains positive. Thus, we expect that the amplitudes of the two oscillators become synchronized as well. We describe here a particular regime appearing at this transition, namely the lag synchronization (LS) [9]. LS appears as a coincidence of *shifted in time* states of two systems, $\mathbf{x}_1(t+\tau_0) = \mathbf{x}_2(t)$.

To characterize LS, we introduce a similarity function S as an averaged over time difference between the variables x_1 and x_2 (with mean values being subtracted) taken with the time shift τ

$$S(\tau) = \left(\frac{\langle (x_2(t+\tau) - x_1(t))^2 \rangle}{(<x_1^2(t)><x_2^2(t)>)^{1/2}} \right)^{1/2}, \qquad (4)$$

and search for its minimum $\sigma = \min_\tau S(\tau)$. If the two signals x_1 and x_2 are independent and their variancies are approximately equal, it yelds $S(\tau) \sim 1$ for all τ. If $x_1(t) = x_2(t)$, as in the case of complete synchronization, $S(\tau)$ reaches its minimum $\sigma = 0$ for $\tau = 0$. Below we describe a nontrivial case, when the similarity function $S(\tau)$ has a minimum for a non-zero time shift τ, meaning the existence of a time lag between the two interacting systems.

First, we follow the transition to phase synchronization in system (3) for $e = 0.165$, $f = 0.2$, $c = 10$. The parameters $\omega_0 = 0.97$ and $\Delta = 0.02$ are chosen by trial in such a way that the appearance of large windows of periodic behavior is avoided. The calculation of the average frequencies $\Omega_{1,2}$ allows us to find the first transition at $\varepsilon = \varepsilon_p \approx 0.036$ to the frequency entrainment $\Omega_1 = \Omega_2 = \Omega$ (Figure 3). Due to the high coherence of the Rössler attractor, the phase difference in the synchronous regime is bounded and oscillates around some mean value $\delta\phi = <\phi_1(t) - \phi_2(t)> \neq 0$. For stronger coupling $\varepsilon = \varepsilon_l \approx 0.14$ we observe a new transition to lag synchronization (see the σ vs ε curve in Figure 3). In Figure 4 we show numerically obtained similarity functions for the range from relatively weak to strong coupling. For weak coupling $\varepsilon < \varepsilon_p$ (curves 1,2), $S \sim 1.5$ and practically does not depend on τ, as expected for independent signals. For intermediate coupling strength

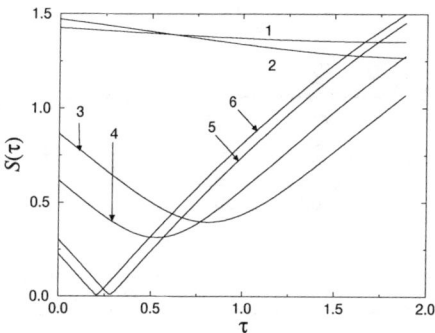

FIGURE 4. The similarity function $S(\tau)$ (4) for different values of coupling strength ε (1: $\varepsilon = 0.01$, 2: $\varepsilon = 0.015$, 3: $\varepsilon = 0.05$, 4: $\varepsilon = 0.075$, 5: $\varepsilon = 0.15$, 6: $\varepsilon = 0.2$). With the increase of coupling a minimum appears indicating the existence of a certain phase shift between interacting systems (curves 3 and 4). In the regime of lag synchronization (curves 5 and 6) the minimum is extremely small.

$\varepsilon_p < \varepsilon < \varepsilon_l$, a minimum of $S(\tau)$ appears (curves 3,4) indicating the existence of some characteristic time shift τ_0 between x_1 and x_2. This shift is related to the phase difference as $\tau_0 = \delta\phi/\Omega$. Note that in this regime the amplitudes are uncorrelated, so the value of $S(\tau_0)$ is relatively large. Further increase of coupling makes at $\varepsilon \approx \varepsilon_l$ this minimum very sharp (curves 5,6) and practically equal to zero. It means that the states of the systems become identical, but shifted in time with respect to each other. It is important that calculations of $S(0)$, i.e. the comparison of x_1 and x_2 without time shift, reveal no transition at $\varepsilon = \varepsilon_l$. For larger couplings $\varepsilon > \varepsilon_l$, the time lag τ_0 continuously decreases, but no further qualitative transitions are observed.

The second transition to LS is also reflected in the Lyapunov spectrum (Figure 3). It corresponds to the change of the sign by the second positive Lyapunov exponent, but does not exactly coincide with it due to intermittency discussed below. This means that the relation appears not only between the phases, but also between the amplitudes. The phase shift remains, and therefore a time lag between the signals x_1 and x_2 is observed.

Very much alike the transition to complete synchronization, the transition to lag synchronization is extremely sensitive to small perturbations. Respectively, it also occurs via intermittency, which can be interpreted as modulational ("on–off") one and appeares when one of the LE crosses zero. Even when the second LE is negative, the local instability can lead to bursts of non-synchronous behavior [20], see Figure 5. Due to this intermittency, σ gradually decreases in the region $0.11 < \varepsilon < 0.14$, until these local instabilities disappear.

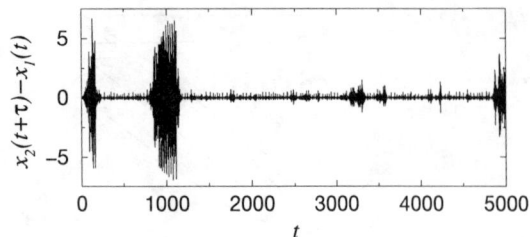

FIGURE 5. The time series $x_2(t+\tau) - x_1(t)$ in the intermittent region $\varepsilon = 0.13$, $\tau = 0.32$. The bursts can be viewed as the excursions from the low–dimensional "synchronous" attractor.

REFERENCES

1. H. Fujisaka and T. Yamada, *Prog. Theor. Phys.* **69**, 32 (1983).
2. A. S. Pikovsky, *Z. Physik B* **55**, 149 (1984).
3. L. M. Pecora and T. L. Carroll, *Phys. Rev. Lett.* **64**, 821 (1990).
4. M. de Sousa Vieira, A. J. Lichtenberg, and M. A. Lieberman, *Int. J. of Bifurcation and Chaos* **1**, 691 (1991).
5. N. F. Rulkov, M. M. Sushchik, L. S. Tsimring, and H. D. I. Abarbanel, *Phys. Rev. E* **51**, 980 (1995).
6. H. D. I. Abarbanel, N. F. Rulkov, and M. M. Suschik, *Phys. Rev. E* **53**, 4528 (1996).
7. L. Kocarev and U. Parlitz, *Phys. Rev. Lett.* **76**, 1816 (1996).
8. M. Rosenblum, A. Pikovsky, and J. Kurths, *Phys. Rev. Lett.* **76**, 1804 (1996).
9. A. Pikovsky, M. Rosenblum, G. Osipov, and J. Kurths, *Physica D* **104**, 219 (1997).
10. A. Pikovsky, M. Rosenblum, and J. Kurths, *Europhys. Lett.* **34**, 165 (1996).
11. G. Osipov, A. Pikovsky, M. Rosenblum, and J. Kurths, *Phys. Rev. E* **55**, 2353 (1997).
12. A. S. Pikovsky, *Sov. J. Commun. Technol. Electron.* **30**, 85 (1985).
13. U. Parlitz, L. Junge, W. Lauterborn, and L. Kocarev, *Phys. Rev. E.* **54**, 2115 (1996).
14. M. G. Rosenblum, G. I. Firsov, R. A. Kuuz, and B. Pompe, in *Nonlinear Analysis of Physiological Data*, Springer Series for Synergetics, edited by H. Kantz, J. Kurths, and G. Mayer-Kress, Berlin: Springer, 1997.
15. M. Rosenblum, A. Pikovsky, and J. Kurths, *Phys. Rev. Lett.* **78**, 4193 (1997).
16. O. E. Rössler, *Phys. Lett. A* **57**, 397 (1976).
17. T. L. Carroll and L. M. Pecora, *IEEE Trans. Circ. and Systems* **38**, 453 (1991).
18. L. M. Pecora and T. L. Carroll, *Phys. Rev. A* **44**, 2374 (1991).
19. A. Pikovsky et al., *Phys. Rev. Lett.* **79**, (1997).
20. A. S. Pikovsky and P. Grassberger, *J. Phys. A* **24**, 4587 (1991).

Cycles from Short Time Series

Stuart Allie and Alistair Mees

Centre for Applied Dynamics and Optimization, University of Western Australia, Nedlands, 6907, Australia

Abstract. We present an algorithm for finding low-order cycles of chaotic maps from possibly very short time series. No information about the map other than the time series is used. The method finds all the periodic points of a piecewise linear approximation of the map. We present an example showing the effectiveness of the method for the Henon map, including a "cycle expansion" calculation of the Hausdorff dimension.

METHOD

We present a method for finding low-order cycles of chaotic maps from short time series. Our method is intended for use when the map is unknown and the only information available is a short time series. We find all the exact periodic points of a given order for a piecewise linear approximation to the map.

The partition for the piecewise linear approximation is defined by triangulating the data [1]. If the data is contaminated by noise, we construct a triangulation model from the data. Methods to do this are described in [2,3]. A triangulation is a collection of n-simplices in R^n where the intersection of two simplices is either the empty set or an m-simplex, $m < n$. (A n-simplex is the closed convex hull of $n+1$ points — the vertices of the simplex — in R^n.)

For a point x in a simplex σ with vertices $v_o, \ldots v_n$, we say that $\lambda_i = \lambda_i(x)$, $i = 0, 1, \ldots, n$, are the *barycentric coordinates* for x with respect to v_i, if

$$x = \sum_i \lambda_i v_i \quad \text{and} \quad 1 = \sum_i \lambda_i.$$

A piecewise linear approximation to $f : R^n \to R^n$ is then given by

$$\hat{f}(x) = \sum_i \lambda_i f(v_i).$$

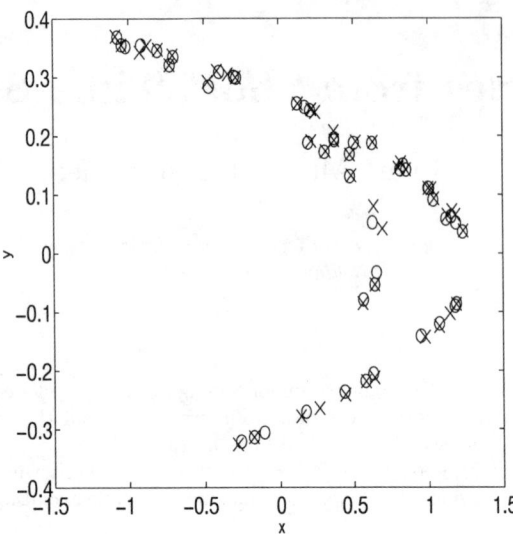

FIGURE 1. Approximate periodic points up to period 7 for the Henon map (circles) and the true points (crosses), calculated using 50 data points.

We note [1] that $\hat{f}$ is continuous in x; if f is C^2, then $\hat{f}(x) = f(x) + O(\Delta^2)$, where $\Delta = \text{diameter}(\sigma)$, and if f is affine, then $\hat{f} = f$.

We approximate iterates of f by iterating $\hat{f}$ rather than trying to approximate f^k directly. The point here is that we can model the map itself reasonably well, but not higher iterates of the map, especially for chaotic systems.

If the time series in R^n is a scalar time series embedded using a time-delay embedding [5,6], then $n-1$ components of the map f will be linear, and our approximation is exact for these components.

We construct a "transition matrix," T, on our collection of simplices where

$$T_{ij} = \begin{cases} 1 & \text{if } f(\sigma^i) \cap \sigma^j \neq \emptyset \\ 0 & \text{otherwise} \end{cases}$$

for $i, j = 1, \ldots, N$ and we have N simplices. For each simplex, which we label k_0, we can recursively search T for sequences where

$$T_{k_0,k_1} T_{k_1,k_2} \cdots T_{k_{p-1},k_0} = 1,$$

so that the index sequence $k_0, k_1, \ldots k_{p-1}$ corresponds to a sequence of simplices which *might* contain a period p point. We then construct the following system of equations:

$$\sum_i \lambda_i^{k_0} v_i^{k_0} = \sum_i \lambda_i^{k_{p-1}} f(v_i^{k_{p-1}}), \tag{1}$$

$$\sum_i \lambda_i^{k_j} v_i^{k_j} = \hat{f}^j(x) = \sum_i \lambda_i^{k_j-1} f(v_i^{k_j-1}); \quad j = 1, \ldots, p-1, \qquad (2)$$

with $\sum_i \lambda_i^{k_j} = 1$ and $\lambda_i^{k_j} \geq 0$, $\forall j = 0, \ldots, p$, $\forall i$. This is simply the periodic point equation

$$x = \hat{f}^p(x)$$

in terms of our piecewise linear approximation $\hat{f}$. This is a system of linear equations in $p(n+1)$ unknowns: the $\lambda_i^{k_j}$. In this way we find, by exhaustive search, *all* the cycles of length p for $\hat{f}$.

It is true that, in general, the number of periodic points of the true map is not necessarily equal to that for the piecewise linear approximation. For low order points and sufficient data, equality will hold, although the authors have not yet been able to construct a suitable, and computable, criterion to guarantee this equality. Presently, several heuristics are used to identify false or missing cycles and these heuristics seem to work well in practice.

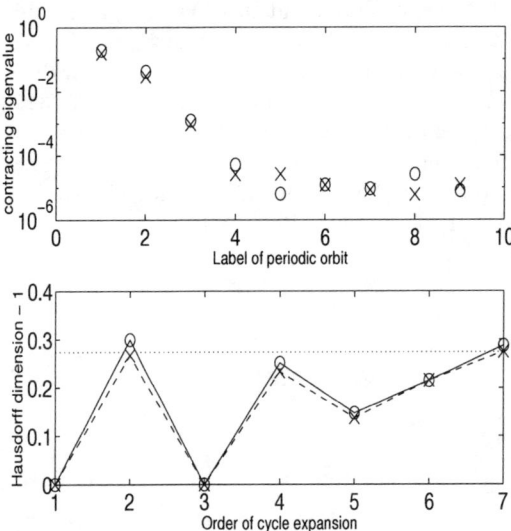

FIGURE 2. Top axes shows the values of Λ_P for the periodic orbits up to period 7 of the Henon map. Circles are the approximate points, crosses are the true values. The Bottom axes shows $D_s = D_h - 1$ for cycle expansions up to order 7. The solid line is for the approximate periodic points, the dashed line is the true map. The dotted line shows the accepted value of $D_s \approx 0.274$.

EXAMPLES

We calculated the approximate periodic points, up to period 7, for the Henon map

$$(x, y) \mapsto (1 - 1.4x^2 + y, 0.3x),$$

using only 50 data points. These periodic points have been used to calculate the Hausdorff dimension using the "cycle expansion" method [4]. Figure 1 shows the true and approximate periodic points up to period 7. Figure 2 shows the values of the stable eigenvalues of each periodic orbit, Λ_P, for both the true map and the approximate map for orbits up to period 7 and D_s =Hausdorff dimension - 1, as a function of the order of the cycle expansion.

More detail and other examples, including the Ikeda map and a chaotic electronic circuit, can be found in [7]. We will present elsewhere a detailed study of the use of cycle expansions to estimate dynamical invariants from short time series.

REFERENCES

1. A. I. Mees, International Journal of Bifurcation and Chaos **1**, 777 (1991).
2. S. P. Allie, A. I. Mees, K. Judd, and D. F. Watson, Phys. Rev. E **55**, 87 (1997).
3. S. P. Allie, A. I. Mees, K. Judd, and D. F. Watson, in *Control and Chaos*, edited by K. Judd, A. I. Mees, K. L. Teo, and T. L. Vincent (Birkhauser, Boston, 1996).
4. R. Artuso, E. Aurell, and P. Cvitanovic, Nonlinearity **3**, 325 (1990).
5. F. Takens, in *Dynamical Systems and Turbulence*, edited by D. A. Rand and L. S. Young (Springer, Berlin, 1981), Vol. 898, pp. 365–381.
6. L. Noakes, International Journal of Bifurcation and Chaos **1**, 867 (1991).
7. S. P. Allie and A. I. Mees, Phys. Rev. E (In press.)

Diagnosing Intermittency

Stephen Hammel [†], John Barnett [†] and Nathan Platt [*]

[†] Code 743, Naval Research and Development,
NCCOSC San Diego California 92152
[*] Institute for Defense Analyses, Alexandria Virginia 22311

Abstract. Many systems are observed to exhibit short bursts of intense activity, followed by periods of quiescence. Sustained alternation between these two qualitatively different states is called intermittency, and we investigate a form known as on-off intermittency. Models for this form of intermittency utilize a master-slave coupling between two simple dynamical systems. Our approach treats the output of the slave system as an observable, and we seek to deduce the character of the background master system which is driving it. For our work this means a statement about the structure of the probability density function for the observed slave variable.

I INTRODUCTION

There are many dynamical systems which exhibit intermittent signals. Such a signal is characterized by an alternation between periods of relative quiescence and periods of large amplitude bursting. This term was originally applied to fluid systems in which point probes in fluid flow produced signals that alternated between laminar flow and turbulent bursts. A particular type of this behavior has been named on-off intermittency [1].

The defining feature of on-off intermittency is the existence of a time-dependent parameter which determines the stability of the system near a bifurcation point. In an on-off intermittent system, there are usually two components: a master driving subsystem, and a slave ("observable") subsystem. As indicated by the names, the slave system depends upon the master system dependent variable, but the master system is independent of the slave system state.

The study of on-off intermittent systems has been motivated by problems that have driven the larger field of applied nonlinear dynamics. One is confronted with a somewhat complex system for which a single scalar variable can be reliably measured. Based upon the behavior of this variable, one would like to be able to reconstruct the underlying system which drives the observed variable. This attempt motivated much of the early work in the method of time delays and delay coordinate embedding.

II A PIECEWISE LINEAR MAP

On-off intermittency has been examined in flows, but for simplicity we will consider only discrete time maps. Consider the map:

$$y_{n+1} = S(y_n) = \begin{cases} S_s(y_n) = ax_n y_n & ; y_n \leq \dfrac{1}{ax_n} \\ S_f(y_n) = \dfrac{ax_n}{ax_n - 1}(1 - y_n) & ; y_n > \dfrac{1}{ax_n} \end{cases} \quad (1)$$

This will be the slave system for which we observe the variable $\{y_n\}$. The master driving system generates $\{x_n\}$, $x_{n+1} = m(x_n)$.

We will examine three different choices for the master system and examine the subsequent effects on the observed signal $\{y_n\}$. The three master maps that we will consider are the dyadic or doubling map, the tent map, and white noise on the interval $(0, 1)$.

$$\text{Dyadic Map: } x_{n+1} = D(x_n) = 2x_n \bmod 1 \quad (2)$$

$$\text{Tent Map: } x_{n+1} = T(x_n) = \begin{cases} 2x_n & ; x_n \leq 1/2 \\ 2 - 2x_n & ; x_n > 1/2 \end{cases} \quad (3)$$

$$\text{White Noise: } x_{n+1} = \text{Unif}(0, 1) \quad (4)$$

It is quite difficult to distinguish the time series generated by the three different drivers. Fig. 1 shows a typical time-series generated by an observed variable y_n for two of the three different drivers (eqns. (2) and (4)). If we observe the behavior of the slave system variable, what can we deduce about the nature of the master driving system? This is the central question which motivates this paper.

One direct measurement of the state of an intermittent system is to determine the probability of a laminar phase of a given length. This is simply a measurement of the quiescent period between active outbursts. A universal asymptotic $-3/2$ power law distribution [2] was described at a *critical* value for a driving parameter (that value at which the system can be expected to sustain intermittent bursting for arbitrarily long periods). This critical value $a_c = e = 2.718\ldots$ The $-3/2$ power law applies to all three of the master driving systems. In addition, each of the three driving maps generates an invariant density which is uniformly distributed on the interval. From this point of view, the three drivers are indistinguishable.

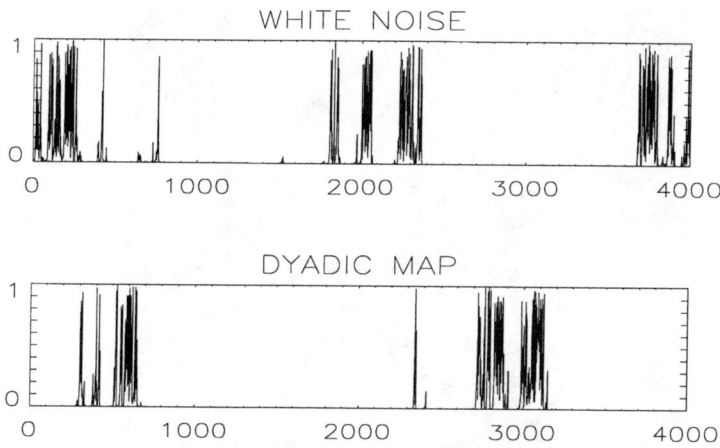

FIGURE 1. Two timeseries of 4000 points generated by eqn(1) with $a = 2.8$. The top series is driven by white noise eqn.(4), and the bottom series is driven by the dyadic (doubling) map, eqn.(2).

However, striking differences in the distribution of laminar phases can be seen when the three different drivers are compared. This is particularly true for parameter values larger than the critical value. Longer laminar phases are substantially more rare for the white noise driver than for both the tent and the dyadic maps. The subsequent sections describe our efforts to understand these differences. One system characteristic is the correlation function, which might be expected to help distinguish the different drivers: $\hat{C}(k) = \lim_{N \to \infty} \sum_{j=1}^{N} (x_j - \langle x \rangle)(x_{j+k} - \langle x \rangle)$. The correlation function for white noise $\hat{C}_{wn}(k) = 0$. But it is interesting to note the correlation function for the dyadic driver $\hat{C}_d(k) = 2^{-k}$ whereas for the tent map, $\hat{C}_t(k) = 0$.

III WHITE NOISE DRIVING

Our approach is to understand the density of the measure for the system (1) above. Ideally we would like to find the invariant density for the system. As a preliminary, we will describe the density induced by the map. Given a particular y_n, how does S transform the uniform measure for $x_n = \text{Unif}(0, 1)$? For this determination we examine the first return map for $\{y_n\}$. A two-dimensional density for y_{k+1} against y_k is shown in fig.2.

To indicate the two different phases of the map, we use S_s for the stretching phase, and S_f for the folding phase. There are two different regions for the density calculation, depending on whether $y_n \leq 1/a$ or $y_n > 1/a$. If $y_n \leq 1/a$,

FIGURE 2. The first return map for white noise driving: y_{n+1} is plotted against y_n, and in the grey-scale plot, regions of greater density are lighter. $\{x_n\}$ is generated by eqn.(4) and $\{y_n\}$ is generated by eqn.(1) with $a = 5$.

then the density is simply $\frac{1}{ay_n}$. To calculate the density for $y_n > 1/a$, note that

$$\frac{1}{\frac{dS_f}{dx}} = \frac{(ax_n - 1)^2}{a(y_n - 1)}$$

Since $S^{-1}(z) = \frac{z}{a(y_n+z-1)}$, the density f is given:

$$f = \frac{1}{y_n}\left|\frac{1}{\frac{dS_f}{dx}}\right|_{x=S^{-1}(z)} = \frac{1}{y_n}\left|\frac{(ax_n-1)^2}{a(y_n-1)}\right|_{x_n=\frac{z}{a(y_n+z-1)}} = \frac{1-y_n}{ay_n(y_n+z-1)^2}$$

The second contribution is given by the normalization factor:

$$\int_{\frac{a(1-y)}{a-1}}^{1} \frac{1-y}{a(y+z-1)^2}dz = \frac{ay-1}{a}.$$

$$f_{\text{left}} = \frac{1}{ay} \quad ; \quad 0 \leq z \leq \frac{a(1-y)}{a-1}$$

$$f_{\text{right}} = \frac{1-y}{ay(y+z-1)^2} \quad ; \quad \frac{a(1-y)}{a-1} \leq z \leq 1$$

Note that the density is singular at both $(y_n, y_{n+1}) = (0,0)$ and $(y_n, y_{n+1}) = (1,0)$

IV THE DYADIC MAP AND THE TENT MAP

An important factor in the distribution of laminar phases for any of the maps is the fact that the three different drivers have very different behavior following the occurrence of a very small value x_k. The next value generated, x_{k+1}, will also be small for both the tent and dyadic maps, whereas the white noise process can produce any value with equal likelihood. This feature creates longer laminar phases for the tent and dyadic map drivers.

The densities generated by both the tent map and the dyadic map are singular at 0. For both cases, the action of the map is simplified: the transformation is of one kind, corresponding to the left half of the interval. Although the first return map for tent map driving is not shown, it is very similar to the first return map for the dyadic map, shown in fig.3. A greater density of points is shown as a lighter shade in the grey-scale plot. The sharp curved boundaries provide a striking contrast with the white noise driver. As noted above, the key to the analysis of both of these drivers is the occurrence of a sequence of small values of x_{n+k} following the appearance of one very small value x_n.

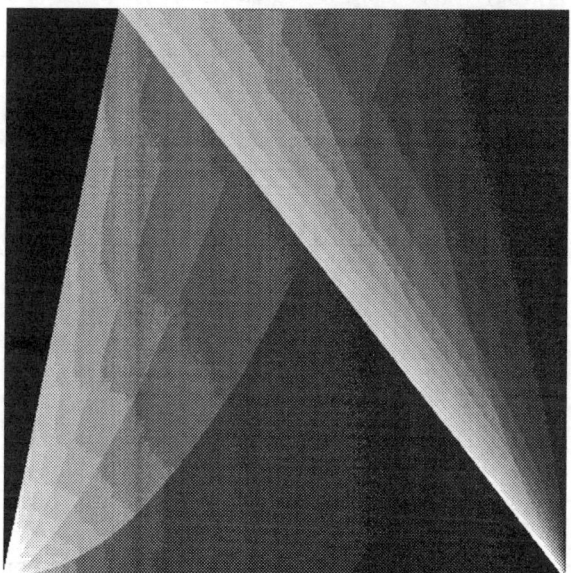

FIGURE 3. The first return map for dyadic map driving. y_{n+1} is plotted against y_n, and in the grey-scale plot, regions of greater density are lighter. $\{y_n\}$ is generated by eqn.(1) with $a = 5$, driven by $\{x_n\}$ generated by eqn.(2)

Hence, for both maps,

$$\begin{aligned} x_{n+1} &= 2x_n \bmod 1 = 2^{n+1} x_0 \bmod 1 \\ y_{n+1} &= ax_n y_n = a^{n+1} \prod_{i=0}^{n} x_i y_0 \end{aligned} \quad (5)$$

Or,
$$y_{n+1} = ax_n \left(a^n \prod_{i=0}^{n-1} x_i y_0 \right) \tag{6}$$

Now, let $A = \log(a), X_n = \log(x_n), Y_n = \log(y_n)$;
$$\begin{aligned} X_{n+1} &= X_n + \log(2) \\ Y_{n+1} &= A + X_n + \left(nA + \sum_{i=0}^{n-1} X_i + Y_0\right) \end{aligned} \tag{7}$$

The action of the map near 0 can be understood with the map in the form (7). Note that as long as $X_n < -A$, ($x_n < 1/a$), we have $X_{n+1} > X_n$, but $Y_{n+1} < Y_n$. This reveals the essential contrast with white noise driving: for $X_n \ll 0$ (equivalently x_n very small), Y_n will first "burrow in" toward 0 before eventually bursting.

Finally, we note that the curved boundaries seen on the first return map are actually a manifestation of the same effect: a sequence of small values of x_k must follow the appearance of a first very small value. Consider $\{(x_n, y_n) : x_n < \epsilon, y_n > 1 - \delta\}$. If $(x_n, y_n) = (\epsilon, 1 - \delta)$, then $(x_{n+1}, y_{n+1}) = (2\epsilon, a\epsilon(1 - \delta))$. Likewise,

$$x_{n+2} = 4\epsilon$$
$$y_{n+2} = 2a^2\epsilon^2(1 - \delta) = \frac{2[a\epsilon(1-\delta)]^2}{1-\delta} = \left(\frac{2}{1-\delta}\right) y_{n+1}^2$$

$$x_{n+3} = 8\epsilon$$
$$y_{n+3} = \frac{8[a\epsilon(1-\delta)]^3}{(1-\delta)^2} = \left(\frac{2^{3/2}}{(1-\delta)^{1/2}}\right)\left(\frac{2^{3/2}[a\epsilon(1-\delta)]^3}{(1-\delta)^{3/2}}\right) = \left(\frac{2^{3/2}}{(1-\delta)^{1/2}}\right) y_{n+2}^{3/2}$$

$$x_{n+4} = 16\epsilon$$
$$y_{n+4} = 2^6 a^4 \epsilon^4 (1-\delta) = \frac{4}{(1-\delta)^{1/3}} \left(2a\epsilon(1-\delta)^{1/3}\right)^4 = \left(\frac{4}{(1-\delta)^{1/3}}\right) y_{n+3}^{4/3}$$

$$x_{n+k+1} = 2^{k+1}\epsilon$$
$$y_{n+k+1} = \left(\frac{2^{\frac{k+1}{2}}}{(1-\delta)^{\frac{1}{k}}}\right) y_{n+k}^{\frac{k+1}{k}}$$

These expressions for $y_{n+2}, y_{n+3}, \ldots, y_{n+k+1}$ and the associated exponents $2, 3/2, \ldots, (k+1)/k$ generate the family of curves visible in fig.3 as sharp steps in the density.

REFERENCES

1. N. Platt, E. A. Spiegel, and C. Tresser, *Phys. Rev. Lett.* **70**, 279 (1993).
2. J. F. Heagy, N. Platt, and S. M. Hammel, *Phys. Rev. E* **49**, 1140 (1994).

Using Delay Differential Equations as Dynamical Classifiers

Michael N. Kremliovsky [†] and James B. Kadtke

Institute for Pure And Applied Physical Sciences
University of California at San Diego
9500 Gilman Dr., La Jolla, CA 92093-0360
[†] *also at RTA Corporation, Washington D.C., 20041-6436*

Abstract. We show how direct estimation of delay differential equations can be used for time series analysis of observed signals corrupted by noise. The specific application we discuss is to detection (identifying the presence of a deterministic signal) and classification (assignment of a particular signal to a known class). These ideas have potentially widespread application to areas such as remote sensing, voice recognition and image processing.

INTRODUCTION

In the last two decades, techniques derived from the theory of nonlinear dynamics have indicated a ground work for novel approaches in signal processing. Until recently, signal processing largely dealt with direct characteristics of the time series under consideration such as amplitude, spectral density, statistical moments, linear regression (ARMA models) and so on. Not surprisingly, this approach can be limited for broadband, noisy, transient signals with nonlinear components. Using such representations, signals produced by the time evolution of the same physical systems can appear dramatically different depending on experimental conditions surrounding its propagation and measurement. Therefore, it can be impossible to detect dynamical similarities or differences by processing a signal $x(t)$ as a direct representation $\{x(t_1), x(t_2), \ldots, x(t_n)\}$.

Contrary to this we propose to detect and classify signals by their dynamical properties [1]. This means that we want to classify the rules of temporal evolution underlying a sample observation, instead of the observation itself. Since we seek to represent the dynamics in a state space, our approach requires estimation of a "rate of change", namely to include the *signal derivative* $dx(t)/dt$ into consideration. This is unusual for conventional signal processing, because the numerical derivative for a noisy observation amplifies the noise, and therefore, often introduces increased errors. However, we have found that via parametric estimation of the signal dynamics, it is sufficient to determine any correlation between the signal measurement and its derivative. Hence, we require only to recover sufficient statistical evidence of the deterministic relations to provide a well-defined hypothesis test.

Specifically, we represent data observations by parametric *dynamical models*, which define the connection between a system state and its rate of evolution under a particular model ansatz. Thus, a D-dimensional dynamical system is described in each instant of time t by its *state vector* $\mathbf{x}(t) = \{x_1(t), x_2(t), \ldots, x_D(t)\}$ in a hypothesized state (phase) space. Correspondingly the time evolution is represented by a *phase trajectory* $\{\mathbf{x}(t) \mid t \in R\}$. In the case of a scalar signal $x(t)$ we can use time-delay embedding $\mathbf{x}(t) = \{x(t), x(t-\tau), \ldots, x(t-(D-1)\tau)\}$. Here, we choose to use first-order **a**-parametric delay differential equations as dynamical models

$$\dot{\mathbf{x}}(t) = F_{\mathbf{a}}[x(t), x(t-\tau), \ldots] \tag{1}$$

for the following reasons:

- these models require an estimate of the first-order derivative only;
- these models are infinite-dimensional in general yet they can describe a complex low-dimensional time evolution;
- these models need only a scalar measurement $x(t)$ to be self-consistent.

At this point we must stress a difference between typical modeling of a time evolution (sometimes called the *Inverse Problem of Dynamical System Reconstruction*) and our application of dynamical models for signal detection/classification. The inverse problem requires one to find an *exact* model, which is capable of reproducing the phase portrait of the system with correct dimensionality and global topological properties. This is an ill-posed problem, strictly speaking, until *all* phase trajectories are measured with infinite precision. In practice, we usually deal with single finite non-ideal measurements. Contrary to the inverse problem, for classification we do not require a model to be exact. Rather, it is only necessary to build enough complexity in the model to be able to describe a signal statistically in terms of its average dynamical properties. In this sense the "flow operator" $\hat{F}[\cdot]$ in the right hand side (RHS) of Eq. (1) is defined by the derivative in the left hand side (LHS), but not necessarily the other way around: i.e. $\dot{x}(t) \Rightarrow F[\cdot]$. Correspondingly, our approach is always applicable since it does not require uniqueness.

DYNAMICAL MODELS AND A FEATURE SPACE

In the order to investigate classification properties of our dynamical models we have first to choose a functional representation for the flow $F[\cdot]$. It is here desirable to have a linear parameter dependence, because in this case the problem of parameter estimation can be reduced to an easily solvable system of linear algebraic equations. The simplest polynomial expansion up to P-th order with $(D-1)$-delays satisfies this requirement ($x(t) \equiv x$, $x(t-\tau) \equiv x_\tau$, $x(t-2\tau) \equiv x_{2\tau}$, ...):

$$\dot{x}(t) = F[\mathbf{x}] \equiv a_0 + a_1 x + a_2 x_\tau + \ldots + a_D x^2 + a_{D+1} x x_\tau + \ldots = $$
$$= \sum_{j=1}^{N} a_j x^{k_1} x_\tau^{k_2} \ldots x_{(D-1)\tau}^{k_D}, \quad (2)$$

where $N = (P+D)!/P!D!$ is the number of monomials and unknown coefficients, $0 \leq k_1 + \ldots + k_D \leq P$. Let us concentrate on the simplest nonlinear case of $P=2$, $D=2$:

$$\dot{x}(t) = a_0 + a_1 x + a_2 x_\tau + a_3 x^2 + a_4 x x_\tau + a_5 x_\tau^2. \quad (3)$$

If $x(t)$ is bounded at ∞ one can set $a_0 = 0$, but this may not be a practical requirement for short time series. Note that we can also re-write Eq. (3) as an integral equation:

$$x(t) = x(0) + a_0 t + a_1 \int_0^t x \, dt' + a_2 \int_0^t x_\tau \, dt' + $$
$$+ a_3 \int_0^t x^2 \, dt' + a_4 \int_0^t x x_\tau \, dt' + a_5 \int_0^t x_\tau^2 \, dt', \quad (4)$$

where integrals can always be defined even if $x(t)$ is a random process. In general, we do not assume that $x(t)$ is in the solution space of Eq. (2), but we will show below that at least one very important case for signal processing, that of polyharmonic processes $x(t) = \sum_i A_i \sin(\omega_i t + \phi_i)$, is a solution of Eq. (3) for certain values of $\mathbf{a}(A, \omega, \phi)$.

Our problem now is to estimate coefficients $\mathbf{a} \equiv \{a_0, \ldots, a_N\}$ in the expansion for a signal $x(t)$ constructed from its discrete measurements $\{x(t_1), x(t_2), \ldots, x(t_n)\}$. Since typically we have many data points to obtain a few parameters ($n \gg N$) it is reasonable to consider an approximate solution minimizing a difference between the LHS and RHS of the Eq. (2). One possibility is to solve Eq. (2) in a least-squares sense, namely to find a set of coefficients $\{a_i\}$ minimizing a cost function

$$Q = (\dot{x} - F[\mathbf{x}, \mathbf{a}])^2. \quad (5)$$

Here, we do not want Q to depend on time (through time dependence of monomials in x), because we are here looking for stationary solutions. Therefore, we must minimize the expectation value:

$$\langle Q \rangle = \lim_{T \to \infty} \frac{1}{T} \int_0^T Q \, dt, \quad (6)$$

where T is the sample window. To do this we need to solve a system of N linear algebraic equations

$$\frac{\partial \langle Q \rangle}{\partial a_i} = 0 \qquad i = 1, \ldots, N. \qquad (7)$$

We can also solve Eq. (2) numerically (with $t = t_1, t_2, \ldots$), for example, using Singular Value Decomposition (SVD) which provides a very robust solution in a least-squares sense [1,6]. In this case, if we estimate the parameters for many observation windows of the signal (n_w windows) we can obtain an ensemble of n_w coefficient vectors $\{\mathbf{a}_j \mid j = 1, \ldots, n_w\}$, which we will call in the following *feature vectors*. Generally, the difference in the distributions of these vectors in a feature space $\mathcal{A}$ allows one to distinguish between dynamical systems belonging to different dynamical classes. Thus, signal classification is reduced to the analysis of feature distributions, via their discrimination, clustering and overlapping in feature space.

For the remainder of this paper, we will concern ourselves with deriving some analytic properties of this dynamical classification scheme. It is highly instructive to examine the feature space partitioning for certain model forms by solving Eq. (7) asymptotically for known signal classes. A simple (though very valuable) signal example is the analysis of harmonic signals. This is a canonical example in signal processing, and thus we would like to find a solution for the case of $x(t) = A\sin(\omega t)$. First of all, we can normalize time such that $t' = \omega t$ and $\tau' = \omega \tau$. Coefficients in the normalized system are then $a_i' = a_i/\omega$. Thus, the information about the time scale of the signal $x(t)$ is contained in the amplitudes of coefficients. We may also re-scale nonlinear coefficients of order p by the factor A^{p-1}. This means that higher absolute amplitudes of the signal ($A \gg 1$) will not have a stronger impact on relative values of nonlinear coefficients. Now, consider Eq. 3. In the case of an infinitely long window, $T \to \infty$, we find that the system of six equations $\partial \langle Q \rangle / \partial a_i =$

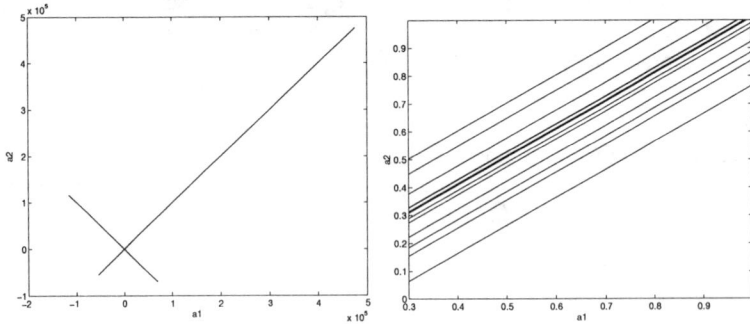

FIGURE 1. In these plots we show the parametrically defined solution (with parameter ω) of Eq. (8) for $\omega > 0$. The fine structure of this transcendental curve is demonstrated in the right plot, representing a 25 fold increase in resolution.

0, $i = 1\ldots6$, can be divided into two non-coupled subsystems, with two ($\{a_1, a_2\}$) and four ($\{a_0, a_3, a_4, a_5\}$) equations/variables, respectively. The first system

$$\frac{\partial \langle Q \rangle}{\partial a_1} = a_1 + a_2 \cos(\tau)$$
$$\frac{\partial \langle Q \rangle}{\partial a_2} = a_2 + a_1 \cos(\tau) + \sin(\tau), \qquad (8)$$

yields solutions $a_1 = \cos(\tau)/\sin(\tau)$ and $a_2 = -1/\sin(\tau)$, while the second system yields: $a_0 = a_3 = a_4 = a_5 = 0$. The resulting transcendental function (a_1, a_2) parametrized by the sine wave frequency $\omega > 0$ is shown in Fig. 1.

To demonstrate this result numerically, we generate artificial data sets consisting of sine waves of various frequencies, and directly estimate the model coefficients with finite window size. These coefficients are shown to lie on the theoretical curve (Fig. 2), in excellent agreement with the analytic results.

Another important signal case is the distribution of features produced by

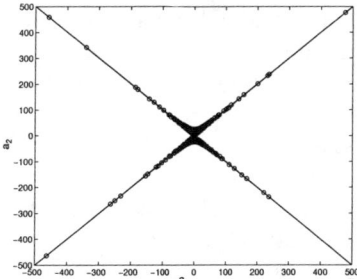

FIGURE 2. Numerical solution in a least-squares sense (indicated by circles) of the Eq. (2) for the class of sine waves with variable frequency $x(t) = \sin(\omega t)$ shows excellent agreement with the analytical solution (solid line; however, see Fig. 1 for the explanation of its fine structure).

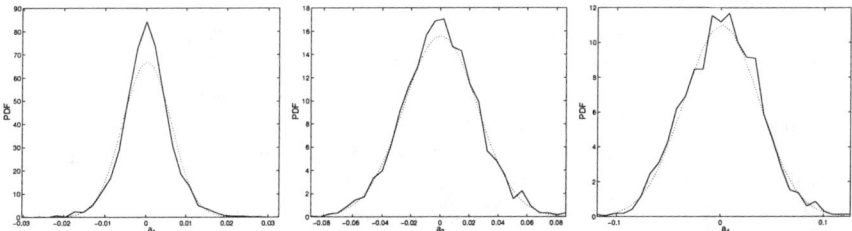

FIGURE 3. Feature PDF (coefficients a_1, a_3 and a_4, from left to right, correspondingly) obtained numerically for pure Gaussian noise (solid curve) in comparison with theoretical Gaussian distributions calculated with the same mean and variance as numerically estimated distributions (dashed curve).

a random Gaussian process with a given autocorrelation $R_{xx}(\tau)$. Here we estimate the features $\mathbf{a}_j$ in windows of finite length T. It is easy to show that

$$E[a_0] = E[a_1] = E[a_2] = E[a_3] = E[a_4] = E[a_5] = 0 , \qquad (9)$$

where $E[\cdot]$ is the expectation value [4]. Thus, the center of the estimated feature distributions must lie at the origin. For large T, it can also be shown analytically that these distributions asymptotically approach normal distributions (Fig. 3); however, the variance is scaled differently for features corresponding to linear and nonlinear terms [4]. Hence, any signal whose features significantly deviate from the origin must contain some correlations and/or a deterministic component. Detectors can therefore be built by measuring this statistical deviation.

SUMMARY

The idea underlying our dynamical detection/classification technique is to estimate the underlying rules of temporal evolution of a system, rather than directly representing its particular time series of observations. This can be done by mapping signals into features, which are parameters of dynamical models consisting of delay differential equations. We showed how the feature space can be analytically partitioned in two distinct important cases: sine waves and a pure Gaussian process. Generally, the features carry statistical information about correlations between states of the system and its rate of evolution; in other words, they can characterize the topology of the underlying state space. Thus, feature distributions can be effectively utilized for detection and classification of dynamical systems.

REFERENCES

1. Kadtke, J. "Classification of Highly Noisy Signals Using Global Dynamical Models" Phys. Lett. A **203** (1995).
2. Kadtke, J., and Kremliovsky, M. AIP Conf. Proc. **375**, 189 (1996).
3. Kremliovsky, M. and Kadtke, J. AIP Conf. Proc. **375**, 205 (1996).
4. Kadtke, J., and Kremliovsky, M. "Estimating Statistics for Detecting Determinism Using Global Dynamical Models", Phys. Lett. A. **229**(2), p.97 (1997).
5. McDonough, R., Whalen, A. "Detection of Signals in Noise", Academic Press, 1995.
6. Press, W.H. et.al. "Numerical Recipes", Cambridge, 509 (1986).
7. Takens, F. In "Dynamical Systems and Turbulence" (Eds. D. Rand and L.S. Young), Springer, 366 (1981).

Weak and Strong Synchronization of Chaos

K. Pyragas

Semiconductor Physics Institute
11 A. Goštauto, LT-2600 Vilnius, Lithuania

Abstract. Some properties of *generalized synchronization* (GS) in unidirectionally coupled chaotic systems are investigated. Depending on the coupling strength, the GS may appear in two different states called the *weak synchronization* (WS) and the *strong synchronization* (SS) characterized by a fractal (nonsmooth) and a smooth synchronization manifold, respectively. Experimentally, the GS and properties of the synchronization manifold can be detected either by an auxiliary response system or by estimating the conditional Lyapunov exponents from observed time series.

Close to the threshold of WS coupled chaotic systems experience on-off intermittency. Unlike the conventional on-off intermittency, where the system dynamics is determined by escape of trajectories from an unstable *smooth* hyperplane, this intermittency is characterized by escape of trajectories from an unstable *fractal* manifold. Despite of this difference the main characteristics of both intermittencies are identical.

INTRODUCTION

Cooperative behavior of coupled chaotic systems has attracted considerable attention lately. Synchronization effects are observed in many physical and biological processes and they are responsible for the transition to low-dimensional attractors in systems with many degrees of freedom. Synchronization of chaos is often understood as a behavior in which two coupled systems exhibit identical chaotic oscillations [1,2]. We refer to this type of synchronization as an *identical synchronization* (IS).

Recently the notion of chaotic synchronization has been generalized for coupled non-identical systems [3,4]. In the case of unidirectionally coupled chaotic systems (master-slave configurations)

$$\dot{X} = F(X), \qquad (1a)$$

$$\dot{Y} = G(Y, X), \qquad (1b)$$

the GS was taken to occur if there exists a map $\Phi : X \to Y$ that takes the trajectories of the attractor in the driving space $X = \{x_1, x_2, \ldots x_d\}$ into the trajectories of the response space $Y = \{y_1, y_2, \ldots y_r\}$ so that $Y(t) = \Phi(X(t))$, and if this map does not depend upon initial conditions of the response system [4]. When Φ differs from identity the detection of the GS in an experiment is a difficult task. One way to recognize the GS is to construct an auxiliary response system Y' identical with Y, link it to the driving system X in the same way as Y is linked to X

$$\dot{Y}' = G(Y', X) \tag{2}$$

and check the existence of the IS between Y and Y' [5]. If such synchronization occurs the response system "forgets" the initial conditions and its asymptotic dynamics is completely determined by the driving system. Geometrically, this implies a collapse of the overall evolution onto a stable synchronization manifold $M = \{(X, Y) : \Phi(X) = Y\}$ in the full state space of two systems $X \oplus Y$, and so leads to a functional relationship between X and Y variables defining the GS [5,6].

The condition of synchronization between Y and Y' is determined by *conditional Lyapunov exponents* [2] $\lambda_1^R \geq \lambda_2^R \geq \ldots \geq \lambda_r^R$ that can be obtained from variational equation of the response system at $\delta X = 0$,

$$\delta \dot{Y} = D_Y G(Y, X) \delta Y, \tag{3}$$

where $D_Y G$ denotes the Jacobean matrix with respect to the Y variable. These exponents define the stability of both the identity manifold $Y' = Y$ in $X \oplus Y \oplus Y'$ state space and the synchronized manifold $Y = \Phi(X)$ in $X \oplus Y$ space [5]. Thus the condition of GS is $\lambda_1^R < 0$.

PROPERTIES OF THE SYNCHRONIZATION MANIFOLD

Note that IS between Y and Y' does not guarantee the smoothness of Φ [7]. The synchronization manifold $M = \{(X, Y) : \Phi(X) = Y\}$ can have a fractal structure. In this case the map Φ will be continuous but not differentiable. Recently, Sauer and Yorke [8] have provided a theorem which shows that only continuously differentiable (C^1) maps preserve the dimension of strange attractors. Thus, for the nonsmooth map Φ, the global dimension of the strange attractor d^G in the whole state space $X \oplus Y$ is larger than the dimension of the driving attractor d^D in X subspace, $d^G > d^D$. This synchronization we define as the WS. For smooth Φ with the degree of smoothness C^1 or higher, these two dimensions are equal in magnitude, $d^G = d^D$. This type of GS is defined as the SS. Obviously, the IS is a particular case of the SS. Generally the threshold of SS can be estimated from the Kaplan-Yorke conjecture [9].

The response system does not have an effect on the global Lyapunov dimension $(d_\lambda^G = d_\lambda^D)$ at the condition [7] $\lambda_1^R < \lambda_m^D$ where m is the minimal integer for which $\sum_{j=1}^m \lambda_j^D < 0$ and $\lambda_1^D \geq \lambda_2^D \geq \ldots \geq \lambda_d^D$ are the Lyapunov exponents of the driving system.

Numerical Example

To illustrate the properties of GS let us consider a simple example of two unidirectionally coupled identical one-dimensional maps:

$$x_{n+1} = f(x_n), \quad (4a)$$
$$y_{n+1} = f(y_n) + k\{f(x_n) - f(y_n)\} \quad (4b)$$

with $f(x) = 4x(1-x)$. Here Eqs.(4a) and (4b) describe the driving and response systems, respectively, k is the coupling strength. At $0 < k < 1$, the coupling term in Eq.(4b) preserves the global stability of the response system since $0 < (1-k)f(y_n) + kf(x_n) < 1$ at any x_n and y_n lying in the interval $[0, 1]$. To observe the GS, we consider an auxiliary response system

$$y'_{n+1} = f(y'_n)) + k\{f(x_n) - f(y'_n)\}. \quad (5)$$

identical with the original response system Eq.(4b), but having different initial condition than that of system (4b). We emphasize that this system does not influence the dynamics of the original response and driving subsystems described by Eqs.(4). It serves only to detect the properties of the system (4).

At any coupling strength k, Eqs.(4) have an invariant manifold $y = x$ and hence admit the IS, which in this case is equivalent to the SS. The thresholds of WS and SS are determined by two different Lyapunov exponents, namely, the conditional Lyapunov exponent λ^R defining the stability of the invariant manifold $y' = y$ in an extended three-dimensional state space $\{x, y, y'\}$, and the transverse Lyapunov exponent λ^I of the invariant identity manifold $y = x$ in the $x - y$ plain. The dependence of these exponents on k is shown in Figure 1(a). $\lambda^R(k)$ becomes zero at two characteristic values of the coupling strength $k = k_w \approx 0.3325$ and $k = k_s = 0.5$ corresponding to the thresholds of WS and SS, respectively. Above the last threshold $k > k_s$, the manifold $y = x$ becomes stable and these two exponents coincide, $\lambda^I(k) = \lambda^R(k)$. This region of k corresponds to the SS since the system attractor lies on the *smooth* manifold $y = x$. As an indicator of the IS between x and y we take the rms. deviation $s_{DR} = \sqrt{\langle(x_n - y_n)^2\rangle}$ At $k > k_s$ it vanishes, $s_{DR}(k) = 0$ [Figure 1(b)]. Similarly, as an indicator of the IS between y and y' we take the rms. deviation $s_{RR} = \sqrt{\langle(y'_n - y_n)^2\rangle}$. It vanishes $s_{RR}(k) = 0$ at $k > k_w$.

In the region $k_w < k < k_s$, we have the IS between the original and auxiliary response systems $y_n = y'_n$ and have no IS between the driving and original

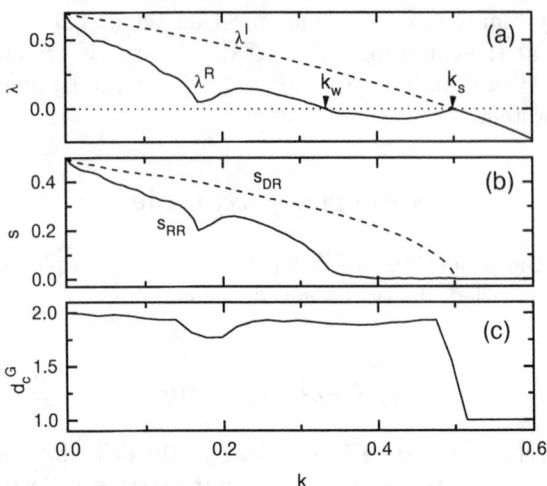

FIGURE 1. (a) Conditional Lyapunov exponent λ^R and transverse Lyapunov exponent λ^I of the invariant identity manifold $y = x$, (b) rms. deviations s_{RR}, s_{DR}, and (c) global correlation dimension d_c^G as functions of coupling strength k.

response systems $y_n \neq x_n$. This corresponds to the WS between the driving x and original response y systems. Here the response system "forgets" its initial conditions, however, the manifold $y = x$ is unstable and the overall dynamics in the $x - y$ plain collapses to another invariant synchronization manifold $M = \{(x, y) : y = \Phi(x)\}$ that has a fractal structure. The IS between y and y' testifies the stability of this manifold. The fractal structure of the manifold is confirmed by calculating the global correlation dimension d_c^G of points placed in $x - y$ plain [Figure 1(c)]. It is between 1 and 2 for $k_w > k < k_s$. The WS shows no evidence in $x - y$ coordinates; the correlation dimension of the strange attractor does not experience any characteristic changes at the threshold of WS. Although the WS can be easily detected with help of an auxiliary response system.

Note, that similar synchronization effects are observed in coupled Henon maps [10] as well as coupled non-identical chaotic flows [7].

Time Series Analysis

Unfortunately, the auxiliary system approach is of limited utility. The method fails for systems whose dynamical equations are not available. If the construction of an auxiliary response system is impossible, the generalized synchronization can be detected by analyzing experimental time series. We have developed an algorithm [10] for estimating the conditional Lyapunov exponents based on two scalar time series one taken from the driving and the other from the response system. This analysis enables one to estimate the domain of GS as well as properties of synchronization manifold without recourse to an auxiliary response system. Moreover, analyzing the two scalar time series one can also estimate the "thickness" of the synchronization manifold which provides a direct test for the smoothness of the map Φ [7]. These algorithms have been tested for various numerical models [10] as well as real experiential systems [11].

ON-OFF INTERMITTENCY

It is well known [12] that on-off intermittency occurs in dynamical systems with certain properties of symmetry. The chaotic attractor of such systems lies on a smooth invariant manifold (usually hyperplain) having lower dimension than the dimension of the full state space. The on-off intermittency is observed just above the blowout bifurcation threshold, when the attractor on the invariant hyperplain becomes a repeller. It appears as a switching of a dynamical variable between two characteristic states. The "off" state is nearly constant and can remain so for very long periods of time. The "on" state is a burst, departing quickly from, and returning quickly to the "off" state.

Here we report an another type of on-off intermittency in which the system may not posses any trivial invariant manifolds [13]. It may occur in any dynamical system consisting of two unidirectionally coupled chaotic subsystems at the threshold of WS.

Let us illustrate this intermittency for the system of coupled logistic maps Eqs. (4). Just below the threshold of WS ($k < k_w$), the conditional Lyapunov exponent becomes positive, $\lambda^R(k) > 0$. It means that the fractal synchronization manifold responsible for WS becomes unstable. Close to the threshold we can expect that the system spends long time in the vicinity of the manifold

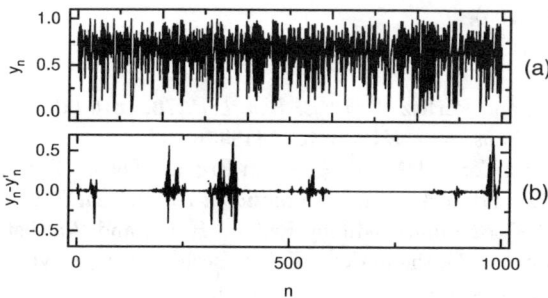

FIGURE 2. Dynamics of (a) the original response system y_n and (b) the difference $y_n - y'_n$ just below the threshold of WS at $k = 0.33$.

and experiences short bursts where the system moves away from this manifold. The expected intermittent behavior is not seen in x_n or y_n dynamics. However, it can be detected with the help of an auxiliary response system. Figure 2 illustrates the dynamics of y_n and the differences $y_n - y'_n$ just below the threshold of WS. The intermittency is not seen in the dynamics of the original response system y_n [Figure 2(a)] and is evident in the signal $y_n - y'_n$ [Figure 2(b)] formed from the difference of the output of the original and auxiliary response systems. Recall that the auxiliary response system y' does not influence the dynamics of the original response y and driving x systems and serves only as an indicator of intermittent behavior in $x - y$ plain which is related to a loss of stability of the fractal synchronization manifold.

To compare this intermittent behavior with the conventional on-off intermittency we calculated the dependence of the mean laminar length τ on the coupling strength k and the distribution of the laminar lengths $P(\tau)$ close to the threshold of WS. We obtained exactly the same power laws $\tau \propto (k_w - k)^{-1}$ and $P(\tau) \propto \tau^{-3/2}$ as those in the conventional on-off intermittency [12].

The identical properties of these two different intermittent processes can be explained as follows. The difference $\delta y_n = y_n - y'_n$ is governed by $\delta y_{n+1} = 4(1-k)(1-2y_n+\delta y_n)\delta y_n$. The properties of intermittent process are determined by small $|\delta y_n| \ll 1$ and we can rewrite this equation as $\delta y_{n+1} = z_n \delta y_n$ where $z_n = 4(1-k)(1-2y_n)$ is a chaotic process determined by Eqs. (4). This is a standard form of a linear map driven with chaotic signal which is considered in the theory of conventional on-off intermittency to derive the above properties [12].

REFERENCES

1. Fujisaka H., and Yamada T., *Prog. Theor. Phys.* **69**, 32 (1983).
2. Pecora L.M., and Carrol T.L., *Phys. Rev. Lett.* **64**, 821 (1990).
3. Afraimovich V.S., Verichev N.N., and Rabinovich M.I., *Radiophys. Quantum Electron.* **29**, 795 (1986).
4. Rulkov N.F., Sushchik M.M., Tsimring L.S., and Abarbanel H.D.I, *Phys. Rev. E* **51**, 980 (1995).
5. Abarbanel H.D.I, Rulkov N.F., and Sushchik M.M, *Phys. Rev. E* **53**, 4528 (1996).
6. Kocarev L., and Parlitz U., *Phys. Rev. Lett.* **76**, 1816 (1996).
7. Pyragas K, *Phys. Rev. E* **54**, R4508 (1996).
8. Sauer T., and Yorke J.A., to appear in Ergodic Theory Dyn. Syst.
9. Kaplan J.L., and Yorke J.A., in *Functional Differential Equations and Approximation of Fixed Points*, edit by Peitgen H.-O., and Walther H.-O., Springer Lecture Notes in Mathematics Vol. 730 Berlin: Springer-Verlag, 1979, p. 204.
10. Pyragas K, *Phys. Rev. E* (1997, in press).
11. Kittel A., Parisi J., and Pyragas K, submitted to *Physica D*.
12. Heagy J.F., Platt N., and Hammel S.M., *Phys. Rev. E* **49**, 1140 (1994).
13. Pyragas K, *Chaos, Solitons and Fractals* (1997, in press).

Extraction of Dynamics from Non-stationary Time Series Data

Liangyue Cao

Department of Mathematics, University of Western Australia, Nedlands, WA 6907, Australia

Abstract. One of the main mechanisms to generate non-stationary data is that the system's environment is always changing with time. It is appropriate to approximate non-stationary time series using the model: $\mathbf{X}_{n+1} = F(\mathbf{X}_n, \mathbf{U}_n)$, where $\mathbf{U}_n$ is the system's environment at the time n. If the $\mathbf{U}_n$ is not observable, we may consider to use the model: $\mathbf{X}_{n+1} = F(\mathbf{X}_n, \hat{\mathbf{U}}_n)$, by somehow learning the function $\hat{\mathbf{U}}_n$ from the available data provided the unknown $\mathbf{U}_n$ is generated from a deterministic system. Several non-stationary time series are tested using the above models. Satisfactory results have been obtained including free-run predictions and bifurcation diagram recovering.

INTRODUCTION

Recently there have been many discussions on predictions of non-stationary time series, e.g., [7,8,10]. To improve the predictions, we may need to know what are the mechanisms to produce the nonstationarity of time series. One of the main mechanisms which is quite obvious and we have understood is that the system's environment is changing with time, or say, there exists external dynamics (driving force) acting on the system. To describe this kind of dynamics, we consider the following system,

$$\begin{aligned} \mathbf{X}_{n+1} &= F(\mathbf{X}_n, \mathbf{U}_n), \\ x_n &= h(\mathbf{X}_n), \end{aligned} \quad (1)$$

where the $\mathbf{U}_n$ is the system's environment or driving force, which is independent of $\mathbf{X}$ but dependent on time n; h is a measurement function.

The system (1) generally produces non-stationary time series $\{x_n\}$. But, when we say "nonstationarity", we may not strictly follow its mathematical definition. More discussions on nonstationarity of time series data could be referred to [6]. In this paper, we are interested in extraction of the dynamics from the time series data generated by (1), where $\mathbf{U}_n$ known and unknown are both considered.

For simplicity we take the $\mathbf{U}_n$ to be scalar and denote it u_n. If u_n can be measured, we fit the following predictive model,

$$x_{n+1} = f(x_n, x_{n-1}, \cdots, x_{n-(d-1)}, u_n, u_{n-1}, \cdots, u_{n-(l-1)}), \tag{2}$$

where d and l are the embedding dimensions, (the time delay τ is assumed to equal 1 in this paper). We will use the *averaged false nearest neighbor* method to find the dimensions d and l [1,5] in our examples, and thereafter we always assume that the d and l have been chosen.

In many practical situations, however, the u_n may not be measured. In this case, we consider the following model instead of (2),

$$x_{n+1} = f(x_n, x_{n-1}, \cdots, x_{n-(d-1)}, \hat{u}_n, \hat{u}_{n-1}, \cdots, \hat{u}_{n-(l-1)}), \tag{3}$$

where $\hat{u}_n$ is *somehow* used to reconstruct the u_n in (2) provided the unknown u_n is generated from a deterministic system. In our early paper [3], we simply set $\hat{u}_n = n$ and $l = 1$ and obtained some satisfactory results. Here $\hat{u}_n$ will be learned during the process of training model.

Using the above two models we test several non-stationary time series including testing free-run predictions and recovering bifurcation diagrams. Next section we briefly introduce our modeling method — wavelet networks. Then we show our numerical simulations.

WAVELET NETWORKS WITH LINEAR TERMS

Our remaining problem is to approximate the function f in (2) and the f and the $\hat{u}$ in (3). We use the method of wavelet networks [2] with linear terms, which has the following approximation form,

$$\hat{f}(\mathbf{y}) = \mathbf{A}\mathbf{y} + b + \sum_{i=1}^{Nb} w_i \psi(\mathbf{D}_i \mathbf{R}_i (\mathbf{y} - \mathbf{c}_i)), \tag{4}$$

where $\mathbf{y} = (y_1, y_2, \cdots, y_m) \in R^m$, $\mathbf{A}\mathbf{y} + b$ is a linear term, $\psi(\mathbf{y}) = \exp(-\mathbf{y}^T \mathbf{y}/2)$, (here we use Gaussian as a base function and one can use wavelet functions to replace it), $\mathbf{c}_i$'s are translation vectors, $\mathbf{D}_i$'s are diagonal matrices built from the dilation vectors, *i.e.*, $\mathbf{D}_i = \text{diag}(d_{1i}, d_{2i}, \cdots, d_{mi})$ and $(d_{1i}, d_{2i}, \cdots, d_{mi})$ is the dilation vector, $\mathbf{R}_i$'s are $m \times m$ matrices, which are used to compensate for the orientation selective nature of the dilations (to make the network more flexible), w_i's are weight coefficients; and Nb is the number of bases used. An algorithm for deriving the fitting parameters of (4) was proposed in [2] and here we only mention that it is a form of back-propagation.

To approximate the f in (2), we just take

$$\mathbf{y}_n = (x_n, x_{n-1}, \cdots, x_{n-(d-1)}, u_n, u_{n-1}, \cdots, u_{n-(l-1)})$$

and derive the parameters in (4) by minimizing the prediction error:

$$\sum_n [x_{n+1} - \mathbf{A}\mathbf{y}_n - b - \sum_{i=1}^{Nb} w_i \psi(\mathbf{D}_i \mathbf{R}_i (\mathbf{y}_n - \mathbf{c}_i))]^2. \quad (5)$$

To approximate the f and the $\hat{u}$ in (3), we first approximate the $\hat{u}$ using polynomials, i.e., let $\hat{u}_n = \sum_{i=1}^{Np} \alpha_i (n/s)^{i-1}$, where s is a scaling factor and we let it equal the length of the time series, and α_i's are the parameters to be determined. Then we take

$$\mathbf{y}_n = (x_n, x_{n-1}, \cdots, x_{n-(d-1)}, \hat{u}_n, \hat{u}_{n-1}, \cdots, \hat{u}_{n-(l-1)})$$

and derive all parameters by minimizing the prediction error (5).

NUMERICAL SIMULATIONS

We tested several artificial time series including time series generated from Hénon map, Ikeda map, and Lorenz equations with driving forces, and some segments of Santa Fe data set B [9]. We have obtained satisfactory results on all of them. Due to the requirement of limit number of pages, in this section we only show our results on the time series from the Ikeda map [4] with driving force. (Note that the Ikeda time series was the most complicated one among the time series we tested; actually the Ikeda attractor is much more complicated than Hénon and Lorenz attractors).

We generate our time series data using the following Ikeda map with driving force,

$$\begin{aligned} x_{n+1} &= 1 + u_n(x_n \cos(\theta) - y_n \sin(\theta)), \\ y_{n+1} &= u_n(x_n \sin(\theta) + y_n \cos(\theta)), \end{aligned} \quad (6)$$

where $\theta = 0.8 - 15/(1 + x_n^2 + y_n^2)$, u_n is the driving force. We test three cases of u_n's varying behavior, which are:

(i) monotonically, i.e., $u_n = 0.0017\,(n-1) - 0.34$;
(ii) torus, i.e., $u_n = 0.296(\sin(n) + 0.3\sin(\pi n + 1) + 0.5\sin(\pi^2 n + 2)) + 0.18$;
(iii) chaotically, i.e., $u_n = 1.04 v_n - 0.34$, $v_n = 4 v_{n-1}(1 - v_{n-1})$;
where $-0.34 \leq u_n \leq 0.7$ for each case.

For each of the three cases, we generate 600 data points and record the x component values, i.e., $x_1, x_2, \cdots, x_{600}$. Shown in Figure 1 are the time series.

Now we assume that the $\{u_n\}$ is unknown. For each case, we use the first 550 data points to fit the predictive model (3) with the embedding dimension $d = 4$ and $l = 1$. We let $Nb = 5 \pm 1$ in the approximation model (4) and $Np = 4 \pm 1$ for the $\hat{u}_n$. Then we test free-run (iterative) predictions on the remaining 50 data points. We also fit a normal predictive model, i.e., without

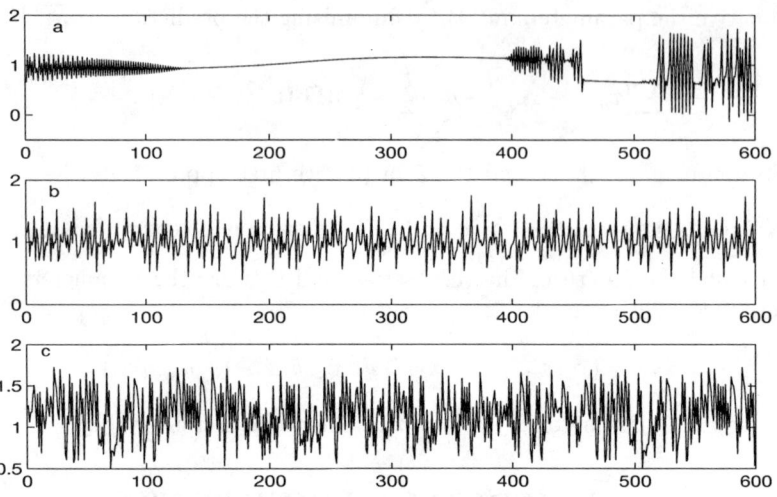

FIGURE 1. Time series of x coordinate of the Ikeda map (6). (a) u_n monotonically increasing. (b) u_n changing in a torus. (c) u_n chaotically changing.

the $\hat{u}_n$ in the model (3), and then test predictions on the same data points. Shown in Figure 2 are the prediction results. One can see that the predictive model (3) performs better than the normal predictive model.

In the following, we turn to test the same time series data as the above, but assume that the $\{u_n\}$ is known; i.e., we have time series data $(x_1, u_1), (x_2, u_2), \cdots, (x_{600}, u_{600})$. We still use the first 550 data points to fit the predictive model (2) with the embedding dimensions $d = 4$ and $l = 1$. We let $Nb = 5 \pm 1$ in the approximation model (4). Then we test free-run predictions on the remaining 50 data points, where only initial state values $x_{547}, x_{548}, x_{549}, x_{550}$ and $u_n, n = 551, 552, \cdots, 600$ are known. Shown in Figure 3 are the prediction results. We can see that the prediction results are much better than those with unknown u_n shown in Figure 2.

We make a few remarks on our results above. 1) when we used higher embedding dimensions the predictions did not improve significantly; 2) our method did not strongly depend on the number of data points which we used to fit the predictive models, provided the driving forces come from deterministic systems with nice smooth properties; 3) in practical tests the predictions with the driving forces changing in a torus are much better than those with the driving forces monotonically increasing; but we need to investigate further if it is true in principle; it may be related to the recurrence of the driving forces.

As a final test, we would like to see how the models we have fitted can capture the dynamics of the underlying system. Once we get the predictive model (2), we can analyze its bifurcation behavior with regarding the u_n as a bifurcation parameter, for example, let $u_n = \beta$ for all n, we can get the

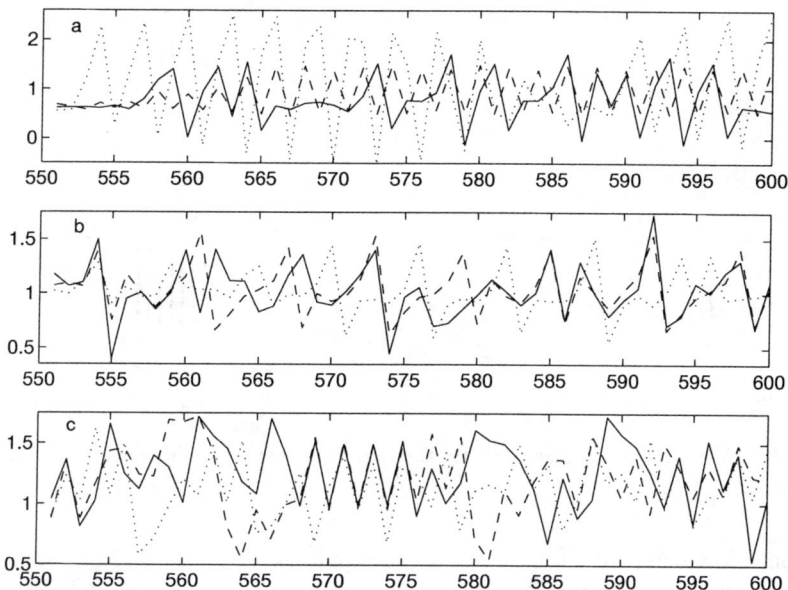

FIGURE 2. Solid line is the actual data, dashed line is the predicted data with the predictive model (3), and dotted line is the predicted data with the normal predictive model. (a) u_n monotonically increasing. (b) u_n changing in a torus. (c) u_n chaotically changing.

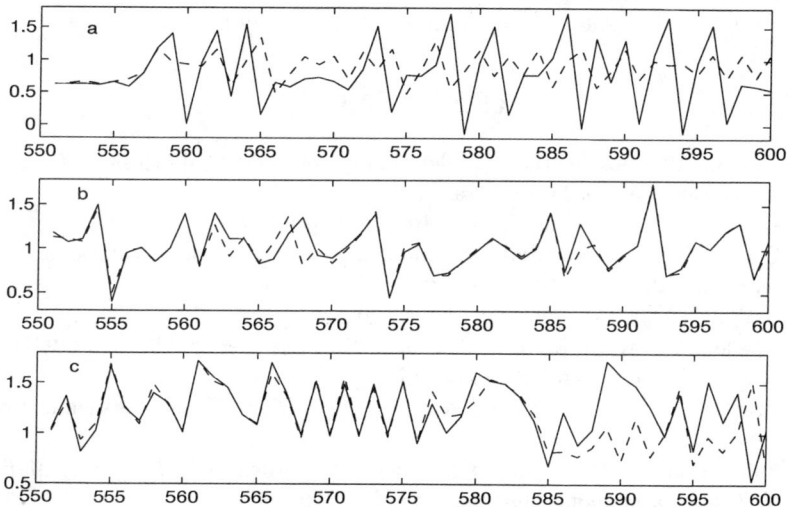

FIGURE 3. Solid line is the actual data, and dashed line is the predicted data with the predictive model (2). (a) u_n monotonically increasing. (b) u_n changing in a torus. (c) u_n chaotically changing.

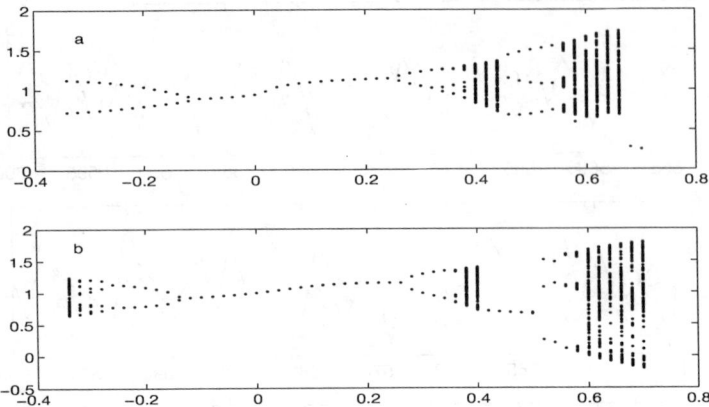

FIGURE 4. Bifurcation diagrams. (a) the recovered bifurcation diagram. (b) the actual bifurcation diagram.

asymptotic behavior of x in (2) at this parameter value β. Now we let β change, we can get the bifurcation diagram of the predictive model. We take the case of parameter changing chaotically as an example. Shown in Figure 4a is the bifurcation diagram generated from the predictive model (2) we fitted, where the parameter β change from -0.34 to 0.7. As a comparison, the actual bifurcation diagram from the Ikeda map (6) is shown in Figure 4b. We can see that the two bifurcation diagrams are almost same. It implies that our models can well capture the dynamics of the underlying system.

REFERENCES

1. Cao, L., Practical method for determining embedding dimension of a scalar time series, *Physica* **D**, in press.
2. Cao, L., Hong, Y., Fang, H., and He, G., *Physica* **D85**, 225 (1995).
3. Cao, L., Mees, A., and Judd, K., Modeling and predicting non-stationary time series, *Int. J. Bif. and Chaos*, in press.
4. Ikeda, K., *Opt. Commun.* **30**, 257 (1979).
5. Kennel, M., Brown, R., and Abarbanel, H., *Phys. Rev. A* **45**, 3403 (1992).
6. Manuca, R., and Savit, R., *Physica* **D99**, 134 (1996).
7. Priestley, M., *Nonlinear and Non-stationary Time Series*, Academic Press, 1988.
8. Subba Rao, T., *Developments in Time Series Analysis in Honour of Maurice B. Priestley*, London: Chapman and Hall, 1993
9. Weigend, A., and Gershenfeld, N.A., *Time Series Prediction: Forecasting the Future and Understanding the Past*, Addison-Wesley, NM, 1994.
10. Weigend, A., Mangeas, M., and Srivasta, A., *Int. J. Neural Systems* **6**, 373 (1995).

The Identification of Time-Delay Systems

M. J. Bünner†, Th. Meyer, A. Kittel, and J. Parisi

Abteilung für Energie- und Halbleiterphysik,
Fachbereich Physik, Universität Oldenburg,
D-26111 Oldenburg, Germany,
† Max-Planck-Institut für Physik komplexer Systeme,
D-01187 Dresden, Germany. [1]

Abstract.
We present a method for the identification of both, scalar and nonscalar time-delay systems. If the dynamics of the system investigated is governed by a time-delay induced instability, the method allows to determine the delay time. In a second step, the time-delay differential equation can be recovered from the time series. The dynamics is not required to be settled on its attractor, which also makes transient motion accessible to the analysis. If the motion actually takes place on a chaotic attractor, the applicability of the method does not depend on the dimensionality of the chaotic attractor. For demonstration, we analyze time series, which are obtained with the help of the numerical integration of a two-dimensional time-delay differential equation.
P.A.C.S.: 05.45.+b

Nonlinear time-delay differential equations have been widely proposed to account for the observed oscillatory, chaotic or hyperchaotic motion of dynamical systems. It has been shown with the help of numerical techniques that scalar time-delay differential equations are able to exhibit high-dimensional chaotic attractors with many positive Lyapunov exponents [1]- [5]. This finding has been supported by experiments [6]- [11] More recently, we introduced a time series analysis method for the identification of scalar time-delay systems [12]- [16]. The method does not put any restriction on the dimensionality of the dynamics analyzed, opening up a door towards the time series analysis of high-dimensional chaotic motion in time-delay systems. Furthermore, the method does not require the motion to be settled on its attractor. In this paper, we present a generalization of the time series analysis method proposed to the

---

[1]) Author for correspondence: Martin J. Bünner, Max Planck-Institut für Physik komplexer Systeme, D-01187 Dresden, Germany, e-mail: buenner@mpipks-dresden.mpg.de

case of nonscalar time-delay systems.

We consider an N-dimensional time-delay differential equation

$$\dot{\vec{y}}_0(t) = \vec{h}(\vec{y}_0(t), \vec{y}_{\tau_0}(t)), \qquad \vec{y}_{\tau_0}(t) = \vec{y}_0(t - \tau_0), \qquad (1)$$

with the initial condition $\vec{y}_0(t) = \vec{y}_i(t)$ for $-\tau_0 \leq t \leq 0$. The state of the system is uniquely defined by N functions on an interval of length τ_0. Therefore, the phase space of system (1) is infinite dimensional. The trajectory in the infinite dimensional phase space can be recovered from the time series $\vec{y}_0(t)$ without loss of information by just taking functions of length τ_0 out of the time series.

It is the specific property of time-delay systems that only a restricted number of coordinates are correlated via the time-evolution equation (1), namely, the $2N$ coordinates $(\vec{y}_0(t), \vec{y}_{\tau_0}(t))$ taken from the phase space and one of its time derivatives $\dot{\vec{y}}_0(t)$. This is interpreted in Ref. [17] as a consequence of the idea that the phase space of time-delay systems can be divided into a linear subsystem and a localized nonlinearity. Therefore, it is not necessary to analyze the dynamics in the infinite dimensional phase space, to verify the existence of an underlying time-delay system. It is sufficient to show the existence of correlations according to Eq. (1). To this end, we analyze the dynamics in a $3N$-dimensional space, which is spanned by the coordinates $(\vec{y}_0, \vec{y}_{\tau_0}, \dot{\vec{y}}_0)$. The dynamics of a time-delay system in the $3N$-dimensional space is restricted to a $2N$-dimensional hypersurface, which is given by the time-evolution equation (1).

Here, we simply test for the existence of such a hypersurface for a given time series. Starting with N scalar time series, $\vec{y}_0(t)$, which have been taken from the system to be investigated, we hypothesize, at first, that the dynamics is governed by an N-dimensional time-delay system

$$\dot{\vec{y}}_0(t) = \vec{h}_r(\vec{y}_0(t), \vec{y}_\tau(t)), \qquad \vec{y}_\tau(t) = \vec{y}_0(t - \tau), \qquad (2)$$

with an unknown function $\vec{h}_r$ and an unknown delay time τ, both of which will be determined in the subsequent analysis, if the ansatz (2) turns out to be successful. Then, we take the values of $(\vec{y}_0, \vec{y}_\tau)$ and $\dot{\vec{y}}_0$ from the time series and analyze their dynamics in the $3N$-dimensional space, which is spanned by the coordinates $(\vec{y}_0, \vec{y}_\tau, \dot{\vec{y}}_0)$. If the coordinates of the trajectory $(\vec{y}_0(t), \vec{y}_\tau(t), \dot{\vec{y}}_0(t))$ are correlated via equation (2), the hypothesis that the system is governed by a time-delay equation with the delay time τ has been verified. If the projected trajectory $(\vec{y}_0, \vec{y}_\tau, \dot{\vec{y}}_0)$ does not fulfill condition (2), the hypothesis has to be rejected. This gives a criterion to determine the delay time from the time series. Additionally, the time-delay differential equation can be constructed by analyzing the functional relationship (2), which gives the function $\vec{h}_r$. Obviously, the only requirement remains that the system dynamics fulfills the time-evolution equation (2), which is true for all kinds of transient motion as well as for the motion on chaotic or hyperchaotic attractors of arbitrary dimension. Therefore, the method permits to analyze high-dimensional chaotic dynamics of time-retarded systems.

The applicability of the method is demonstrated with the help of a computer experiment. We consider the two-dimensional time-delay differential equation, which has been chosen to serve its demonstrational purpose best:

$$\dot{u} = -v + f(u_{\tau_0}), \qquad (3)$$
$$\dot{v} = g_1(u) + g_2(v),$$

with the initial condition: $u(t) = u_i(t), -\tau_0 \le t \le 0$, and $v(t = 0) = v_i$. The functions f and g_1, g_2 are given by:

$$f(u_{\tau_0}) = \frac{a u_{\tau_0}}{1 + u_{\tau_0}^{10}}, \qquad (4)$$

$$g_1(u) + g_2(v) = -\frac{1}{T}(v - u). \qquad (5)$$

Equation (3) has some similarity with the Mackey-Glass system [18]. The dependence of $\dot{u}$ on the time-delayed value u_{τ_0} is the same as it is the case in the Mackey-Glass system. But while the dependence of $\dot{u}$ on u induces exponential relaxations in the Mackey-Glass system, in (3) it is similar to a damped oscillator.

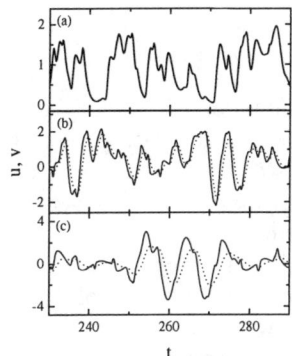

FIGURE 1. Time series of the system (3) obtained with the help of numerical integration $(a = 3.00, \tau_0 = 20.00)$ for different values of T: (a) $T = 0.10$; (b) $T = 0.60$; (c) $T = 1.90$.

We present three time series of u and v for different T in Fig. 1. In Fig. 1(a), it is clearly seen how the variable v, for $T = 0.10$, follows the variable u and the dynamics of u resembles that of a Mackey-Glass system. The values of (u, v) are positive for all times. We mention at this point that system (3) is invariant under the transformation $(u, v) \to (-u, -v)$. Therefore, there exists another attractor with negative values of (u, v). In Fig. 1(b) and Fig. 1(c), we

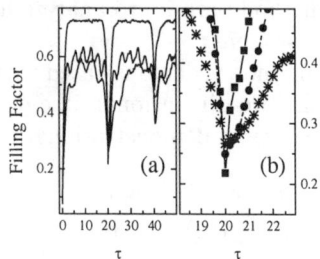

FIGURE 2. (a) Filling factor of the two time series $(u(t), v(t))$ of the system (3) with $a = 3.00, \tau_0 = 20.00$ for different values of T ($T = 0.10, 0.60, 1.90$; the values of T are indicated in the figure). We used $100,000$ data points for the filling factor analysis, which were taken out of a time series of $1,000,000$ data points. In (b) a blow-up of the filling factor in the vicinity of the delay time $\tau = \tau_0$ is shown ($squares : T = 0.10; circles : T = 0.60; stars : T = 1.90$).

observe that the two coexisting attractors are merged. Variable v no longer follows variable u, but it develops an independent dynamics. In these cases, the system reveals its nonscalar nature.

For the identification, we chose a restrictive ansatz for demonstrational purposes,

$$\dot{u} = -v + f_r(u_\tau), \tag{6}$$
$$\dot{v} = g_{r,1}(u) + g_{r,2}(v). \tag{7}$$

The delay time τ and the functions f_r and $g_{r,1}, g_{r,2}$ are yet unknown and will be determined in the following. At first, we perform a filling factor analysis [19]. The two time series $(u(t), v(t))$ are projected to a five-dimensional space which is spanned by the coordinates $(u, u_\tau, \dot{u}, v, \dot{v})$. The five-dimensional space is covered with equally sized hypercubes and the number of hypercubes which has been visited by the trajectory is counted under variation of τ. The results are presented in Fig. 2 for different values of T. The minimum in the filling factor for $\tau = \tau_0 = 20.00$ is well detected for all values of T.

Then, the existence of the functional relationship (2) has to be shown, in order to verify the underlying time-delay induced instability. The special form of ansatz (6)-(7) together with the time-evolution equations (3) allows for a convenient way of proving the existence of the function (2). We emphasize, though, that, in general, it is expected to be more troublesome. We apply an intersection with the help of the condition $v(t^i) = 0$. If the nonscalar ansatz (6)-(7) is successful, the values $\dot{u}^i = \dot{u}(t^i)$ and $u^i_{\tau_0} = u_{\tau_0}(t^i)$ have to be correlated via $\dot{u}^i = f_r(u^i_{\tau_0})$. Plotting $\dot{u}^i$ versus $u^i_{\tau_0}$ as is shown in Fig. 3(a), the existence of the smooth function f_r is verified. In the next step, the functional

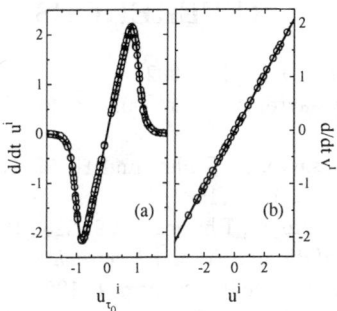

FIGURE 3. Recovery of the time-delay differential equation from the time series:(a) Comparison of the data points $(\dot{u}^i, u^i_{\tau_0})$, which are shown as open circles, with the function f (line) in eq. (4). (b) Comparison of the data points $(\dot{v}^i, u^i)$, which are shown as open circles, with the function g_1 (line) in eq. (5).

relationship between $\dot{v}^i$ and u^i is investigated. We use the same intersection condition $v(t^i) = 0$ as above. According to the nonscalar ansatz (6)-(7), the coordinates of the intersection points are correlated via $\dot{v}^i = g_{r,1}(u^i)$ (see Fig. 3(b)). Plotting $\dot{v}^i$ versus u^i yields the function $g_{r,1}(u^i)$. The recovery of the function $g_{r,2}(v^i)$ is to be done in the same spirit and is not shown here. From a practical point of view it is interesting whether the identification of a nonscalar time-delay system can be accomplished with only a scalar variable available for the analysis. This is discussed elsewhere [19].

In conclusion, we have presented a generalization of a recently proposed method for the identification of scalar time-delay systems by analyzing the time series. The method is generalized in the way that it can be applied to nonscalar time-delay systems. We have shown that an N-dimensional time-delay system can be identified with the help of N time series.

We emphasize that we only require the motion to obey the time-delay differential equation. The motion is not required to be located in certain parts of phase space. If the dynamical system possesses coexisting attractors, the method can be applied to motions on every coexisting attractor. Moreover, the method is also applicable to transient motions. If the motion is on a chaotic or hyperchaotic attractor, the applicability of the analysis method does neither depend on the dimensionality nor on the number of positive Lyapunov exponents of the chaotic or hyperchaotic attractor. Therefore, we find that the present method might open up a door towards the time series analysis of high-dimensional chaotic motion in time-delay systems.

We thankfully acknowledge valuable discussions with J. Poinke, O. E. Rössler, the ENGADYN group, A. Politi and H. Kantz. Financial support of the Deutsche Forschungsgemeinschaft is acknowledged.

REFERENCES

1. J. D. Farmer, Physica D **4**, 366 (1982).
2. M. Le Berre, E. Ressayre, A. Tallet, and H. M. Gibbs, Phys. Rev. Lett. **56**, 274 (1986).
3. M. Le Berre, E. Ressayre, A. Tallet, and H. M. Gibbs, D. L. Kaplan, M. H. Rose, Phys. Rev. A **35**, 4020 (1987).
4. K. Ikeda and K Matsumoto, Physica D **29**, 223 (1987).
5. M. Schanz, *Zur Analytik und Numerik zeitlich verzögerter synergetischer Systeme*, Ph. D. Thesis, Universität Stuttgart, 1997.
6. H. M. Gibbs, F. A. Hopf, D. L. Kaplan, and R. L. Shoemaker, Phys. Rev. Lett. **46**, 474 (1981).
7. R. T. Arecchi, G. Giacomelli, A. Lapucci, and R. Meucci, Phys. Rev. A **43**, 4997 (1991).
8. I. Fischer, O. Hess, W. Elsäßer, and E. O. Göbel, Phys. Rev. Lett. **73**, 2188 (1994).
9. G. Giacomelli, R. Meucci, A. Politi, F. T. Arecchi, Phys. Rev. Lett. **73** (1994) 1099.
10. A. Namajunas, K. Pyragas, A. Tamasevicius, Phys. Lett. A **201**, 42 (1995).
11. G. Giacomelli, A. Politi, Phys. Rev. Lett. **76** (1996) 2686.
12. M. J. Bünner, M. Popp, Th. Meyer, A. Kittel, U. Rau, and J. Parisi, Phys. Lett. A **211**, 345 (1996).
13. M. J. Bünner, Th. Meyer, A. Kittel, J. Parisi, in *Nonlinear Physics of Complex Systems*, J. Parisi, S. C. Müller, W. Zimmermann (Eds.), Springer-Verlag, Berlin, 1996, p. 229.
14. M. J. Bünner, M. Popp, Th. Meyer, A. Kittel, and J. Parisi, Phys. Rev. E **54** (1996) R3082.
15. M. J. Bünner, *Die Identifikation von zeitretardierten Systemen mittels Zeitreihen-Analyse*, Ph. D. thesis, Carl von Ossietzky Universität Oldenburg, 1997.
16. H. Voss, J. Kurths, submitted to Phys. Lett. A.
17. M. J. Bünner, Th. Meyer, A. Kittel, J. Parisi, submitted to Z. für Naturforsch. a.
18. M. C. Mackey and L. Glass, Science **197**, 287 (1977).
19. M. J. Bünner, Th. Meyer, A. Kittel, J. Parisi, submitted to Phys. Rev. E.

Synchronizing Hyperchaotic Circuits

Arūnas Tamaševičius*, Antanas Čenys*, Audrius Namajūnas*,
Gytis Mykolaitis* and Erik Lindberg[†]

*Semiconductor Physics Institute, A.Goštauto 11, Vilnius LT 2600, Lithuania
[†] Technical University of Denmark, Lyngby DK 2800, Denmark

Abstract. With regard to possible applications to secure communications, the applicability of synchronizing hyperchaotic circuits with a single dynamical variable is discussed. Several specific examples are reviewed, including the fourth-order circuits with two positive Lyapunov exponents as well as the oscillator with a delay line characterized by multiple positive Lyapunov exponents.

INTRODUCTION

Synchronization of chaotic systems has attracted in this decade much attention (see [1-5] and references therein). The most comprehensive bibliography on this topic has been compiled by Guanrong Chen [6]. The last updated version of the bibliography includes more than 750 entries.

Chaotic synchronization is believed to have interesting applications in secure communications [7-9]. However, nearly all investigations in this field deal with low-dimensional chaotic systems characterized by only one positive Lyapunov exponent. Meanwhile, masking signals with comparatively simple chaos does not ensure sufficient security. In some cases extracting of the messages can be performed by means of common signal processing techniques [10].

For enhanced security hyperchaotic systems with two or more positive Lyapunov exponents seem to be promising. It was, however, believed that hyperchaos could not be synchronized by a single variable and one needed to transmit as many variables as there were positive Lyapunov exponents. If this was the case, the very use of hyperchaos would seem rather doubtful since most communication systems operate with just one signal.

Fortunately, several advanced methods have been recently described suggesting either scalar transmitted signal techniques [11,12] or even simpler strategy [13] employing a single dynamical variable to synchronize hyperchaotic systems.

The objective of this paper is to demonstrate that a large number of hyperchaotic circuits with more than one positive Lyapunov exponents described in literature so far can be synchronized with only one variable.

SINGLE DYNAMICAL VARIABLE STRATEGY

General remarks

The advantage of the scalar transmitted signal technique [12] is evident. From a practical point of view, however, this approach leads to some inconvenience since it requires direct access to all or at least several dynamical variables in the transmitter as well as in the receiver. In addition, large sets of parameters should be adjusted empirically for a specific hyperchaotic system.

Let us recall here that considering synchronization of hyperchaotic systems we regard possible applications to secure communications. In communications one deals, as a rule, with artificial systems including specially designed electronic circuits. In other words, we are not restricted to either existing natural oscillators or to mathematical models, such as the hyperchaotic Rössler equations considered in [12]. Instead, we can make free use of any hyperchaotic circuit that admits simple one-variable synchronization technique [13].

Given a hyperchaotic oscillator as a transmitter

$$d\vec{v}^*/dt = \vec{F}(\vec{v}^*), \tag{1}$$

where $\vec{v}^* = (v_1^*, v_2^*, ...v_s^*, ...v_m^*)$ is an m-dimensional ($m \geq 4$) state vector and v_s^* is a selected accessible variable, in some cases the receiver can be synchronized by adding the one-variable signal (shown in braces $\{...\}$) to the corresponding input:

$$d\vec{v}'/dt = \vec{F}'(\vec{v}),$$
$$dv_s/dt = F_s(\vec{v}) - \{k(v_s - v_s^*)\}. \tag{2}$$

Here $\vec{v}' = (v_1, v_2, ...v_{m-1})$, so that $\vec{v} = (\vec{v}', v_s)$ is a corresponding m-dimensional state vector of the receiver, and k is a control parameter. The asterisks "*" denote the variables attributed to the transmitter.

Specific examples

In this subsection four specific examples are presented to illustrate the performance of the method. The examples include three fourth-order hyperchaotic circuits with two positive Lyapunov exponents also a delay line circuit with multiple positive Lyapunov exponents. The following notations are used:

$x = U_{C1}/U_0$, $y = \rho I_{L1}/U_0$, $z = U_{C2}/U_0$, $w = \rho I_{L2}/U_0$, $t = t/\tau$,

$\rho = \sqrt{L_1/C_1}$, $\tau = \sqrt{L_1 C_1}$, $a = \rho/R$, $b = R/\rho$, $\varepsilon = C_2/C_1$, $\mu = L_2/L_1$.

Hyperchaotic circuit with a combined parallel-series resonance loop

The circuit shown in Figure 1 has been first proposed in [14], later investigated numerically [15] and studied experimentally [16]. General purpose diode can be used for the nonlinear element G.

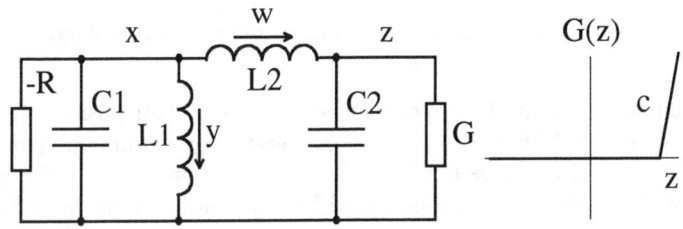

FIGURE 1. Hyperchaotic circuit and I/V characteristic of the nonlinear element.

The synchronization properties of the circuit has been examined in [17] both numerically and experimentally. The dynamics of the receiver is described by the set of equations

$$dx/dt = ax - y - w - \{k(x - x^*)\},$$
$$dy/dt = x,$$
$$\varepsilon dz/dt = w - c(z - 1)H(z - 1), \qquad (3)$$
$$\mu dw/dt = x - z.$$

Here $H(u)$ is the Heaviside function; $H(u < 0)=0$, $H(u \geq 0)=1$. The circuit has been confirmed to be a hyperchaotic one with two positive Lyapunov exponents, $\lambda_1 = 0.44$, $\lambda_2 = 0.15$ at $a = 0.6$, $\varepsilon \approx 0.3$, $\mu = 0.2$, $c \to \infty$ [14]. Hyperchaos has been found to exist over wide range of the parameter a, e.g. $0.25 \leq a \leq 0.8$ with the other parameters fixed at $c = 10$, $\varepsilon \approx \mu \approx 0.3$ [15,16].

The receiver can be synchronized to the transmitter, $x = x^*$ at $k \gtrsim 0.3$ [17]. It is interesting to note that robust synchronization can be achieved via other variables, y or z as well.

Related experiments with a passive nonlinear resonator used for the receiver and containing no negative resistor -R have been described in [18].

Hyperchaotic circuit with two parallel resonance loops

The circuit contains two parallel LC loops (see Figure 2) coupled by means of a diode [19]. In contrast to the previous circuit this oscillator includes a two-variable nonlinearity $G(x, z) = c(x - z - 1)H(x - z - 1)$.

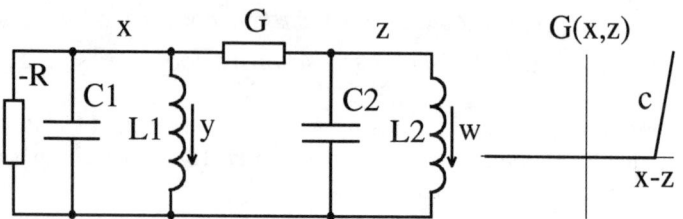

FIGURE 2. Hyperchaotic circuit and I/V characteristic of the nonlinear element.

The oscillator exhibits hyperchaotic behavior with two positive Lyapunov exponents in the certain ranges of the parameter a for a common values of the others, $c = 40$, and $\varepsilon \approx \mu \approx 0.3$ [19], e.g. $\lambda_1 = 0.12$ and $\lambda_2 = 0.07$. Synchronization of the circuit has been observed both numerically in the equations

$$\begin{aligned}
dx/dt &= ax - y - c(x - z - 1)H(x - z - 1) - \{k(x - x^*)\}, \\
dy/dt &= x, \\
\varepsilon dz/dt &= -w + c(x - z - 1)H(x - z - 1), \\
\mu dw/dt &= z.
\end{aligned} \qquad (4)$$

via x and experimentally by adding an appropriate control current to the capacitor C1. However, in contrast to the first example it seems that synchronization can not be achieved via other variables.

Matsumoto-Chua-Kobayashi hyperchaotic circuit

The first experimental observation of hyperchaotic oscillations has been described by Matsumoto, Chua and Kobayashi [20] in a circuit shown in Figure 3. In addition to the linear negative resistor -R the circuit contains the second, N-type negative resistor implemented by means of a negative impedance converter shunted with two diodes [20].

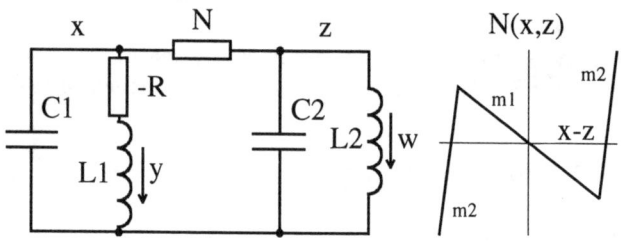

FIGURE 3. Hyperchaotic circuit and I/V characteristic of the nonlinear element.

So, in contrast to the two above circuits this one contains two active elements. The equations for the oscillator can be given in the form

$$dx/dt = -y - N(x,z)$$
$$dy/dt = x + by - \{k(y - y^*)\},$$
$$\varepsilon dz/dt = -w + N(x,z), \quad (5)$$
$$\mu dw/dt = z,$$

where $N(x,z)$ is a two-variable three-segment function, thus more complicated one than $G(z)$ or $G(x,z)$ in the above examples. The authors of the paper [20] found two positive Lyapunov exponents, $\lambda_1 = 0.24$ and $\lambda_2 = 0.06$.

The synchronization phenomenon of this circuit has been studied numerically in [13]. Though the circuit is rather complicated and involves two active elements it can be still synchronized via one particular variable, namely y. The largest transversal Lyapunov exponent, calculated for $x = x^*$, $y = y^*$, $z = z^*$, $w = w^*$ becomes negative at $k \gtrsim 0.6$ [13] indicating synchronism between the transmitter and the receiver. Synchronization, however, can not be achieved via other variable, x, z or w.

Hyperchaotic circuit with delay line

The circuit presented in Figure 4 is an electronic analog [21] of the physiological Mackey-Glass system [22].

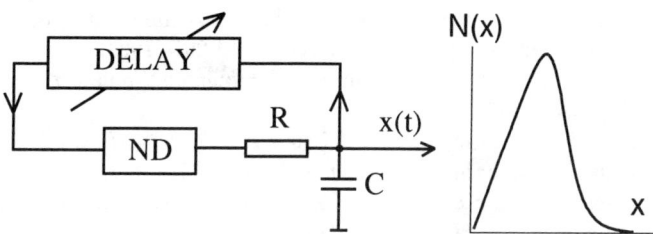

FIGURE 4. Hyperchaotic circuit and I/V characteristic of the nonlinear element.

The dynamics is given by the following delay differential equation

$$dx/dt = -x + N(x,\tau) - \{k(x - x^*)\}, \quad N(x,\tau) = \frac{2x(t-\tau)}{1 + x^{10}(t-\tau)}. \quad (6)$$

Here $x = U_C/U_0$, where U_0 is the stationary output value derived from $N(U_0) = U_0$, $t = t/RC$, and $\tau = T_{del}/RC$ is the dimensionless time and the dimensionless delay, respectively. Though the circuit is described by one-variable equation, it can be considered as an infinite-dimensional one [22] because of the time delay τ. The number of positive Lyapunov exponents increases with the parameter τ. For example, at $\tau=6$ there are three positive Lyapunov exponents, $\lambda_1 = 0.05$, $\lambda_2 = 0.033$, and $\lambda_3 = 0.01$ [23]. Nevertheless, the receiver can be synchronized via single variable, as given in Eq. (6) provided $k > 0.65$. The number of positive Lyapunov exponents can be increased further by means of the delay τ [22], meanwhile synchronization is still easily achieved with only one variable x as shown both numerically and experimentally in [23].

REFERENCES

1. Pecora L.M., and Carroll T.L., *Phys. Rev. Lett.* **64**, 821 (1990).
2. Pecora L.M., and Carroll T.L., *Phys. Rev. A* **44**, 2374 (1991).
3. Carroll T.L., and Pecora L.M., *Int. J. Bifurcation and Chaos* **2** 659 (1992).
4. Lai Y.-C., and Grebogi C., *Phys. Rev. E* **50** 1894 (1994).
5. Kocarev L., and Parlitz U., *Phys. Rev. Lett.* **76** 1816 (1996).
6. Chen G., *Control and synchronization of chaotic systems (a bibliography)*, Houston: ECE Dept., University of Houston, TX, 1996, available from ftp: "ftp.egr.uh.edu/pub/TeX/chaos.tex".
7. Kocarev Lj., Halle K., Eckert K., Chua L.O., and Parlitz U., *Int. J. Bifurcation and Chaos* **2** 709 (1992).
8. Cuomo K.M., and Oppenheim, *Phys. Rev. Lett.* **71** 65 (1993).
9. Li H.-J., and Chern J.-L., *Phys. Lett. A* **206** 217 (1995).
10. Perez G., and Cerdeira H.A., *Phys. Rev. Lett.* **74** 1970 (1995).
11. Kocarev L., and Parlitz U., *Phys. Rev. Lett.* **74** 5028 (1995).
12. Peng J.-H., Ding E.J., Ding M., and Yang W., *Phys. Rev. Lett.* **76** 904 (1996).
13. Tamaševičius A., and Čenys A., *Phys. Rev. E* **55** 297 (1997).
14. Nishio Y., Mori S., and Saito T., In: *Int. Seminar on Nonlinear Circuits and Systems*, Moscow, 1992, pp. 60-69.
15. Tamaševičius A., Namajūnas A., and Čenys A., *Electron. Lett.* **32** 957 (1996).
16. Tamaševičius A., Namajūnas A., Čenys A., and Mykolaitis G., In: *4th Int. Workshop on Nonlinear Dynamics of Electronic Systems*, Seville: Centro Nacional de Microelectronica, 1996, pp. 363-368.
17. Tamaševičius A., Mykolaitis G., Čenys A., and Namajūnas A., *Electron. Lett.* **32** 1536 (1996).
18. Tamaševičius A.,Čenys A., and Mykolaitis G., *Electron. Lett.* **32** 2029 (1996).
19. Tamaševičius A.,Čenys A., Mykolaitis G., Namajūnas A., and Lindberg E., *Electron. Lett.* **33** 542 (1997).
20. Matsumoto T., Chua L.O., and Kobayashi K., *IEEE Trans. Circ. and Systems I* **33** 1143 (1986).
21. Namajūnas A., Pyragas K., and Tamaševičius A., *Phys. Lett. A* **201** 42 (1995).
22. Farmer D., *Physica D* **4** 366 (1982).
23. Namajūnas A., Tamaševičius A., and Čenys A., *Electronics and Electrical Engineering*, Kaunas, in press (1997).

Solving inverse problems of identification type by optimal control methods

Suzanne Lenhart*, Vladimir Protopopescu[†] and Jiongmin Yong[‡]

* Mathematics Department, University of Tennessee, Knoxville, TN, 37996-1300
[†] Oak Ridge National Laboratory, Oak Ridge, TN, 37831-6364
[‡] Mathematics Department, Fudan University, Shanghai, 200433, China

Abstract. Inverse problems of identification type for nonlinear equations are considered within the framework of optimal control theory. The rigorous solution of any particular problem depends on the functional setting, type of equation, and unknown quantity (or quantities) to be determined. Here we present only the general articulations of the formalism. Compared to classical regularization methods (e.g. Tikhonov coupled with optimization schemes), our approach presents several advantages, namely: (i) a systematic procedure to solve inverse problems of identification type; (ii) an explicit expression for the approximations of the solution; and (iii) a convenient numerical solution of these approximations.

DIRECT PROBLEM

We consider a system described by the state function, u, which satisfies the abstract operator state equation with data F:

$$A(u(\xi); F(\xi)) = 0. \qquad (1)$$

The state function and the data may be scalar or vector and the operator A is, in general, nonlinear. The independent variables of the problem, ξ, which in the evolution problem contain the time, t, take values in the domain $\Omega \subset \mathbb{R}^n$, with sufficiently regular boundary, $\partial\Omega$.

In order to fix ideas, we assume that the operator A is a second order partial differential operator, but the present formalism is much more general, including equations of higher order, ordinary differential equations, hybrid systems, etc. The formulation applies as well to elliptic systems (u being independent of time) and hyperbolic (one redefines the state functions and rewrite the equation as a first order evolution system).

If the system (1) accurately describes a realistic physical situation, the direct problem is well-posed, i.e. the data F (parameters, functions, coefficients, sources, initial and boundary values, etc.) determine uniquely a regular solution u [1–3].

INVERSE PROBLEM

In contrast, the inverse problem which consists of determining a part of F from the solution u, is - in general - ill-posed [4]. In fact, in almost all the practical situations, the solution is never completely known and one must deal with only partial or imprecise knowledge of it [5].

Formally, the inverse problem corresponding to the direct problem (1) is given below:

"Given partial (and perhaps noisy) observations, $B\tilde{u}$, of the true solution $\tilde{u}$, in a subdomain Ω' of the phase space, $\Omega' \subset \Omega$, and a (2) known part, f_1, of the data F, we seek to determine the unknown part of the data, f."

This inverse problem corresponds to parameter identification in a model with known structure involving some unknown data [6]. The standard method of solving these identification problems is based on Tikhonov's regularization. In this approach, one constructs - starting from actual observations, $B\tilde{u}$, - a cost function:

$$J_\beta(f) = \frac{1}{2}\|Bu - B\tilde{u}\|^2 + \frac{\beta}{2}\|f\|^2, \quad \beta > 0 \qquad (3)$$

where u is the solution of (1) for the data $F = (f_1, f)$. The exact form of the cost functional and the types of norms involved depend on the concrete problem that has to be solved. The quadratic form used here makes the illustration clearer and simpler. Tikhonov's approach and its variants seek to minimize the functional $J_\beta(f)$ - for a fixed β - over the set of unknown data. It assures - in principle - that the observation of the model solution obtained with f, Bu, will approach the actual observation, $B\tilde{u}$. The minimization is carried out by using an optimization program. The functions f_β which achieve the absolute minimization of $J_\beta(f)$ represent the approximate solution of the inverse problem (2). The disadvantages of this approach are:
- when β is very small, the problem is unstable;
- when β is very large, the solution is not accurate;
- there is no systematic procedure for finding the absolute minimum;
- there are no systematic means to control the approximations;
- the optimization program is implemented by "trial and error" which makes the numerical calculations difficult and inefficient.

SOLUTION OF THE INVERSE PROBLEM BY OPTIMAL CONTROL METHODS

To eliminate most of the disadvantages above, we propose a new approach to the inverse problem of identification, based on optimal control for operator equations as developed by J.-L. Lions [7–9].

Our idea is to consider a family of functionals (3) for $\beta \geq 0$. For each β strictly positive one considers the unknownn data, f, as a control which belongs to a certain bounded set, $\mathcal{F}$; the control has to be adjusted - always remaining in the set - in order to minimize the functional $J_\beta(f)$. The minimum of the cost functional over f is attained at the optimal control, $f = f_\beta$:

$$J_\beta(f_\beta) = \inf_{f \in \mathcal{F}} J_\beta(f). \tag{4}$$

Letting the sequence of β tend toward zero, one can verify that the sequence f_β converge in an appropriate sense to an element of the control set, $f^* \in \mathcal{F}$ (see Section 5). This element represents the solution - perhaps non-unique - of the inverse problem (2).

The idea to use the techniques of optimal control to approximate in a systematic manner the solution of an inverse identification problem is new. Indeed, as mentioned above, the majority of the traditional approaches [4–6] couple Tikhonov's regularization with an optimization algorithm. Recently, Puel et Yamamoto [10,11] have obtained uniqueness and stability results and reconstruction algorithms for the identification of sources in a wave equation. Their work is based on the results of exact controllability and uniqueness of the Hilbert uniqueness method (HUM) obtained by J.-L. Lions [12–14]. These results are limited to linear problems and moreover the identication is realized for a special type of observations. Finally, the other approaches which use the control theory - not necessarily optimal - for solving inverse problesm are limited to a particular cases where: (i) the equation is specified [15]; (ii) the control problem is the linear quadratic type and can be treated by Riccati equations [16]; or (iii) there is a bilinear control of a concrete type [17,18].

SOLUTION OF OPTIMAL CONTROL PROBLEM

To prove that there exists an optimal control f_β which minimizes (3) one first establishes the needed existence, uniqueness, and a priori estimates for the solution u of the direct problem (1) and uses a minimizing sequences argument. To characterize the optimal control, one procedes as follows:

- One shows that the function $f \to u(f)$ is differentiable. In general, there exists a directional derivative (evaluated at the optimal control, f_β):

$$\left.\frac{\partial u}{\partial f}\right|_{f_\beta,\text{ direction }\ell} = \psi \qquad (5)$$

which is the solution of the linear problem:

$$L(\psi; f_1, f_\beta, \ell, u_\beta) = 0 \qquad (6)$$

corresponding to problem (1). The operator L is linear in ψ and depends nonlinearly on f_β and on $u_\beta = u(f_\beta)$.

- By differentiating $J_\beta(f)$ and using the estimates on u and ψ, one obtains the characterization of the optimal control $f_\beta \in \mathcal{F}$ that minimizes J_β over $\mathcal{F}$.

- In deriving this characterization, one constructs the adjoint system corresponding to (1), for the adjoint function p:

$$L^*(p; f_1, f_\beta, u_\beta) = 0 \qquad (7)$$

and the resulting characterization for f_β in terms of state and adjoint variables:

$$f_\beta = G(u_\beta, p). \qquad (8)$$

- Replacing f_β in (1) and (7) by the characterization (8), one obtains the optimality system (OS):

$$\begin{aligned} A(u_\beta; f_1, G(u_\beta, p)) &= 0 \\ L^*(p; f_1, G(u_\beta, p), u_\beta) &= 0. \end{aligned} \qquad (9)$$

- The uniqueness of the solution of the OS guarantees the uniqueness of the optimal control, but - in general - the OS admits more than one solution, and then one has to analyze the uniqueness of f_β by other methods.

- Finally, the solution (the solutions) of the OS are calculated numerically by applying an iterative algorithm for two point boundary value problem (TPBVP).

PASSAGE TO THE LIMIT AND SOLUTION OF THE INVERSE PROBLEM

The cost functional (3) is an approximation in the sense that the second term is artificial - as always in Tikhonov regularization. By letting β tend toward zero one expects to obtain a control f_0 which minimizes only the difference between the actual observation and the solution of (1) with f replaced by f_0.

We assume that the inverse problem (2) has a solution, i.e. there exists $f^* \in \mathcal{F}$ such that $u^* = u(f^*)$ satisfies $Bu^* = B\tilde{u}$ a.e. in Ω'. Then we can prove that there exists f_0 (not necessarily equal to f^*) such that on a (sub)sequence $\beta \to 0$, we have

$$f_\beta \to f_0$$
$$u_\beta = u(f_\beta) \to u_0 \tag{10}$$

and
$$B u_0 = B \tilde{u} \quad \text{a.e. in } \Omega'.$$

The generic idea of justifying the result is the following:
With $f^* \in \mathcal{F}$, one has $J_\beta(f_\beta) \leq J_\beta(f^*)$, i.e.,

$$\frac{1}{2}\|Bu_\beta - B\tilde{u}\|^2 + \frac{\beta}{2}\|f_\beta\|^2 \leq \frac{\beta}{2}\|f^*\|^2. \tag{11}$$

By utilizing the a priori estimates independent of β obtained on u_β one has on a subsequence of $\beta \to 0$, that f_β goes toward f_0 and u_β goes toward u_0 (with convergence of appropriate derivatives); moreover, $u_0 = u(f_0)$. Letting β tend to 0 in (11), one also obtains $Bu_0 = B\tilde{u}$. Thus, in the observation domain the solution u_0 is close (ideally identical) to the actual measured values and so the control f_0 is identified.

Remarks. The condition required to ensure (10) is equivalent to the observability condition used in HUM. This result is formal in the sense that the topologies of the limits are not specified.

TWO EXAMPLES

We applied the approach described above to solve two concrete identification problems motivated by underwater, geophysical, and seismic explorations.

1. *Identification of the reflexion coefficient on a part of the boundary.* [17] For $\Gamma \subset \mathbb{R}^2$, $T > 0$, and $w \in C(\Gamma)$, $w < 0$, we consider the spatio-temporal domain $\Omega = D \times (0, T)$, where $D = \{(x, y, z) | (x, y) \in \Gamma,\ w(x, y) < z < 0\}$. We divide the boundary of the domain D in three disjoint parts:

$$(\partial D)_1 = \{(x, y, z) | (x, y) \in \partial\Gamma,\ w(x, y) < z < 0\}$$
$$(\partial D)_2 = \Gamma \times \{z = 0\}$$
$$(\partial D)_3 = \{(x, y, w(x, y)) | (x, y) \in \Gamma\}$$

and consider the acoustic wave equation:

$$\begin{aligned}
u_{tt} - \nabla(E\nabla u) &= f_1, & &\text{in } \Omega \\
u &= 0 & &\text{on } (\partial D)_1 \times (0,T) \\
\frac{\partial u}{\partial \nu} &= 0 & &\text{on } (\partial D)_2 \times (0,T) \\
\frac{\partial u}{\partial \nu} + fu &= 0 & &\text{on } (\partial D)_3 \times (0,T) \\
u &= g_1,\ u_t = g_2 & &\text{in } D \times \{0\}.
\end{aligned} \qquad (12)$$

From observations $\tilde{u}$ of the solution of the system (12) effected in a subdomain $\Omega' \subset \Omega$ we try to identify the reflection coefficient f in a boundary condition of Robin type. As before, we consider that the control set is bounded, $0 \leq f \leq M$, and we look for a solution in $L^2(0,T;H^1(\Omega))$; the approximation f_β minimizes

$$J_\beta(f) = \frac{1}{2}\int_{\Omega'}(u-\tilde{u})^2 + \frac{\beta}{2}\int_\Gamma f^2.$$

We obtain the explicit characterization

$$f_\beta(x,y) = \min\left\{\left(\frac{1}{\beta}\int_0^T u_\beta p(x,y,w(x,y),t)dt\right)^+, M\right\} \qquad (13)$$

where the superscript $+$ denotes the positive part of the function and u_β is the solution of (12) for $f = f_\beta$ and p is the solution of the adjoint system:

$$\begin{aligned}
p_{tt} - \nabla(E\nabla p) &= (u_\beta - \tilde{u})\chi_{\Omega'} & &\text{in } \Omega \\
p &= 0 & &\text{on } (\partial D)_1 \times (0,T) \\
\frac{\partial p}{\partial \nu} &= 0 & &\text{on } (\partial D)_2 \times (0,T) \\
\frac{\partial p}{\partial \nu} + fu &= 0 & &\text{on } (\partial D)_3 \times (0,T) \\
p &= p_t = 0 & &\text{in } D \times \{T\}.
\end{aligned} \qquad (14)$$

2. *Identification of the potential for the wave equation* [18]. We describe the propagation of the acoustic wave equations in a spatial domain $D \subset \mathbb{R}^3$ by the system

$$\begin{aligned}
u_{tt} &= \Delta u + fu + f_1 & &\text{in } \Omega = D \times (0,T) \\
u &= 0 & &\text{on } \partial D \times (0,T) \\
u &= g_1,\ u_t = g_2 & &\text{in } D \times \{0\}.
\end{aligned} \qquad (15)$$

From observations $\tilde{u}$ effected in $\Omega' \subset \Omega$ we try to determine the potential f. The control set is bounded, $|f| \leq M$, and we look for a solution in $L^2(0,T;H_0^1(\Omega))$; the approximation f_β minimizes

$$J_\beta(f) = \frac{1}{2}\int_{\Omega'}(u-\tilde{u})^2 + \frac{\beta}{2}\int_\Omega f^2.$$

We obtain the characterization

$$f_\beta = \max\left(-M, \min\left(\frac{-u_\beta p}{\beta}, M\right)\right) \tag{16}$$

where u_β is the solution of the system (15) for $f = f_\beta$ and p solves the adjoint problem:

$$\begin{aligned} p_{tt} - \Delta p &= f_\beta p - (u_\beta - \tilde{u})\chi_{\Omega'} & &\text{in } \Omega \\ p &= 0 & &\text{on } \partial D \times (0,T) \\ p &= p_t = 0 & &\text{in } D \times \{T\}. \end{aligned} \tag{17}$$

Remark. Both examples contain bilinear control problems.

ACKNOWLEDGMENTS

S.L. and V.P. acknowledge partial support from DOE'S Office of BES under contact No. AC05-96OR22464 with Lockheed Martin Energy Research Corporation. J.Y. is supported in part by the NNSF of China, the Chinese State Education Commission Science Foundation, and the Trans-Century Training Programme Foundation for Talents of the State Education Commission of China.

REFERENCES

1. J.-L. Lions, *Équations Différentielles Opératorielles et Problèmes aux Limites*, (Springer Verlag, Berlin, 1961).
2. J.-L. Lions and E. Magenes, *Non-Homogeneous Boundary Value Problems*, vols. 1-3, (Springer Verlag, New-York, 1972).
3. L. C. Evans, *Partial Differential Equations*, vols. 3A-3B (Berkeley Mathematics Lecture Notes, Berkeley, 1993).
4. C. W. Groetsch, *Inverse Problems in the Mathematical Sciences*, (Vieweg, Braunschweig, Wiesbaden, 1993).
5. I. J. D. Craig and C. Brown, *Inverse Problems in Astronomy*, (Adam Hilger Ltd., Bristol and Boston, 1986).
6. H. T. Banks et K. Kunish, *Estimation Techniques for Distributed Parameter Systems*, (Birkhäuser, Boston, 1989).
7. J.-L. Lions, *Optimal Control of Systems Governed by Partial Differential Equations*, (Springer Verlag, New York, 1971).
8. J.-L. Lions, *Some Methods in the Mathematical Analysis of Systems and Their Control*, (Science Press, Beijing & Gordon and Breach, New York, 1981).
9. J.-L. Lions, *Controllabilité Exacte*, vols. 1-2, (Masson, Paris, 1988).
10. J.-P. Puel and M. Yamamoto, C. R. Acad. Sci. Paris **320**, série I, 1171 (1995).
11. M. Yamamoto, Inverse Problems **11**, 481 (1995).
12. J.-L. Lions, C. R. Acad. Sci. Paris **302**, série I, 471 (1986).
13. J.-L. Lions, C. R. Acad. Sci. Paris **307**, série I, 865 (1988).

14. J.-L. Lions, SIAM Review **30**, 1 (1988).
15. D. L. Russell, SIAM J. Control and Opt. **4**, 276 (1966).
16. I. Lasiecka and R. Triggiani, Differential and Algebraic Riccati Equations with Applications to Boundary/Point Control Problems: Continuous Theory and Approximation Theory, in *Lecture Notes in Control and Information Sciences*, vol. 164 (Springer Verlag, New York, 1991).
17. S. Lenhart, V. Protopopescu, and J. Yong, Applicable Analysis (*to appear*).
18. M. Liang, preprint, University of Tennessee, 1997.

DYNAMICAL CONTROL AND
SPATIO-TEMPORAL PHENOMENA

Karhunen-Loeve Mode Control in Chaotic Reaction-Diffusion System

Ioana Triandaf and Ira B. Schwartz

Naval Research Laboratory,
Special Project for Nonlinear Science, Code 6700.3
Plasma Physics Division
Washington DC 20375-5000

Abstract. We introduce a chaos control method which stabilizes unstable states of a spatio-temporal process based on analyzing the dynamics of the main coherent structure in the data represented by the highest energy Karhunen-Loeve mode. The problem is then reduced to the application of embedding techniques to the control of a time series given by the amplitude of a dominant spatial mode. Control is applied to a reaction-diffusion process where we stabilize an unstable orbit inside a chaotic regime. By perturbing the boundary condition, stabilization occurs as the result of a travelling wave propagating through the bulk media.

INTRODUCTION

In this paper, we are interested in stabilizing unstable states and eliminating chaos in a spatio-temporal process by using the natural dynamics of the system; i.e., we wish to control chaos based on exploiting stable manifolds already present in the data. This type of control is based on a geometric model constructed directly from data obtained by well-known embedding techniques [1,2]. The geometric model contains important information about the flow such as attractors, stable and unstable manifolds, and replaces the equations of state with an approximate model of phase space. The control amounts to finding suitable states in the given data and using parameter adjustments to maintain the system on a desired state by directing the process to stable manifolds that can be identified from data. This approach to chaos control was initiated by Ott, Grebogi and Yorke [3].

We apply the algorithm to the following reaction-diffusion system:

$$\frac{\partial u}{\partial t} = D\frac{\partial^2 u}{\partial x^2} + \frac{1}{\epsilon}[v - f(u)]$$

$$\frac{\partial v}{\partial t} = D\frac{\partial^2 v}{\partial x^2} - u + \alpha, \quad x \in [0,1],$$

(1)

subject to Dirichlet boundary conditions.

Parameters ϵ and α are assumed positive and fixed. The choice of the Van der Pol-like reaction term determines the type of patterns that form. It is a phenomenological model which reproduces the spatio-temporal two-front bursting patterns observed for a chlorite-iodide reaction in the Couette flow reactor. For our method we will consider the two-front pattern with two diffusion dominated regions at the boundaries and chaotic bursting occuring in the middle region (Fig.1). The nonlinear term we consider is given by $f(u) = u^2 + u^3$. The remarkable feature of this chaotic pattern is that it couldn't be observed in the absence of diffusion, and the chemical reaction itself would evolve in a steady manner [4].

KARHUNEN-LOEVE DECOMPOSITION

The Karhunen-Loeve procedure applies to a discretized spatio-temporal pattern, say a solution of (1), $\mathbf{u}(x,t) = (u(x,t), v(x,t))$, given in terms of a computational spatial grid $\mathbf{x} = (x_1, \ldots, x_p)$, and at discrete intervals in time $\{t_n\}$: $\{\mathbf{u}^n(\mathbf{x})\} = \{\mathbf{u}(\mathbf{x}, t_n)\}_{n=1,M}$.

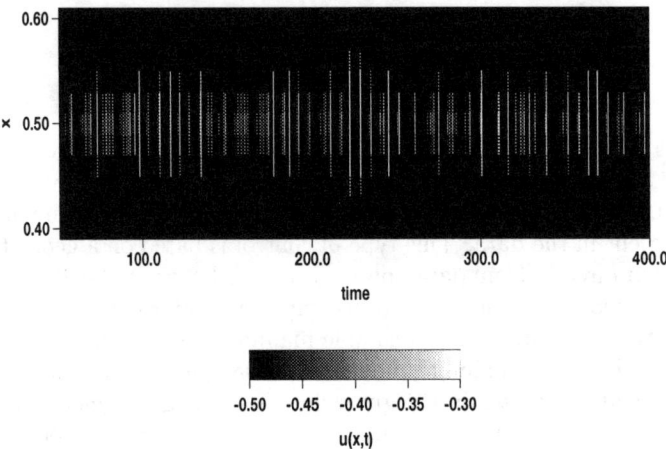

FIGURE 1. Chaotic spatio-temporal pattern for the u-variable at $D = 0.032249, \alpha = 0.1$, $\epsilon = 0.1$. The image is enhanced to show the spatio-temporal dependence of the boundary between the reaction and diffusion regimes.

The KL modes are the orthonormal eigenfunctions of the autocorrelation matrix: $K(x,x') = <\mathbf{u}(x,t)\mathbf{u}(x',t)>$, where the brackets stand for time average and the vector product is the dyadic product. The field **u** may be expanded as:

$$\mathbf{u}(x,t) = \sum_n a_n(t)\psi_n(x), \qquad (2)$$

where the amplitudes of the KL modes are orthogonal in time: $<a_n(t)a_m(t)> = \lambda_n \delta_{nm}$. We performed KL decomposition on a chaotic solution of the system (1). The pattern for the u-variable is shown in Fig. 1.

THE KL MODE CONTROL AlGORITHM

The algorithm addresses spatio-temporal data such as a solution of the reaction-diffusion system (1), or in general any spatio-temporal pattern obtained from experimental measurements. As shown in [5] the appropriate sampling in space is always an important issue for control. In the method we propose here, we eliminate this problem by taking as our time series the amplitude of the highest energy KL mode and use it for analyzing the dynamics. This time series contains information on the statistically most representative spatial structure. This is the term $a_1(t)$ in the expansion (2). By construction, the highest energy KL mode is the one structure which best approximates the spatio-temporal pattern in L^2 norm. For the system (1) the highest energy KL mode contains 90% of the energy. Heuristically, that means that about 90% of the data, in time, lies close to this spatial mode. By using this approach of

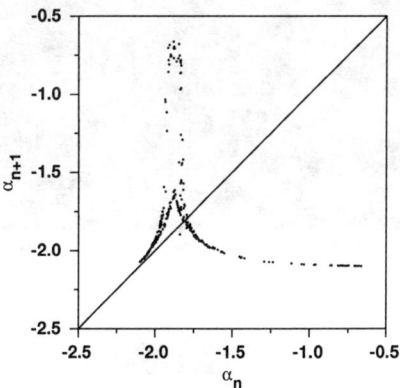

FIGURE 2. First return map of the maxima of $\alpha(t)$, the amplitude of the first KL mode (dimensionless units).

sampling the dynamics, we essentially approximate **u** by the first term in the expansion (2).

We further reduce the dimension of the sampled dynamics by sampling the chosen time series discretely, for example at the successive maxima. This gives a map of the form:

$$\alpha_{n+1} = F(\alpha_n, p) \tag{3}$$

where p is the control parameter and α_n denote the successive maxima of $a_1(t)$. The control parameter will be taken to be one of the Dirichlet boundary conditions. So we control by adjusting the feed rate at one of the boundaries. The map F is shown in Fig.2 where we identify a period one unstable orbit in the dynamics. The data in Fig.2 constitutes the geometric model we use for control. (Notice that for the period one fixed point, this representation is sufficient to capture the relevant dynamics.)

As our linear control algorithm, we used the OGY control method, which consists in changing the parameter at each iteration of the map F, so that the next iterate will fall on the stable manifold of the orbit we want to stabilize. The orbit to stabilize corresponds to a period one saddle of the map F, which we denote by α_0; i.e., $\alpha_0 = F(\alpha_0, p)$. This would correspond to a periodic function $\alpha(t)$ and a time-periodic pattern of the original solution of (u, v) of (1). ¿From data it is possible to determine the eigenvalues and eigenvectors of an orbit, for example by using a least square fit for data near the desired orbit. For a two-dimensional map F, with stable and unstable eigenvalues λ_s and λ_u, the OGY technique requires the parameter change:

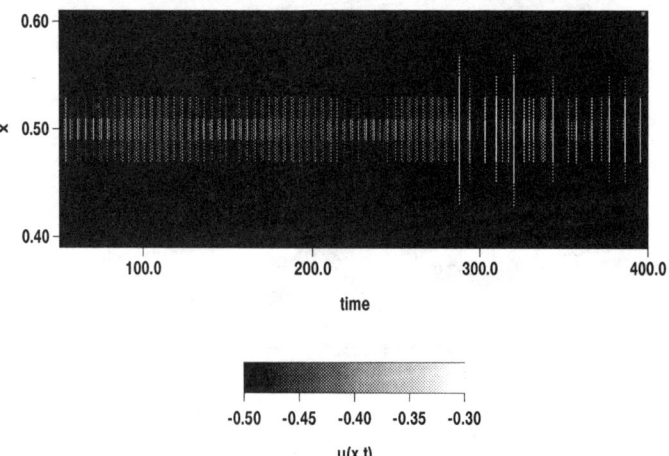

FIGURE 3. Stabilized pattern obtained by using KL control algorithm. The control is interrupted after 250 successive bursts and the chaotic pattern returns.

$$\delta p_n \equiv \frac{\lambda_u \cdot \xi \cdot \mathbf{f}_u}{(\lambda_u - 1)\mathbf{g} \cdot \mathbf{f}_u}, \quad (4)$$

where ξ is the distance from the current iterate to the orbit chosen for control, f_u is the contravariant vector corresponding to the unstable direction and the vector g is the derivative of the unstable state u_0 with respect to the parameter p, which can be determined from data by knowing the orbit at two nearby values of p.

Notice that although the dynamics and control representations in Eqs. 3 and 4 are discrete, the parameter fluctuations at the boundary are not pulses similar to δ functions. Rather, the boundary condition is turned on at one maxima, and left on until the next maxima, at which point it is adjusted based on Eq. (4).

The stabilized pattern is shown in Fig. 3. After 250 iterates we release the control and the chaotic pattern is reestabilished.

In Fig.4 we show the corresponding parameter fluctuations.

DISCUSSION

The way control affects the solution can be understood by looking at a single pulse in the control parameter (i.e. one of the boundary conditions) which is kept on for an entire cycle. We plotted in Fig. 5 the difference u^*, between the solution with a single pulse and the chaotic solution without a pulse. We notice the pulse generates a travelling wave which propagates inside the domain for the u-variable. This behavior is distinct from the case where only diffusion is present (See [8] for further details.).

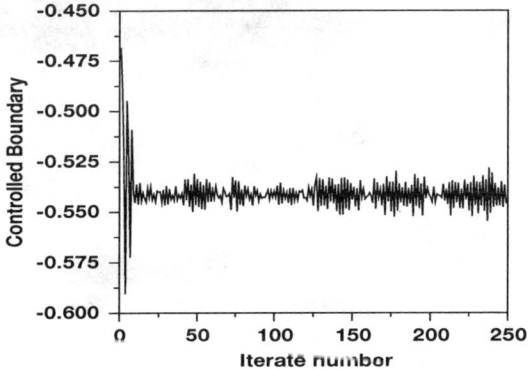

FIGURE 4. Adjustments in the feed rate used to obtain the controlled pattern in Fig.3.

Finally, we note that the method we present is more robust than previous methods [6,7] in that it doesn't require special placement of time series measurements and control adjustments are one order of magnitude smaller.

REFERENCES

1. Broomhead, D.S. and King, G.P., "Extracting qualitative dynamics from experimental data", *Physica D* 20, (1986), 217-236.
2. Takens, F.,"Detecting Strange Attractors in Turbulence" in *Dynamical Systems and Turbulence* edited by D.A. Rand and L.-S. Young Lecture Notes in Mathematics 898, Springer-Verlag, Berlin (1981), 366-381.
3. Ott, Edward C., Grebogi, Celso, and Yorke, James A. *Phys. Rev. Lett.* 64, 1196 (1990).
4. Arneodo, A. and Elezgaray, J., in *Spatial Inhomogeneities and Transient Behavior in Chemical Kinetics*, edited by Gray,P., Nicolis,G., Baras,F., Borckmans,P. and Scott,S.K. (Manchester University, Manchester, 1990),p.415.
5. Qin,F., Wolf,E.E., Chang,H.-C. , *Proceedings of the American Control Conference*, San Francisco, June 1993. F. Qin, E. E. Wolfe, and H.-C. Chang, *PRL* vol. 72, 1459 (1994).
6. Schwartz,Ira and Triandaf,Ioana *Phys. Rev E*, 50, 2548 (1994).
7. Pierre,Th., Bonhomme,G. and Atipo,A., *Phys. Rev.Lett.*,vol 76, 2290 (1996).
8. Triandaf,Ioana and Schwartz,Ira B., "Karhunen- Loeve mode control of chaos in a reaction-diffusion process", *Phys. Rev. E*, in press, 1997.

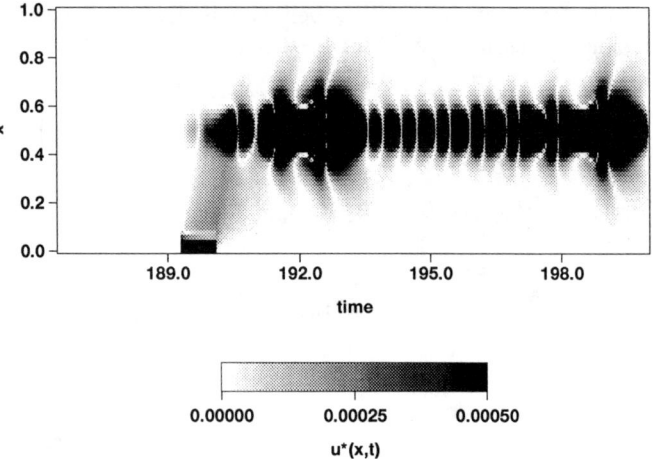

FIGURE 5. The spatio-temporal pattern generated when taking the difference u^*, between the chaotic solution in Fig.1 and the same chaotic solution with a pulse at the boundary during a single cycle.

Stability of bound states of pulses in the Ginzburg-Landau equations

Vsevolod V. Afanasjev[1]

School of Electrical Enineering, University of New South Wales, Sydney 2052
Australia

Boris A. Malomed[2]

Department of Interdisciplinary Studies
Faculty of Engineering, Tel Aviv University, Tel Aviv, Israel

Abstract. We consider bound states (BS) of quasi-soliton pulses in the quintic Ginzburg-Landau (GL) equation and in the driven damped nonlinear Schrödinger equation. Using the perturbation theory, we derive dynamical systems describing the interaction between weakly overlapping pulses in both models and formation of bound states. While all BS's in the GL model are unstable, one of them has very weak instability, so it can be considered stable for applications. For the damped driven model, we demonstrate the existence of fully stable BS's, provided that the amplitude of the driving field exceeds a very low threshold.

I INTRODUCTION

Various forms of the Ginzburg-Landau equations and solitary-pulse (SP) solutions to them is a topic that has been attracting a great deal of attention, see, e.g., [1–3]. This is stimulated both by physical applications, that extend from nonlinear fiber optics to travelling-wave convection in binary fluids, and by the interest to fundamental dynamical properties of models based on the GL equations.

One of the interesting properties for GL pulses is the possibility to form bound states, which have been predicted analytically [3] and demonstrated numerically [4]. However, the BS stability remains a controversial problem. Numerical simulations [4], performed only for a strongly perturbed model,

---

[1] e-mail: afanasjev@cse.unsw.edu.au
[2] e-mail: malomed@eng.tau.ac.il

have produced unstable BS, while the analytical results [3] (formally valid for weakly perturbed systems only) predicted stable BS.

A different but related model, viz., the driven damped nonlinear Schrödinger (NLS) equation (see, e.g., [5]) also has SP which can form a BS. The first numerical results specially aimed at studing the BS's in this model were reported in [6]. It was found in [6] that the BS's exist indeed with a separation between the bound pulses that is in fairly good agreement with the analytical predictions, even in the case when the parameters assumed to be small in the analysis were actually not so small. However, the numerical study was limited to in-phase solitons.

The objective of the present paper is to advance understanding of the BS stability in two most important models, viz., the quintic GL and driven damped NLS equations, in the case when they may be treated as perturbations of the NLS equation. This case is quite realistic for the applications (at least) to the optical fibers and, simultaneously, it admits a consistent analysis based on the perturbation theory.

II THE QUINTIC GINZBURG-LANDAU MODEL

We take the quintic GL equation in the standard form that emphasizes its proximity to the unperturbed NLS limit:

$$iu_z + \frac{1}{2}u_{\tau\tau} + |u|^2 u = iP[u], \qquad (1)$$

where $P[u] = -\alpha u + \beta u_{\tau\tau} + \gamma |u|^2 u - \Gamma |u|^4 u$. Here we are using the "fiber" notation, i.e., z and τ are the propagation distance and the so-called reduced time. All the parameters α, β, γ, and Γ are assumed to be positive. They account for the linear losses, spectral filtering (or diffusion in other physical contexts), nonlinear gain, and stabilizing higher-order nonlinear losses, respectively.

The SP solutions are assumed to be close to the soliton of NLS with an amplitude η, $u = \eta \,\text{sech}\,[\eta(\tau - T)] \exp[i(\eta^2 z/2 + \phi)]$, T and ϕ being arbitrary constants. The perturbation theory can be applied provided that the dimensionless parameters α, β, γ, and Γ are all small. In this case, the amplitude of the stable pulse is [1] $\eta^2 = (16\Gamma)^{-1}\left[5(2\gamma - \beta) + \sqrt{25(2\gamma - \beta)^2 - 480\alpha\Gamma}\right]$. The small dissipative perturbations make the asymptotic form of the soliton far from its center oscillating, $u \approx 2\eta \exp(-\eta|\tau| + i\chi|\tau|)$, $\chi = \alpha\eta^{-1} + \beta\eta$.

The next step is to consider the interaction between two weakly overlapping pulses with equal amplitudes. We will introduce the normalized propagation distance $x \equiv 2\sqrt{2}\eta^2 z$, the normalized separation between the SP's $r \equiv \eta(T_1 - T_2)$, and the phase difference between them $\psi \equiv \psi_1 - \psi_2$, where the subscripts 1 and 2 mark the positions and phases of the two interacting pulses. Regarding the overlapping between the solitons as another small perturbation, one can derive a system of effective evolution equations for r and ψ:

$$\frac{d^2r}{dx^2} + \frac{\sqrt{2}}{3}\beta\frac{dr}{dx} + e^{-r}[\cos(br) + b\,\sin(br)]\cos\psi = 0, \qquad (2)$$

$$\frac{d^2\psi}{dx^2} + \frac{\lambda}{\sqrt{2}}\frac{d\psi}{dx} - e^{-r}\cos(br)\sin\psi = 0, \qquad (3)$$

where the four original control parameters combine into three final ones: β and

$$\lambda \equiv \frac{1}{15}\sqrt{25(2\gamma - \beta)^2 - 480\alpha\Gamma}, \quad b \equiv \frac{-\alpha + \beta\eta^2}{\eta^2}. \qquad (4)$$

Equations (2) and (3) may be regarded as equations of motion for a mechanical system with two degrees of freedom, in the presence of friction, in the potential $U(r,\psi) = -e^{-r}\cos(br)\cos\psi$, which has a set of local extrema at

$$br_0 = \tan^{-1}b + \frac{\pi}{2}(1+2n),\ \psi_0 = \pi m, \qquad (5)$$

where $n = 0, 1, 2, ...$, $m = \pm 1, \pm 2, ...$. A peculiarity of the system (2) and (3) is that, while the effective mass corresponding to the degree of freedom r is $+1$, that for ψ is -1. The negative effective mass drastically changes the stability of the FP's. In particular, due to this fact, all the local extrema (5) are actually *saddles*.

Equations (2) and (3) also have the second set of the FP's,

$$br_0 = \frac{\pi}{2}(1+2n),\ \psi_0 = \frac{\pi}{2}(1+2m). \qquad (6)$$

Comparing the FP's (5) and (6), we notice that, for coinciding values of the integer n, they have nearly equal separation r between the bound pulses, but the relative phase ψ differs by $\pi/2$. The stability analysis of the FP's (6) reveals that there are two relatively large negative eigenvalues corresponding to rapidly decaying perturbations [as well as in the case of FP (5)] and two exponentially small complex eigenvalues

$$\sigma_{1,2} = \pm ib\sqrt{\frac{3}{\beta\lambda}}e^{-r_0} + \frac{3}{2}\left(\frac{b}{\beta\lambda}\right)^2\left(\sqrt{2}\beta + \frac{3\lambda}{\sqrt{2}}\right)e^{-2r_0}. \qquad (7)$$

Obviously, the FP (6) is an unstable spiral. Thus we obtain two types of unstable BS's in the quintic GL model: Depending on the phase difference

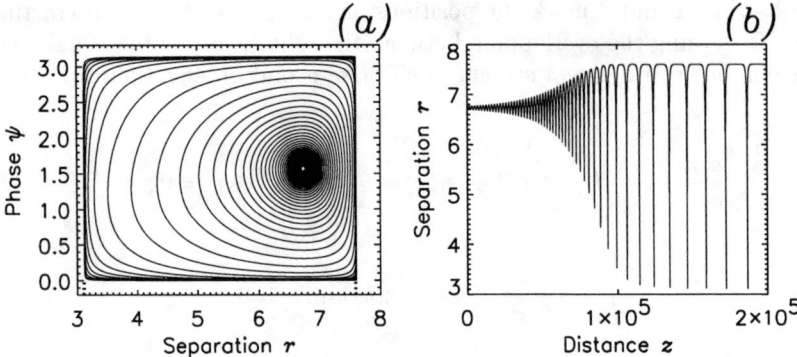

FIGURE 1. (a) An example of a dynamical trajectory of the full four-dimensional system (2) and (3) in projection onto the plane (r, ψ). The trajectory is unwinding around the fixed point (6) with $n = 1$. The parameters are $b = 0.7$, $\beta = 0.525$, $\lambda = 0.35$. (b) The same trajectory versus the distance z.

between the SP's, their BS should be unstable as a saddle or as a spiral. Precisely this was observed in the recent numerical experiments performed at *non small* values of the perturbation parameters (and for the opposite sign in front of the dispersion term) [4].

Returning to the perturbative analysis, we notice that the real part of the eigenvalue (7), accounting for the instability of the spiral, is proportional to the *square* of the exponentially small factor e^{-r_0}. Thus the instability of the spiral is extremely weak and one may interpret this FP, provided that the underlying perturbation parameters are small indeed, as a practically stable BS of the pulses.

To check the correctness of this picture, we performed numerical simulations of the system (2) and (3). As an illustration, in Fig. 1 we display a projection of the four-dimensional dynamical trajectory onto the plane (r, ψ). This trajectory pertains to the case $\lambda = \frac{1}{2}b$, $\beta = \frac{3}{4}b$ [in this case the dynamical system (2) and (3) coincides with that for the interacting SP's governed by the *cubic* GL equation, so this case is of additional interest], and $b = 0.7$. The FP was taken as per Eq. (6) with $n = 1$. Notice that the numerical values of the perturbation parameters are not really small in this case; nevertheless, the trajectory, exactly as it is predicted by the reduced system, is slowly unwinding around the FP, filling the interior of the separatrix-grid cell, and finally the motion practically stops when the trajectory gets very close to the boundaries of the cell.

III THE DRIVEN DAMPED MODEL

This model is based on the equation [5]

$$iu_t + \frac{1}{2}u_{xx} + |u|^2 u = -i\alpha u + \epsilon\, e^{i\Omega t}, \tag{8}$$

where we have switched to the traditional (non-fiber-optics) notation, though this model has some optical applications too. It is well known that this model supports stable SP solution (existing above a continuous wave (cw) background supported by the drive in competition with the friction) [5]. Far from the center of the pulse, its asymptotic form is $u(x,t) \approx 2\eta \exp(i\Omega t - \eta|x| + ik|x| + i\psi)$, where the soliton's amplitude η is related to the driving frequency by the relation $\eta = \sqrt{2\Omega}$, and ψ is a phase constant. The wave number k, because of which the soliton's tail is oscillatory and thus gives rise to an effective interaction potential with local minima, is $k = \alpha/\eta$ [3].

Combining the results of [7] and [3], it is straightforward to derive a system of equations describing the interaction of two weakly overlapping pulses in the model (8). The form of the equations simplifies in terms of the variables

$$2\sqrt{2}\eta^2 t \equiv \tau,\ \eta\Delta \equiv r,\ \alpha/\eta^2 \equiv b,\ \pi\epsilon/2\eta^3 \equiv E, \tag{9}$$

where Δ is the separation between the centers of the two pulses. The eventual form of the dynamical system is

$$\frac{d^2\psi_j}{d\tau^2} + \sqrt{2}b\frac{d\psi_j}{d\tau} + \frac{1}{2}(-1)^{j-1}e^{-r}\cos(br)\sin(\psi_2 - \psi_1) + \frac{b}{4} + \frac{E}{4}\sin\psi_j = 0, \tag{10}$$

$$\frac{d^2 r}{d\tau^2} + e^{-r}\left[\cos(br) + b\,\sin(br)\right]\cos(\psi_2 - \psi_1) = 0, \tag{11}$$

where j takes values 1 and 2, ψ_j being the phase constants of the two pulses.

The system (10) and (11) has the FP's

$$br_0 = \frac{\pi}{2} + \tan^{-1} b + \pi n,\ n = 1, 2, 3, \ldots \tag{12}$$

$$\psi_1 = \psi_2 = -\sin^{-1}(b/E);\ \psi_1 = \psi_2 = -\pi + \sin^{-1}(b/E); \tag{13}$$

which are similar to the FP's (5) considered in section II. In what follows, the common values of $\psi_1 = \psi_2$ at the FP will be denoted as ψ_0. Note that this FP exists if $|b/E| \le 1$, which is a well-known threshold condition [5], which we will assume to be satisfied.

Stability analysis of the FP is straightforward. First of all, the soliton must be stable in isolation, which implies a well-known fact: Out of the two FP's in Eq. (13), one should take the one with $E \cos\psi_0 > 0$ [5]. Next, the perturbations of the separation and phase decouple in the linearized equations governing evolution of the small perturbations around the FP and in order to provide for its stability against the perturbation of the separation, one should take $n = 1$ in Eq. (12), which is known too [6]. After this, a remaining previously unexplored issue is an accurate analysis of the stability against phase perturbations. Technically, it is quite easy, and leads to the final result: The phase perturbations do not produce instability provided that

$$\frac{1}{4} E^2 \cos^2 \psi_0 > e^{-2r_0}. \tag{14}$$

The meaning of the condition (14) is obvious: phase locking of both pulses to the external drive is able to suppress the phase instability that rendered the FP's (5), considered in section II, unstable. Therefore, condition (14) is not satisfied in the absence of the drive ($E = 0$), but if the drive is present, it is very easy to satisfy this condition, as its right-hand side is exponentially small, while the left-hand side is not.

IV CONCLUSION

In this work we have made an effort to clarify an important issue that has remained controversial, namely, stability of bound states of pulses in the quintic GL equation and in the driven damped NLS model, both of which are well known to support stable isolated pulses. We have found that in the quintic GL equation two types of bound states are possible, saddles and spirals. While both of them are unstable, the instability of the spirals is extremely weak. In the damped damped model, the situation is essentially simpler. Using the description in terms of the dynamical system, we have demonstrated that the fixed point, corresponding to the pair of pulses stably locked to the driving force, can easily become stable, provided that the drive's amplitude exceeds a very low threshold value. This stable bound state was observed earlier in direct simulations of the driven damped model.

REFERENCES

1. Malomed, B.A., *Physica D* **24**, 155 (1987).
2. van Saarloos, W., and Hohenberg, P.C., *Phys. Rev. Lett.* **64**, 749 (1990).
3. Malomed, B.A., *Phys. Rev. A* **44**, 6954 (1991); *Phys. Rev. E* **47**, 2874 (1993).
4. Afanasjev, V.V., and Akhmediev, N., *Opt. Lett.* **20**, 1970 (1995); *Phys. Rev. E* **53**, 6471 (1996).
5. Kaup, J.D., and Newell, A.C., *Proc. Roy. Soc. London, Ser. A* **361**, 413 (1978).
6. Cai, D., Bishop, A.R., Grønbech-Jensen, N., and Malomed, B.A., *Phys. Rev. E* **49**, 1677 (1994).
7. Karpman, V.I., and Solov'ev, V.V., *Physica D* **3**, 487 (1981).

Controlling symmetric vortex configurations

Á. Péntek[1], J. B. Kadtke[1], and Z. Toroczkai[2,3]

[1]*Institute for Pure and Applied Physical Sciences, University of California at San Diego, 9500 Gilman Dr., La Jolla, CA 92093-0360*
[2]*Department of Physics, Virginia Polytechnic Institute and State University, Blacksburg, VA 24061-0435*
[3]*Institute for Theoretical Physics, Eötvös University, Puskin utca 5-7, H-1088 Budapest, Hungary*

Abstract. A high-dimensional chaos control algorithm is applied to stabilize unstable orbits of symmetric point-vortex configurations. We discuss possible applications to an experimental plasma system.

INTRODUCTION

Hamiltonian dynamics of point vortices has attracted interest for many years [1–5] due to, e.g. its central role in understanding vortex dynamics in superfluid helium [6], superconductors and dynamics of non-neutral plasma filaments [7]. The point vortex dynamics often captures most of the qualitative features of the evolution of vorticity in such flows. Our recent results [8,9] also show that these simple Hamiltonian models provide a good formulation to develop control algorithms for more complex fluid flows.

In this paper we apply a previously developed high-dimensional control algorithm [10] to stabilize unstable symmetric states of three point vortices inside a circular geometry. The flow model is based on Hamiltonian point vortex dynamics. The control algorithm is a modified Ott, Grebogi, Yorke (OGY) [11] scheme which can be implemented in higher dimensional systems, e.g. with phase space dimension larger than 3. We compare the model point-vortex dynamics with the evolution of vorticity in a viscous flow, and then demonstrate the method's effectiveness using numerical examples that model an experimental system of confined non-neutral plasma.

To fix a simple framework for our discussion, we restrict ourselves to one of the simplest symmetric configurations that lead to chaotic dynamics, namely the dynamics of three identical collinear vortices. Fig. 1 shows a typical time-evolution of three identical vorticity distributions with Gaussian profile in a viscous flow evolved using a Navier-Stokes simulation. For time scales much shorter than the viscous one, the vortices essentially follow a Hamiltonian dynamics (Fig. 1(a-i)). Since the collinear state is unstable, a simple instability develops that leads to an exchange of vortices. One extremal vortex changes position with the central one. On time scales comparable with the viscous one, the initially sharp Gaussian profile starts to diffuse, and the size of the vortices increases (see Fig. 1). Because during the vortex exchange process the vortices pass close to each other, they will merge (Fig 1(k-l)). This is essentially an inviscid process, however this process obviously cannot be modeled by Hamiltonian vortex dynamics. In several experimental applications we are

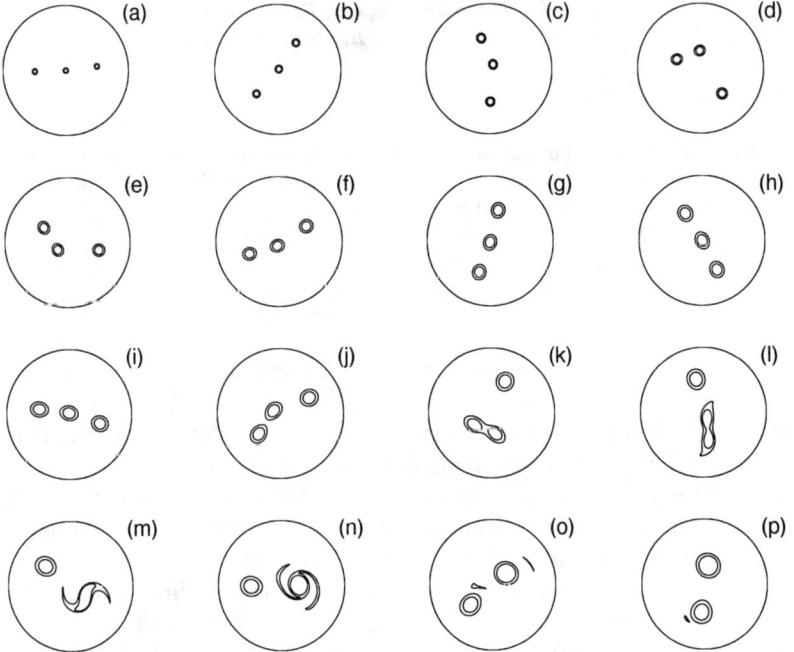

FIGURE 1. Time evolution of three collinear identical vortices in a viscous flow. Lines of constant vorticity are shown at 0.33 and 0.1 of the maximum vorticity

interested in, however, the Reynolds number is extremely high and the viscous time scale is than on the order of several thousands of rotation times. Hence, for a long time the Hamiltonian dynamics is the only relevant process that governs the evolution of vorticity. In the following we will be interested only in this part of the vorticity evolution.

To develop the control scheme, we first briefly review the formulation of point vortex dynamics. Let $z = x + iy$ denote the complex coordinate of a single vortex of circulation Γ. The effect of the boundary of the circular domain with radius R is simply accounted for by image charges at R^2/z^*. The discussion can be further simplified by rendering the equations-of-motion dimensionless with the substitution: $z \longrightarrow z/R$, and $t \longrightarrow 2\pi t/\Gamma R^2$, and by performing a transformation into a reference frame co-rotating with the vortex system with uniform angular velocity ω. The equations of motion then read as

$$\dot{z}_k^* = -\frac{i}{2}\left[\sum_{j\neq k}^{3}\frac{1}{z_k - z_j} + \sum_{j}^{3}\frac{z_j^*}{1 - z_j^* z_k} - 2\omega z_k^*\right], \quad k = 1, 2, 3. \quad (1)$$

The main advantage of this co-rotating frame is that some of the periodic orbits in the 6-D phase space reduce to a single fixed point (with proper choice of ω).

THE PERTURBATIONS

To control the vortex dynamics we require non-symmetric and non-homogeneous perturbations to provide sufficient "spatial orthogonality" of the modes [10]. We have initially introduced simple perturbations applied only at the cylinder surface: namely, a uniform distribution of sources and sinks with variable strength $\cos(k\theta)$. Although such a perturbation may be difficult to implement in a real fluid experiment, it is easily realized in the plasma analog of the vortex dynamics [7], where similar perturbations can be generated by external electric fields. For $k = 1, 2,$ and 3, this leads to a uniform, a quadrupole, and a sextapole field, respectively. Since all these fields are highly symmetric, none of them alone can effectively control the vortex dynamics. However, their linear combinations, resulting in strongly asymmet-

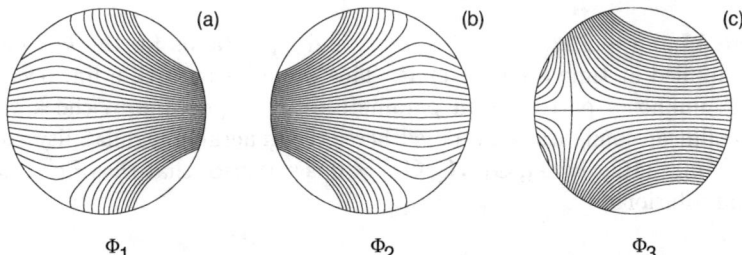

FIGURE 2. Perturbation fields used to control the vortex dynamics: (a) $\dot{z}^* = \Phi_1 (1 + z + z^2)$, (b) $\dot{z}^* = \Phi_2 (-1 + z - z^2)$, and (c) $\dot{z}^* = \Phi_3 (1 + z - z^2)$. The strength of these fields Φ_1, Φ_2, Φ_3 are computed by the control algorithm.

ric and inhomogeneous fields, proved to be a successful choice that satisfies the controllability conditions. Here, we have introduced three linearly independent external fields as shown in Fig. 2(a-c). Using these, the perturbed equations of motion of the kth vortex can be written:

$$\dot{z}_k^* = -\frac{i}{2}\left[\sum_{j\neq k}^3 \frac{1}{z_k - z_j} + \sum_j^3 \frac{z_j^*}{1 - z_j^* z_k} - 2\omega z_k^*\right]$$
$$+\Phi_1\left(1 + z_k + z_k^2\right) + \Phi_2\left(-1 + z_k - z_k^2\right) + \Phi_3\left(1 + z_k - z_k^2\right) \quad (2)$$

THE CONTROL ALGORITHM

Here we aim to modify the dynamics of the fully-viscid continuous fluid using a control model based on the point vortex dynamics. Hence, the model governing equations (2) are known, and the high-dimensional control is straightforward to calculate [10]. In the case of an experimental system, the method can be implemented in an identical fashion, using the Jacobian and the perturbation matrix obtained from experimental time-series [12].

Let $\dot{\mathbf{r}} = \mathbf{A}(\mathbf{r}, \mathbf{\Phi})$ represent the 6-dimensional unperturbed dynamical system of Eq. (1). To achieve control, we adopt the approach of OGY, i.e. we require that the trajectory lies on the stable manifold of the fixed point after a time $2\Delta t$ [10]:

$$\delta \mathbf{r}(t + 2\Delta t) = \alpha \,|\delta \mathbf{r}(t)|\, \mathbf{e}_s, \quad (3)$$

where $\mathbf{e}_s$ is one of the stable eigenvectors of the unstable fixed point. Eq (3) then can be written explicitly as

$$\mathbf{MG}_1\,\delta\Phi_1^{(1)} + \mathbf{MG}_2\,\delta\Phi_2^{(1)} + \mathbf{MG}_3\,\delta\Phi_3^{(1)} + \mathbf{G}_1\,\delta\Phi_1^{(2)} + \mathbf{G}_2\,\delta\Phi_2^{(2)} + \mathbf{G}_3\,\delta\Phi_3^{(2)} =$$
$$= \frac{1}{\Delta t}\left(\alpha\,|\delta\mathbf{r}(t)|\,\mathbf{e}_s - \mathbf{M}^2\delta\mathbf{r}(t)\right) \quad (4)$$

where $\mathbf{M} \equiv \mathbf{1} + \mathbf{J}\Delta t$, $\mathbf{G}_k = \partial \mathbf{A}/\partial \Phi_k$, and $\Phi_k^{(p)}$ stands for the kth perturbation applied at time $t + (p-1)\Delta t$. This provides us with 6 linear equations and 6 unknown perturbation parameters. It is possible to check that, for the collinear configuration studied here, the general controllability condition $\det\{\mathbf{MG}_1; \mathbf{MG}_2; \mathbf{MG}_3; \mathbf{G}_1; \mathbf{G}_2; \mathbf{G}_3\} \neq 0$ is satisfied, that thus Eq. 4 has nontrivial solutions.

NUMERICAL EXPERIMENTS

To illustrate the above control scheme, we actively stabilized the simplest unstable periodic orbits of the three vortex system inside a cylinder, the collinear state. The angular velocity of the configuration is $\omega =$

$(3+a^4)/(1-a^4)/(4a^2)$. Here a is the radial distance of the non-central vortices from the center of the cylinder. In the co-rotating frame, the periodic orbit becomes a single fixed point. Analytic calculation of the Jacobian at $a = 0.5$ reveals the existence of four real eigenvalues, $\lambda_{1,2} = \pm 4.7565$ and $\lambda_{3,4} = 0$, and two purely complex ones, $\lambda_{5,6} = 1.6027i$. The stable eigendirection $\mathbf{e}_s$ is then obtained as the eigenvector associated with the single negative real eigenvalue.

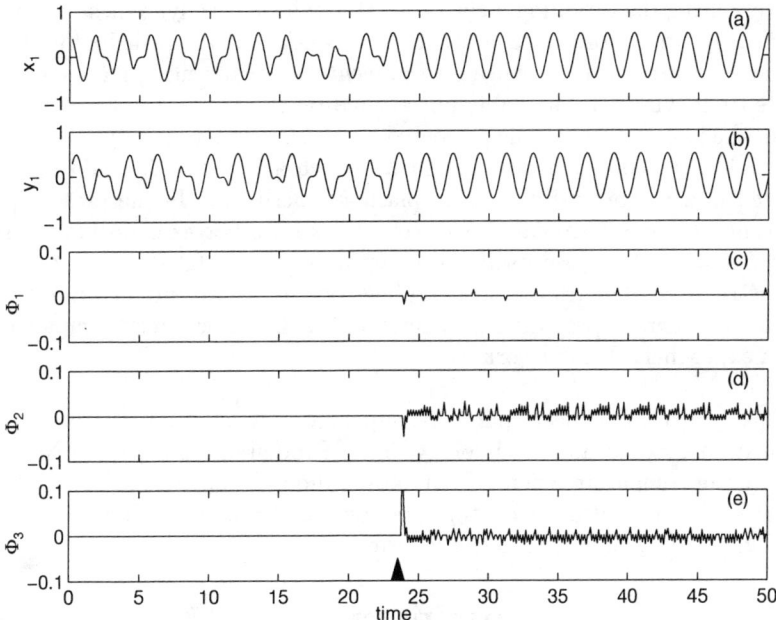

FIGURE 3. Vortex trajectories in the lab frame and the applied perturbations. $\Delta t = 0.1$ and $\alpha = 0.1$ has been used.

For the numerical simulation we have started the system close to the unstable fixed point, to avoid the complication of introducing a targeting algorithm. To demonstrate the control Fig. 3(a-b) shows the time evolution of a single coordinate pair (x_1, y_1) of the vortex system in the collinear case, at first without control (up to mark). Then, after the dynamics reaches a small neighborhood of the desired fixed point, the controller is switched on, and the trajectory is stabilized. Fig. 3(c-e) also shows the time evolution of the required perturbations. Clearly this demonstrates that the algorithm can effectively control the dynamics with only tiny perturbations applied on the boundary.

In a realistic Navier-Stokes simulation our algorithm is effective on time scales significantly shorter than the viscous one ($Re^{1/2}$). This is attractive since this time scale can be rather large in some plasma systems, where the viscosity is small and there are no boundary-layer effects due to the free-slip conditions on the boundary. Even in these circumstances, it should be noted that there is an additional non-viscous effect that is not present in the Hamiltonian model, i.e. the vortex merger due to the finite size of the vortices. Therefore, effective control can be reached only with concentrated vortices that are far from each other, typically with $2\rho/d \gtrsim 1.8$, where d is the distance between vortices and ρ the vortex radius computed according to Ref. [14]. Although this condition can be easily fulfilled while the controller is active, during the targeting algorithm or the full chaotic dynamics the vortices may easily come close to each other. In this context, the principal usefulness of such a control scheme will be to *prevent* the vortex merger on short time scales from symmetric unstable initial conditions.

For the numerical example presented here, i.e. three-vortex dynamics inside a cylinder, the controller was designed in such a way that it could be implemented in a magnetically-confined plasma experiment. In the present paper the controller was formulated using exactly known vortex coordinates. However, the control scheme itself can be entirely reformulated in a phase space reconstructed purely from wall signal (e.g. boundary pressure or voltage) measurements, thereby providing an algorithm which can be directly implemented in an experimental framework.

This project has been partially supported by a grant from the U.S.-Hungarian Science and Technology Joint Fund under Project JFNo. 501, and by the Hungarian Science Foundation under Grant Nos. OTKA T17493, F17166. ZT has also been sponsored by the National Science Foundation through the Division of Material Research.

REFERENCES

1. E. A. Novikov, and Yu. B. Sedov, Sov. Phys. - JETP **48**, 440 (1978).
2. H. Aref, Phys. Fluids **22**, 393 (1979); Ann. Rev. Fluid Mech. **15**, 345 (1983).
3. J. B. Kadtke, *PhD Thesis*, Brown University, (1987).
4. A. Pentek, T. Tel, and Z. Toroczkai, J. Phys. A **28**, 2191 (1995); Fractals **3**, 33 (1995).
5. T. H. Havelock, Phyl. Mag. S. 7 **11**, 617 (1931).
6. P. H. Roberts, and R. J. Donnelly, Ann. Rev. Fluid Mech. **6**, 179 (1974).
7. K. S. Fine, A. C. Cass, W. G. Flynn, and F. Driscoll, Phys. Rev. Lett. **75**, 3277 (1995).
8. J. B. Kadtke, A. Pentek, and G. Pedrizzetti, Phys. Lett. A **204**, 108 (1995).
9. A. Pentek, J. B. Kadtke, and G. Pedrizzetti, *Controlled capture of vortices in open viscous flows*, preprint (1996).
10. A. Pentek, J. B. Kadtke, and Z. Toroczkai, Phys. Lett. A **224**, 85 (1996).
11. E. Ott, C. Grebogi, and J. A. Yorke, Phys. Rev. Lett. **64**, 1196 (1990);F. J. Romeiras, C. Grebogi, E. Ott, and W. P. Dayawansa, Physica D **58**, 165 (1992).

12. T. Shinbrot, C. Grebogi, E. Ott, and J. A. Yorke, Nature **363**, 411 (1993).
13. Z. Toroczkai, Phys. Lett. A **190** 71 (1994); B. Sass, and Z. Toroczkai, J. Phys. A **29**, 3545 (1996).
14. K. S. Fine, C. F. Driscoll, J. H. Malmberg, and T. B. Mitchell, Phys. Rev. Lett. **67**, 588 (1991).

Quantifiers for spatio-temporal bifurcations in coupled map lattices

Nandini Chatterjee[1] and Neelima Gupte[2]

[1] *Department of Physics, University of Pune, Pune 411 007, INDIA*
[2] *Department of Physics, Indian Institute of Technology, Madras, Chennai 600 036, INDIA*

Abstract. The bifurcation behaviour of spatially extended systems shows interesting features which are as yet poorly understood. We analyse spatio-temporal bifurcations in coupled map lattices which maybe classified as purely spatial or spatio-temporal in nature. We construct quantifiers, which can detect all types of bifurcation behaviour. We demonstrate the utility of our quantifiers in the context of spatially or temporally periodic behaviour in a system of coupled sine circle maps. The evolution of spatial period two initial conditions inside the temporal period one tongue shows bifurcations which are purely spatial and spatio-temporal in nature which are successfully picked up by our quantifiers.

INTRODUCTION

The study of the spatio-temporal bifurcation behaviour that arises in spatially extended systems is necessary for the understanding of phenomena which arise in a variety of systems in nature and in the laboratory. Bifurcation phenomena can be seen in experiments of cellular patterns [1,2], in Josephson junction arrays [3], and in coupled oscillator systems [4].

We study spatio-temporal bifurcations in a system of coupled sine circle maps. Such systems have proved to be analytically and computationally tractable and are yet able to capture generic behavior observed in spatially extended systems which possess oscillatory behavior [5]. The spatial extent of the system provides a multiplicity of attractors and consequently a variety of bifurcations, which can be purely spatial, purely temporal or spatio-temporal in nature. Bifurcations of all these types have been observed in experimental systems, purely spatial as seen in experiments on cellular patterns [1,2], purely temporal of the in-phase to splay kind as seen in Josephson junction arrays [3], and spatiotemporal as observed in coupled oscillators [4].

The specific model under study is a coupled map lattice of sine-circle maps with nearest neighbour diffusive symmetric normalized coupling and periodic boundary conditions defined by the evolution equation

$$\theta_{t+1}(i) = (1-\epsilon)f(\theta_t(i)) + \frac{\epsilon}{2}f(\theta_t(i+1)) + \frac{\epsilon}{2}f(\theta_t(i-1)) \; mod\, 1 \quad (1)$$

where $\theta_t(i)$ is the angular variable associated with the ith site, at time t and $f(\theta_t(i)) = \theta_t(i) + \Omega - \frac{K}{2\pi}\sin(2\pi\theta_t(i))$, the familiar single circle map. The parameters Ω and K, where Ω is the natural frequency at $K = 0$ and K, the strength of the nonlinearity are taken to be uniform at each site and ϵ which lies between 0 and 1 is the strength of the coupling parameter. Earlier studies of this system show that the regions of stability of the spatially synchronised (spatial period one) solutions of different temporal periods corresponding to winding numbers $\frac{P}{Q}$ form a set of Arnold tongues in the $\Omega - \epsilon - K$ space as seen in Fig. 1 [8].

The $\frac{P}{Q} = \frac{0}{1}$ tongue corresponds to the temporal period one case. If the system is evolved with spatial period 2 initial conditions [7], then three different kinds of solutions can be observed in the $\frac{P}{Q} = \frac{0}{1}$ tongue and $\frac{P}{Q} = \frac{1}{1}$ tongue of Fig.1. The phenomena in the $\frac{0}{1}$ tongue can also be observed in the $\frac{1}{1}$ tongue and can be similarly analysed (See Fig.2). In each tongue, the first solution is a spatially synchronised, temporal fixed point or a spatial period one temporal period one (SP1TP1) solution, which can be seen in the major portion of the tongue in the dark banded portion of Fig. 2. The second solution is the spatial period 2, temporal period 1 (SP2TP1) solution as can be seen in the triangular region with vertical hashes at the bottom of the plot in Fig. 2. Further, a spatial period 2 temporal period 2 (SP2TP2) solution of travelling wave type can be seen in the triangular region with horizontal hashes at the top of the plot. Thus the SP1TP1 or the synchronised solution temporal period one solution undergoes a pure spatial bifurcation to a SP2TP1 solution in the lower region of parameter space, and a spatio-temporal bifurcation to a SP2TP2 or travelling wave solution in the upper part of the parameter space. We need to construct a quantity which can detect each one of these bifurcations. We set up new sum and difference variables and recast the equation of evolution as given by Eq. 1 in a convenient form. The stability matrix constructed in terms of these new variables can detect all the bifurcations observed including the purely spatial bifurcation.

CONSTRUCTING THE QUANTIFIERS

Consider a spatially periodic solution of spatial period k in a lattice of kN sites. For such a solution, at any time t, the value of the variable at the ith lattice site is the same as the value at the $(i+k)$th site. Thus, the difference between the variable values of the ith and the $(i+k)$th lattice site, approaches zero for all such pairs of neighbours. Further for homogeneous values of map

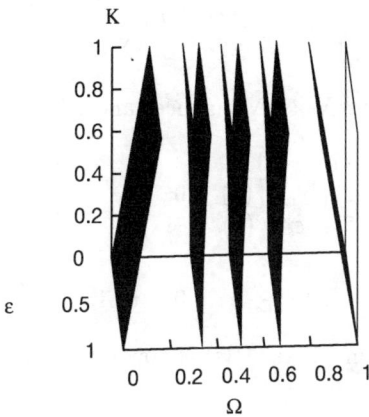

FIGURE 1. Arnold tongues in the $\Omega-\epsilon-K$ space for the spatially synchronized solutions. The temporal periods shown from left to right correspond to $\frac{P}{Q} = \frac{0}{1}, \frac{1}{3}, \frac{1}{2}, \frac{2}{3}$, and $\frac{1}{1}$.

parameters, the sum of the ith and $(i+k)$th site is also a constant. So we construct sum and difference variables as follows

$$a_t^k(i) = \theta_t(i) - \theta_t(i+k) \qquad (2a)$$
$$b_t^k(i) = \theta_t(i) + \theta_t(i+k) \qquad (2b)$$

Using Eq.1 we obtain the equations of evolution for the coupled sine circle map lattice as defined above in terms of $a_t^k(i)$ and $b_t^k(i)$ as

$$a_{t+1}^k(i) = (1-\epsilon)g^a\left(a_t^k(i), b_t^k(i)\right)$$
$$+ \frac{\epsilon}{2}\left(g^a(a_t^k(i+1), b_t^k(i+1)) + g^a(a_t^k(i-1), b_t^k(i-1))\right) \qquad (3a)$$
$$b_{t+1}^k(i) = (1-\epsilon)g^b\left(a_t^k(i), b_t^k(i)\right)$$
$$+ \frac{\epsilon}{2}\left(g^b(a_t^k(i+1), b_t^k(i+1)) + g^b(a_t^k(i-1), b_t^k(i-1))\right) + 2\Omega \mod 2 \qquad (3b)$$

Here, $g^a(a_t^k(i), b_t^k(i)) = a_t^k(i) - \frac{K}{\pi}\sin(\pi a_t^k(i))\cos(\pi b_t^k(i))$ and $g^b(a_t^k(i), b_t^k(i)) = b_t^k(i) - \frac{K}{\pi}\sin(\pi b_t^k(i))\cos(\pi a_t^k(i))$. It can be easily shown that, $\forall i$, $a_t^k(i) = 0$ and $b_t^k(i) = s_m$ ($m \equiv i \mod k$), where $s_1, s_2, \ldots s_k$ are all distinct *constants*, and are solutions of Eqs. 2a and 2b for a fixed Ω and K. To study the stability of any spatially periodic solution with spatial period k we need to examine the eigenvalues of the linear stability matrix. The stability condition for a spatially periodic solution with (spatial period k) is given by the largest eigen-value $|\lambda^{largest}| \leq 1$.

We expand Eqs. 2a and 2b about $a_t^k(i) = 0$ and $b_t^k(i) = s_m, m : 1, 2 \ldots k$ distinct constants, upto the linear order to obtain the linear stability matrix J_t^{2kN}

$$J_t^{2kN} = \begin{pmatrix} A_t'^{kN} & B_t'^{kN} \\ B_t'^{kN} & A_t'^{kN} \end{pmatrix} \qquad (4)$$

where $A_t'^{kN}$ and $B_t'^{kN}$ are $kN \times kN$ matrices and $A_t'^{kN}$ is given by

$$A_t'^{kN} = \begin{pmatrix} \epsilon_s A_t^k(1) & \epsilon_n A_t^k(2) & 0 & \cdots & 0 & \epsilon_n A_t^k(kN) \\ \epsilon_n A_t^k(1) & \epsilon_s A_t^k(2) & \epsilon_n A_t^k(3) & 0 & \cdots & 0 \\ 0 & \epsilon_n A_t^k(2) & \epsilon_s A_t^k(3) & \cdots & 0 & 0 \\ \vdots & \vdots & \vdots & \vdots & \vdots & \vdots \\ \epsilon_n A_t^k(1) & 0 & \cdots & 0 & \epsilon_n A_t^k(kN-1) & \epsilon_s A_t^k(kN) \end{pmatrix} \qquad (5)$$

$B_t'^{kN}$ has an identical form where $A_t^k(i)$ is now replaced by $B_t^k(i)$, $\epsilon_s = (1 - \epsilon)$ and $\epsilon_n = \frac{\epsilon}{2}$ and

$$A_t^k(i) = \left(1 - K \cos(\pi a_t^k(i)) \cos(\pi b_t^k(i))\right) \qquad (6a)$$

$$B_t^k(i) = \left(K \sin(\pi a_t^k(i)) \sin(\pi b_t^k(i))\right) \qquad (6b)$$

Imposing the conditions $a_t^k(i) = 0$ and $b_t^k(i) = s_m$ where $m : 1, 2 \ldots k$, the stability matrix J_t^{2kN} given by Eq. 4 reduces to a block diagonal form

$$J_t^{2kN} = \begin{pmatrix} M_t^{kN} & 0 \\ 0 & M_t^{kN} \end{pmatrix} \qquad (7)$$

where the blocks M_t^{kN} are identical and are of the form $A_t'^{kN}$ with each, $A_t^k(i)$ now given by $\bar{A}_t^k(i) = (1 - K \cos(\pi s_i))$. It can be shown that if the stability matrix has non-negative entries for a spatial period k solution, the stability analysis for a lattice of kN sites can be reduced to the analysis of k sites [9]. The synchronised and spatial period 2 solutions can thus be set up in terms of the $a_t^k(i)$ and $b_t^k(i)$ variables of Eqs. 2a and 2b for $k = 1$ and $k = 2$ and the corresponding stability matrices M_t^{1N} (which turns out to be circulant [8]) and M_t^{2N} can be set up by expanding about the stable solution. The eigen-values of the two cases can be explicitly obtained using the general results above.

To study the bifurcations in the $\frac{P}{Q} = \frac{0}{1}$ tongue we examine a lattice of $2N$ sites. The solutions of interest are the SP1TP1 (synchronised) solution and the SP2TP1 and SP2TP2 (travelling wave) solutions. We can also impose the conditions for SP2TP1 and SP2TP2 in terms of the $a_t^1(i)$ and $b_t^1(i)$. We write these as $\tilde{a}_t(i)$ and $\tilde{b}_t(i)$ because now all the solutions namely SP2TP1, SP1TP1 and SP2TP2 (travelling wave) can be expressed using the same variables. We invoke the closure condition for various solutions which can be expressed in the following way. For SP2TP1 we obtain $\tilde{a}_{t+1}(i) = \tilde{a}_t(i) = -\tilde{a}_t(i+1) \neq 0$ and $\tilde{b}_{t+1}(i) = \tilde{b}_t(i) = \tilde{b}_t(i+1) = const$. For SP1TP1 of we have the condition $\tilde{a}_{t+1}(i) = \tilde{a}_t(i) = \tilde{a}_t(i+1) = 0$ and $\tilde{b}_{t+1}(i) = \tilde{b}_t(i) = \tilde{b}_t(i+1) = const$. Similarly the travelling wave solution can be expressed as $\tilde{a}_{t+1}(i) = -\tilde{a}_t(i) = \tilde{a}_t(i+1) \neq 0$ and $\tilde{b}_{t+1}(i) = \tilde{b}_t(i) = \tilde{b}_t(i+1) = const$.

The evolution equations in terms of $\tilde{a}_t(i)$ and $\tilde{b}_t(i)$ have exactly the same form as Eqs. 2a and 2b. We now expand about the solution $\tilde{a}_t(i) = const$ and $\tilde{b}_t(i) = const$ to obtain the stability matrix J_t^{2N}. In this case, although the matrix has negative entries, the analysis for $2N$ lattice sites can be reduced to that of two lattice sites [9]. This basic matrix whose largest eigenvalue is a signature of all spatial, temporal and spatio-temporal transitions is given by

$$M_t^{tw}(1) = \begin{pmatrix} (1-\epsilon)(\tilde{A}_t'(1) + \tilde{B}_t'(1)) & \epsilon(\tilde{A}_t'(1) - \tilde{B}_t'(1)) \\ \epsilon(\tilde{A}_t'(1) + \tilde{B}_t'(1)) & (1-\epsilon)(\tilde{A}_t'(1) - \tilde{B}_t'(1)) \end{pmatrix} \quad (8)$$

Here $\tilde{A}_t'(1)$ and $\tilde{B}_t'(1)$ have the same form as Eqs. 6a and 6b where $a_t^k(i) = \tilde{a}_t(1)$ and $b_t^k(i) = \tilde{b}_t(1)$. The largest eigenvalue $\tilde{\lambda}$ is given by

$$\tilde{\lambda} = (1-\epsilon)(1 - K \cos \pi \tilde{a}_t(1) \cos \pi \tilde{b}_t(1))$$
$$+ \sqrt{\epsilon^2 (1 - K \cos \pi \tilde{a}_t(1) \cos \pi \tilde{b}_t(1))^2 + (1 - 2\epsilon)(K \sin \pi \tilde{a}_t(1) \sin \pi \tilde{b}_t(1))^2} \quad (9)$$

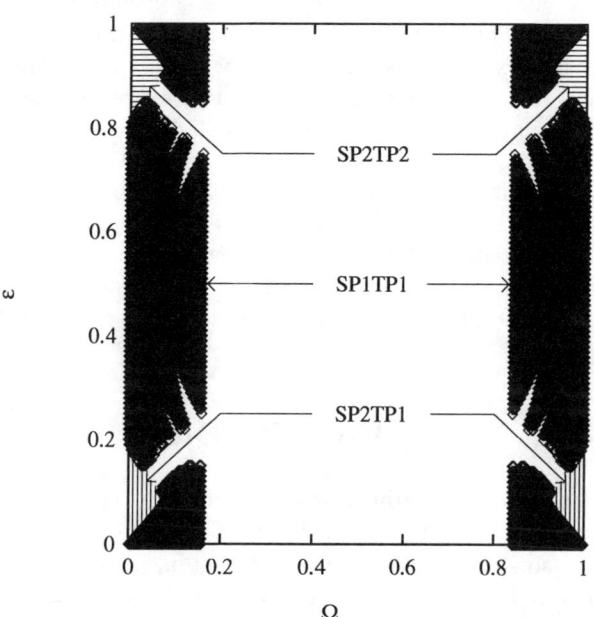

FIGURE 2. This is a 2-D slice of the 3-D plot of the Arnold tongues at $K = 1$. Inside the $\frac{P}{Q} = \frac{0}{1}$ and $\frac{P}{Q} = \frac{1}{1}$ tongues at the left and right of the plot (the only two tongues shown here), the vertical hashed regions show the regions of stability of the SP2TP1 solution, the horizontal hashed triangular regions that of the SP2TP2 solution while the dark band shows the SP1TP1 stable region.

TABLE 1. Eigenvalues at bifurcation points for $\Omega = 0$ and $K = 1$. The underlying periodicity of the new solution is detected via the closure conditions (See text).

ϵ value at $\Omega = 0.0$	M_t^2	M_t^1	M_t^{tw}	Change of solution	Type of Bifurcation
0.01	< 1	> 1	< 1	SP1TP1 to SP2TP1	Pure spatial
0.195	< 1	< 1	> 1	SP2TP1 to SP1TP1	Pure spatial
0.805	< 1	> 1	< 1	SP1TP1 to SP2TP2	Spatio-temporal
0.99	< 1	< 1	> 1	SP2TP2 to SP1TP1	Spatio-temporal

Using this and the condition for closure we can obtain the widths of the ϵ interval for which stable solutions are obtained.

We demonstrate this for the SP2TP1 solution for $\Omega = 0$ and $K = 1$. Using the closure condition and the fact that $0 \leq \theta_t(i) \leq 1$ we find that the solution is given by $\tilde{b}_t(1) = 1$ and $\tilde{a}_t(i)$ is arbitrary. The largest eigenvalue is given by $\tilde{\lambda} = 1 + \cos(\pi \tilde{a}_t(1))$. For the edge of the interval $\tilde{a}_t(i) = 0.5$. Solving for ϵ using the evolution equation for $\tilde{a}_t(i)$ we obtain the edge of the SP1TP1 solution as $\epsilon = \frac{1}{2+\pi}$ which is 0.1904 a result also obtained numerically. Numerical analysis for the other values of Ω shows the regions of stability for the three solutions.

From the stability of the eigenvalues as indicated in table I, we can conclude that the M_t^2 matrix is unable to detect the pure spatial bifurcation. On the other hand, the M_t^1 matrix is able to detect everytime the synchronized solution with temporal period one goes unstable but is unable to indicate the spatial or temporal nature of the new solution. However the matrix M_t^{tw} not only detects the spatio-temporal bifurcation but the pure spatial bifurcation too. Further the closure conditions which are in a neat form are able to indicate the spatio-temporal nature of the new solution.

Table I also gives the obtained bifurcation points for $\Omega = 0.0$ and $K = 1$ and we find that the analytic and numerical edges for ϵ for $\Omega = 0$ and $K = 1$ solution match well.

CONCLUSIONS

In conclusion, we have identified two kinds of bifurcations namely the purely spatial and the spatiotemporal one in the $\frac{P}{Q} = \frac{0}{1}, \frac{1}{1}$ tongues of a coupled map lattice of sine circle maps. We have set up new sum and difference variables and recast the equations of evolution. The stability matrix set up in terms of these new variables is able to capture the signature for all the bifurcations observed including the purely spatial bifurcation. To identify the type of solution however it is necessary to invoke the closure conditions. Such bifurcations have been observed in various experimental situations like Josephson junction arrays where the in-phase to splay phase bifurcation can be seen [3] and the synchronized to rotating wave behaviour seen in rings of coupled chaotic oscillators [4]. We also hope this analysis will be useful in the understanding and analysis of other kinds of bifurcation behavior seen in spatially extended systems.

REFERENCES

1. Coullet P. and Iooss G., *Phys. Rev. Lett.* **64**, 866 (1990)
2. Faivre G. *et al*, *Euro. Phys. Lett.* **9**, 779 (1989).
3. Wiesenfeld K. and Swift J., *Phys. Rev. E* **51**, 1020, (1995).
4. Matias M., Perez-Munuzuri V., Lorenzo M., Marino I. and Perez-Villar V., *Phys. Rev. Lett.* **78**, 219 (1997).
5. Kaneko K., *Theory and Applications of Coupled Map Lattices* , edited by K. Kaneko (John Wiley, England, 1993) and references therein.
6. Bohr T., Bak P., and Jensen M.H., Phys.Rev. A **30** 1960 (1984).
7. The initial conditions considered are of the kind $(\theta_t(i) + \theta_t(i+1)) = 1$.
8. Chatterjee N. and Gupte N., *Phys. Rev. E* **3** 4457 (1996), *Physica A* **224**, 422-432 (1996).
9. Chatterjee N. and Gupte N., unpublished.

Reconstruction of a set of differential equations modelling an experimental homoclinic chaos in the Belousov-Zhabotinskii reaction.

C. Letellier, J. Maquet, L. Le Sceller and G. Gouesbet

LESP - UMR 6614 - CORIA, Place Emile Blondel, BP 8
76131 Mont Saint-Aignan Cedex, France, E-Mail : letellie@coria.fr

F. Argoul and A. Arnéodo.

Centre de Recherche Paul Pascal, Avenue Schweitzer, 33600 Pessac, France

Abstract. We analyze chemical experimental time series by means of recent tools from nonlinear dynamics. More specifically, experiments on the Belousov-Zhabotinskii reaction [1] in a continuous flow reactor reveal a spiralling strange attractor which arises from a (global) homoclinic bifurcation. By using a global vector field reconstruction method, a set of ordinary differential equations is obtained from the measurements of the time dependence of [CeIV]. We show that a tridimensional space is sufficient to embed the behavior of the BZ reaction as suggested by previous works and that the reconstructed model allows us to exhibit topological properties which are not clearly evidenced from the experimental data. We investigate topological properties from the reconstructed set of ODEs and compare them with properties of the phase portrait directly reconstructed from the data and of the phase portrait associated with a 3D model proposed by Richetti *et al.*

I EXPERIMENTS

The Belousov-Zhabotinskii reaction is performed in an open continuously stirred tank reactor (27 ml volume) fed by the three feed lines [2] :
1) first line : $[NaBrO_3] = 7.5 \; 10^{-3}$ mol.l^{-1} $[H_2SO_4]$=1.5N
2) second line : $[CH_2(COOH)_2] = 0.15$ mol.l^{-1} $[H_2SO_4]$=1.5N
3) third line : $[Ce_2(SO_4)_3] = 5 \; 10^{-4}$ mol.l^{-1} $[H_2SO_4]$=1.5N

The three fluxes are maintained nearly equal, using a peristaltic pump. The flow rate through the reactor (φ) plays the role of the control parameter ; it is fixed to be equal to 0.130 ml.mn^{-1}. The dynamics is monitored by measuring the absorbance of the solution at 360 nm ; the stirring rate is 600 rpm while the temperature is regulated to 41°C. The time dependence of [CeIV] is recorded and a part of the time series is displayed in FIG. 1.

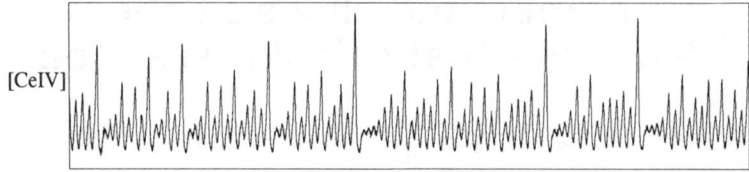

FIGURE 1. *Time evolution of [CeIV] in the case of homoclinic chaos evidenced in the BZ experiment.*

II PHASE SPACE RECONSTRUCTION

A Directly from the experimental data

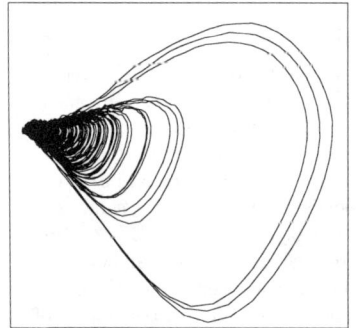 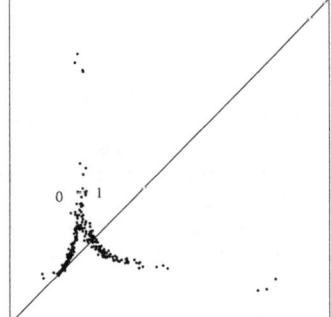

FIGURE 2. *Phase space projection by using derivative coordinates from the experimental time series.*

FIGURE 3. *First-return map computed from the phase space reconstruction from the experimental time series.*

The homoclinic reinjection process in the neighborhood of an underlying saddle focus has been evidenced [2,3]. Such a large amplitude relaxation followed by small amplitude oscillations is also evidenced by the phase space directly reconstructed from the experimental data (FIG. 2). Nevertheless, the three large amplitude oscillations are located at the beginning of the recorded time series and we conjecture that they represent a transient behaviour since they are not observed in the remaining part of the data (approximatively 86 000 points). The dimension of the phase space is expected to be equal to 3 as shown by Mindlin *et al* [4] under other operating conditions or by Richetti *et al* [5]. Consequently, a phase space spanned by the time series $X(t) = [Ce^{4+}(t)]$ and its first two time derivatives, $Y = \dot{X}$ and $Z = \ddot{X}$, is used to investigate the topological properties of the underlying dynamics. A first-return map to a Poincar section is computed and two not well visited monotonic branches are exhibited (FIG. 3). Each monotonic branche is labelled by an integer allowing us to encode periodic orbits by symbolic sequences. Moreover, the topological characterization is not easy to achieve since Moreover, the topological characterization is not easy to achieve since there is no hole in the middle of the attractor. there is no hole in the middle of the attractor.

Such a configuration, always associated with a homoclinic situation, implies that no safe Poincar section may be identified. Furthermore, an inadequate embedding is unfortunately obtained from the time series since many crossings occur in the neighborhood of the fixed point where the noise contamination degrades the observability of the dynamical structure, that is, it is difficult to count oriented crossings required for estimating linking numbers between couples of periodic orbits [6,7].

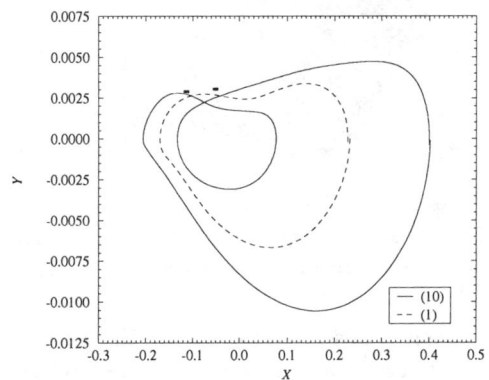

FIGURE 4. *Plane projection of a couple of periodic orbits encoded by (10) and (1), respectively. The associated linking number is equal to the half sum of the oriented crossings, i.e. $L(10,1) = -1$.*

In addition, the trajectory does not visit very well the monotonic branches of the first-return map (FIG. 3), leading to many difficulties to extract the population of periodic orbits. Indeed, we are just able to compute a small number of linking numbers as exemplified in (FIG. 4). Here, each oriented crossing is counted as being equal to ± 1 (according to the usual convention). The linking number is equal to the half sum of the oriented crossings in a regular plane projection of a couple of periodic orbits.

B Reconstructed model

We now attempt a global vector field reconstruction starting from the time evolution of [CeIV]. As we use successive time derivatives of the experimental time series [8,9], the set of reconstructed ODEs may be written as :

$$\begin{cases} \dot{X} = Y \\ \dot{Y} = Z \\ \dot{Z} = F_s(X,Y,Z) \end{cases} \quad (1)$$

where F_s is a multivariate polynomial function to be estimated from the data. After removal of a small drift observed on the data and having applied a slight smoothing, an estimated function $\tilde{F}_s$ (with 56 monomials) is found. A plane projection of the trajectory obtained by integrating the reconstructed set of ODEs is displayed in FIG. 5, superimposed to the phase portrait directly reconstructed from the experimental data, exhibiting a favourable comparison.

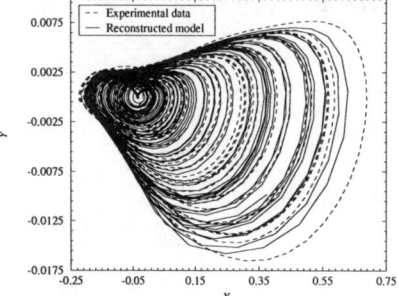

FIGURE 5. *Phase space projection obtained by integrating the reconstructed set of ODEs.*

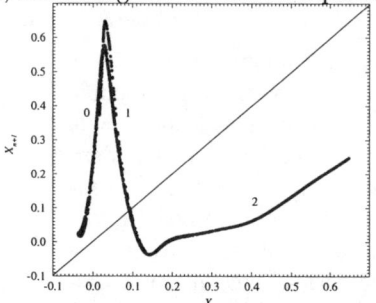

FIGURE 6. *First-return map obtained by integrating the reconstructed set of ODEs.*

Since it is possible to integrate the reconstructed model during a long time, we now possess a large amount of pseudo-periods allowing an easy extraction of periodic orbits. By using a Poincar section, a first-return map is computed and three monotonic branches are clearly evidenced (FIG. 6) while the first-return map associated with the experimental data only exhibits two monotonic branches (FIG. 3). The attractor generated by the reconstructed model may therefore be splitted into three strips whose topological properties are different (FIG. 8).

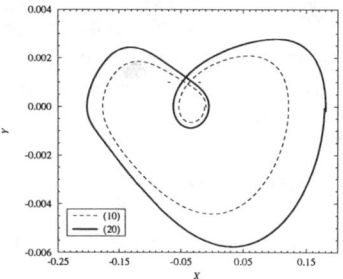

FIGURE 7. *Couple of periodic orbits encoded by (210) and (10), respectively. The linking number $L(210, 10)$ is found to be equal to -2.*

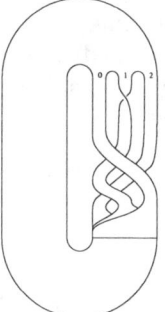

FIGURE 8. *Template of the reconstructed model as a schematic view of the topological properties of the dynamics.*

By counting linking numbers between couples of periodic orbits (for instance, FIG. 7), we found that the template characterizing the reconstructed model is as displayed in FIG. 8. We then observed that all linking numbers between periodic orbits extracted from the experimental data are in agreement with this template. The model is therefore topologically validated.

From the first-return map associated with the reconstructed model, a layered structure on branches 0 and 1 is observed. It may be possible that such a structure could be diffused by noise in the case of the experimental data (FIG. 3). Following works by Fang [10], such a layered structure does not disturb the topological analysis since only low periodic orbits are implied in the topological characterization. In all cases, the branches 0 and 1 are well enough defined.

We have shown that many difficulties to characterize the dynamics starting from the experimental data have been encountered due to i) the external noise perturbations which diffuse the structure of the dynamics, ii) the homoclinic situation implying many oriented crossings in the neighborhood of the fixed point which are perturbed by the noise contamination and iii) the small amount of pseudo-periods (154) forbidding the efficient extraction of periodic orbits. Conversely, the topological characterization of the phase portrait obtained by integrating the reconstructed model is easily achieved since i) the noise has been removed in the reconstructed model (see [9]), ii) the structure of the dynamics is clearly exhibited and iii) a large amount of pseudo-periods (more than 4000) is generated from the model.

C Comparison with a model

We now compare the reconstructed model with a phenomenological model proposed by Richetti et al [5]. This model is of interest since it is quite similar to the model (1) excepted for the third equation which is replaced by

$$\dot{Z} = -\eta Z - \nu Y - \mu X - k_1 X^2 - k_2 Y^2 - k_3 XY - k_4 XZ - k_5 X^2 Z \qquad (2)$$

with the control parameters available from [2]. A plane projection of the phase portrait associated with this phenomenological model is displayed in (FIG. 9). It looks rather different from phase portraits obtained either from the experimental data (FIG. 2) or by integrating the reconstructed model (FIG. 5). Moreover, a first-return map is found with four monotonic branches (FIG. 11). The template of the phenomenological model is found to be different from the one associated with the reconstructed model. Indeed, it is not even compatible with the one extracted from the reconstructed model. In other words, strip 2 of the reconstructed template presents a null local torsion while strip 3 of the phenomenological template has a local torsion equal to -3. In addition, these strips are not linked with the others in the same way. It means that the topological properties are different and, consequently, the dynamical behaviour is different. Although the model by Richetti et al has been validated from the Silnikov point of view, many departures from the data arise. We think that this model should be modified following the reconstructed model.

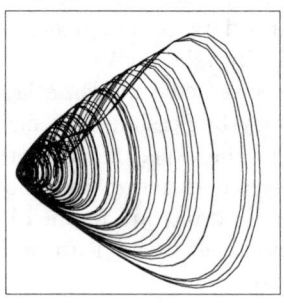

FIGURE 9. *Plane projection of the phase portrait associated with the phenomenological model by Richetti et al.*

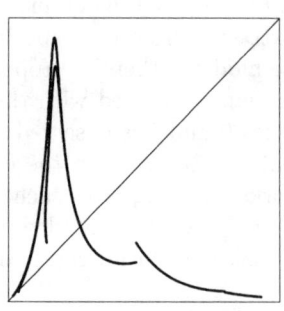

FIGURE 10. *First-return map of the 3D model by Richetti et al.*

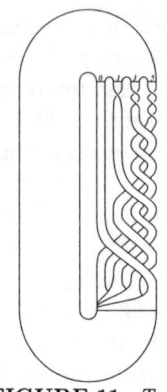

FIGURE 11. *Template of the model by Richetti et al.*

III CONCLUSION

Using a reconstructed model may be very helpful to characterize dynamical behaviours and, particularly, in the case of a homoclinic chaos. Indeed, we showed that topological properties of the dynamics may be evidenced by investigating the phase portrait generated by the reconstructed model while they were diffused by the noise contamination in the experimental data. As the estimated function $\tilde{F}_s$ presents a rather large number of monomials, we plan to use some prescriptions to avoid an overparametrization of the model as proposed by Aguirre and Mendes [11].

REFERENCES

1. *Oscillations and traveling waves in chemical systems*, Ed. R. J. Field & M. Burger., John Wiley & Sons, New York, 1985.
2. Argoul F., Arnéodo A., Richetti P. *Phys. Let. A* **120** (6), 269, (1987).
3. Arnéodo A., Argoul F., Elezgaray J., Richetti P. *Physica D* **62**, 134, (1993).
4. Mindlin G. B., Solari H. G., Natiello M. A., Gilmore R., Hou X. J. *J. Nonlin. Sci.* **1**, 147, (1991).
5. Richetti P., Argoul F. & Arnéodo A. *Phys. Rev. A* **34**, 726, (1986).
6. Tufillaro N. B., Abbott T., Reilly J. *An Experimental Approach to Nonlinear Dynamics and Chaos,* Addison-Wesley, New York, (1992).
7. Letellier C., Dutertre P., Maheu B. *Chaos* 5 (1), 271-282, (1995).
8. Gouesbet G., Letellier C. *Phys. Rev. E* **49** (6), 4955-4972, (1994).
9. Letellier C., Le Sceller L., Dutertre P., Gouesbet G., Fei Z. & Hudson J. L. *J. Phys. Chem.* **99**, 7016-7027, (1995).
10. Fang. H. P. *J. Phys. A* **28**, 3901, (1995).
11. Aguirre L. A. & Mendes E. M. A. M. *Int. J. Bif. & Chaos* **6** (2), 279, (1996).

Multistability and invariants in delay-differential equations

Boualem Mensour and André Longtin

Department of Physics, University of Ottawa
150 Louis Pasteur, Ottawa, Ontario, Canada K1N 6N5

Abstract.
Nonlinear delay-differential equations (DDE's), such as the Mackey-Glass or the Ikeda equations, are strongly multistable when the delay-to-response time ratio is large and have an attractor dimension that increases with the delay. We discuss how properties such as dynamical invariants and multistability can be studied using linear stability analysis and power spectra. As the delay increases, the spectrum converges to an exponentially decaying envelope with a superimposed periodic modulation. Linear stability analysis can predict the position of the peaks of this modulation and quantify multistability. Also, the number of peaks in the power spectrum and the number of linear modes within an inverse characteristic autocorrelation time can be used to estimate the attractor dimension.

INTRODUCTION

Neural and optical delayed-feedback systems can have a large variety of self-oscillatory modes, exhibit bifurcation routes to chaos such as period-doubling, and can also exhibit multistability at large delay. These systems are naturally modeled by nonlinear delay-differential equations (DDE's) in which the time evolution depends not only on the present state but also on states at a given time in the past. DDE's are infinite-dimensional dynamical systems. However, their phase space attractors have a finite fractal dimension proportional to the delay [1], making them useful to study general properties of high-dimensional chaos such as occurs in fluid or optical turbulence.

Multistability is the coexistence of periodic solutions, chaotic solutions, or both [2]. This property makes DDE's important for memory storage purposes [3]. The information can be encoded into stable periodic solutions [4,5] or unstable periodic solutions stabilized by chaos-control techniques [6] using special initial functions.

In this paper, we report how certain properties of DDE's at large delay, such as dynamical invariants and multistability, can be studied using power spectra and linear stability analysis. These numerical techniques are far simpler for estimating dynamical invariants such as the attractor dimension (D) and Lyapunov exponents in comparison with usual methods (such as the correlation integral), especially when the attractors are high-dimensional (e.g. $D > 10$). These are the kinds of attractor dynamics that are useful for storage purposes.

The power spectrum of the Mackey-Glass equation at large delay exhibits a periodic modulation superimposed on an exponentially decaying background, as first observed by Farmer [1]. We found that the peaks of this modulation are associated with the modes of oscillations predicted by linear stability analysis of the dynamics around the fixed point. Also, the number of such modes below a characteristic frequency agrees closely with the Lyapunov dimension of the attractor. Further, the inverse of the decay rate of the spectrum at mid-frequencies agrees with the sum of the positive Lyapunov exponents, and is proportional to the sum of the real parts of the unstable linear modes. Our results suggest that for this class of dynamical systems, connections between spectral properties and dynamical invariants occur at mid-frequencies, rather than in the asymptotic frequency range as suggested in ref. [7]. Detailed results are in [8].

OSCILLATIONS IN THE SPECTRA OF DDE'S

Our study focusses on the Mackey-Glass (MG) delay-differential equation

$$\frac{dx(t)}{dt} = -bx(t) + F(x(t-\tau)) \; ; \quad F(x(t)) = \frac{ax(t)}{1 + x(t)^c}, \tag{1}$$

with parameters $a = 0.2, b = 0.1, c = 10$ [1,7]. As the ratio $R = \tau/\tau_r$ of the delay τ to response time $\tau_r = 1/b$ increases, the chaotic solution loses some of its regularity, as can be seen on going from Fig.1a to Fig.1b. When the delay is small (e.g. $\tau = 25$, as in Fig.1c), the spectrum is broadband with a few broad peaks. There is no clear relationship between the position of these few peaks and the frequency of the linear modes. However, if the delay is increased further to e.g. $\tau = 200$ (Fig.1d), a clear periodic modulation of the broadband background appears. This modulation dies out past $\alpha\delta^{-1}$, where α is a constant that depends on the feedback used ($\alpha \approx 1$ in the MG model), and δ is the autocorrelation time of $F(x(t))$. This time is estimated as the $1/e$ time from the maximal value at lag zero of the central peak in the autocorrelation function (Fig.2a). The number of peaks in the power spectrum increases linearly with the delay. The peaks are no longer visible for $\tau \geq 2000$, since their spacing is proportional to $1/\tau$; thus, past a certain delay, they can no longer be resolved for a given level of spectral frequency resolution.

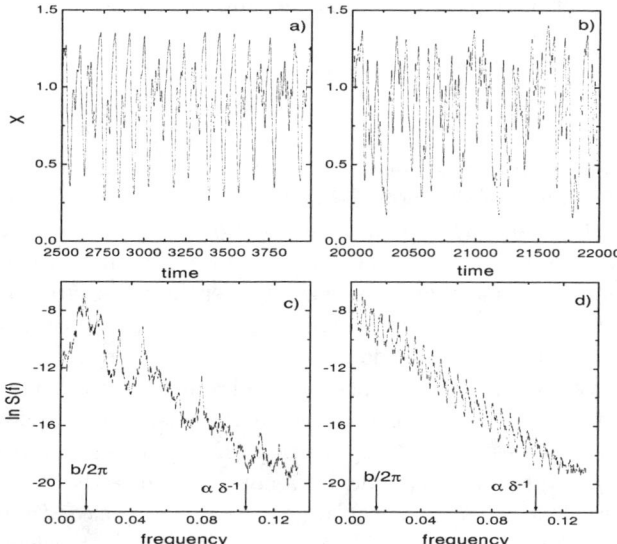

FIGURE 1. a-b) Solutions of Eq.1 for $\tau = 25$, and $\tau = 200$ respectively. c-d) Power spectrum of solutions in (a-b). The spectra are obtained using the FFT algorithm with a Hanning window. The spectra are averaged over 100 consecutive 8192-point time series.

For large delay, the spectra show three different ranges of frequencies: 1) the low frequency range $0 < f < b/2\pi$ with a Lorentzian-type of decay, which contains the most energetic Fourier modes of the chaotic motion (see [9]); 2) the mid-to-high frequency range (or dissipative range) $b/2\pi \leq f \leq \alpha\delta^{-1}$, which has an exponential-type decay $S(f) = Ae^{-\mu^{-1}f}$ (A is a constant and μ^{-1} is the decay rate) with a superimposed modulation; 3) the very high frequency range $f > \alpha\delta^{-1}$ which exhibits a slower decay rate (not shown). As R increases beyond the value 5, the mean decay rate of the background in region (2) stays the same.

LINEAR STABILITY ANALYSIS AND DYNAMICAL INVARIANTS

The frequencies of the modes that compose the periodic modulation in Fig.1d are solutions of the characteristic equation (Eq.2). The latter is obtained by linearizing the DDE around its fixed point x^* and taking trial solutions of the form $u(t) = x(t) - x^* = u_0 e^{st}$:

$$s + b - \beta e^{\theta\tau} = 0, \qquad (2)$$

where $\beta \equiv F'(x^*)$. This equation has an infinite number of roots $s_n = \lambda_n \pm i2\pi f_n$, with f_n satisfying

$$\frac{(n-1/2)}{2\tau} < f_n < \frac{n}{2\tau} \quad n = 1, 3, 5, \cdots \qquad (3)$$

All modes show up as clear peaks in the power spectrum at frequencies corresponding to those of the roots, with the most unstable ones being the dominant ones (Figs.1d, 2b). Some of the stable modes close to the imaginary axis appear also in the power spectrum because of nonlinear coupling to the unstable modes. It is as if all these eigenfrequencies are excited by the chaotic fluctuations, even though they are stable in the absence of these fluctuations. The number of degrees of freedom in the system is directly related to these "spectrally visible" unstable and stable modes of the linearized system.

The one-to-one correspondence between the peaks in the power spectrum and the linear modes is valid only for large R. For lower R values, as in [10] (see also Fig.1c), this correspondence is not exact since the modes are not at the same frequencies as the spectral peaks. It is also clear from our results that most of the energy in the chaotic waveform lies in the low frequencies. This is similar to the case of fluid turbulence, where most of the energy lies in the long wavelength modes.

We also note that the real parts of the roots decrease with increasing delay (Fig.2b). This explains why the peaks in the power spectrum decrease in amplitude with increasing delay. An approximate expression for the frequencies of these modes can be obtained for $\lambda_n \approx 0$:

$$f_n = \frac{n}{2\tau}(1 - \frac{1}{b\tau}) \approx \frac{n}{2\tau} \quad n = 1, 3, 5, \cdots \qquad (4)$$

and for n large, these modes decay as

$$\lambda_n = -\frac{\ln(-2\pi f_n/\beta)}{\tau} \approx -\frac{1}{\tau}\ln n - \frac{1}{\tau}\ln\left(\frac{-\pi}{\beta\tau}\right) \qquad (5)$$

This dependence of the real part of the eigenvalue on the logarithm of the root number has also been found for the Lyapunov exponents [1,9]. The decay rates (as a function of n) of the real part of the roots and of the negative Lyapunov exponents are found to be the same when n becomes large. Thus, these negative Lyapunov exponents can be estimated from the expression for the stable modes (Eq.5), i.e.

$$\lambda^L_{(n+1)/2} = \lambda_n \qquad (6)$$

for n odd and sufficiently large.

These negative modes do not, a priori, provide information on attractor dimension and Kolmogorov-Sinai entropy, since one needs the positive exponents and the first few negative ones to estimate such quantities. Nevertheless,

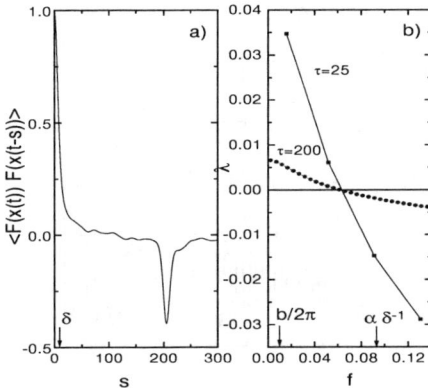

FIGURE 2. a) Autocorrelation function of the feedback term $F(x(t))$ in (Eq.1) for $\tau = 200$. The autocorrelation time δ is found at $1/e$. b) Roots of Eq.2 for $\tau = 25$ and $\tau = 200$. Note that the real part λ of the root (ordinate) is plotted against its imaginary part $\omega/(2\pi) = f$.

this agreement between modes and exponents shows that apparently nonlinear characteristics of DDE's, such as self-oscillation frequencies and negative Lyapunov exponents, are in fact closely related to simple linear characteristics.

We have found [8] that the sum of positive Lyapunov exponents is equal to the inverse decay rate of the power spectra μ in the dissipative range, and is proportional to the sum of the real parts of the unstable modes.

We have also found [8] that the Lyapunov dimension can be estimated as the number of peaks (or the number of roots in Eq.2) within the frequency range $(0, \alpha\delta^{-1})$. For example, in the MG model for $\tau = 200$, we found $\delta \simeq 9.72$ and a dimension $D_L = \alpha\tau/\delta \simeq 21$. The number of peaks in the power spectrum within $(0, \alpha\delta^{-1})$ frequency range is also found to be approximately 21, in good agreement with the Lyapunov dimension found in [1,11].

The Lyapunov dimension is also found to be proportional to the ratio of the (nonzero) lag of the first autocorrelation peak to δ, since this lag corresponds to the delay τ in Eq.1 (see Fig.2a).

Finally, the number of multistable solutions observed for a given delay is found to correspond to the number of unstable modes. This correspondence is obtained when the parameter a of the feedback function is chosen such that the fundamental solution (i.e. the limit cycle past the Hopf bifurcation) is square-wave-like with smooth plateaus free from spurious structure such as spikes. Figure 3 shows the coexistence of three periodic multistable solutions for $a = 0.1376$ and $\tau = 300$.

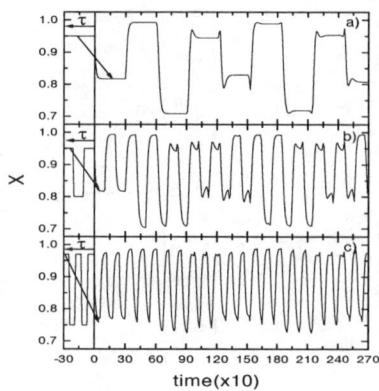

FIGURE 3. Multistable periodic solutions of Eq.1 obtained for $a = 0.1376$ and $\tau = 300$ using piecewise constant initial conditions: a) The fundamental (period-doubled) solution; b) The third harmonic; c) The fifth harmonic.

CONCLUSION

Power spectra and linear stability analysis are relevant to understand multistability and chaos in DDE's. Estimating invariants using such simple techniques is advantageous in comparison with standard algorithms to compute correlation integrals or Lyapunov exponents. Our results hold for the MG and Ikeda systems over a range of delays [8]. Future work will assess their validity for other values of a, b, c in Eq.1.

REFERENCES

1. J.D. Farmer, *Physica* **4D**, 366 (1982).
2. K. Ikeda, K. Kondo, and O. Akimoto, *Phys. Rev. Lett.* **49**, 1467 (1982).
3. K. Ikeda and K. Matsumoto, *Physica* **29D**, 223 (1987).
4. T. Aida and P. Davis, *IEEE J. Quantum Electron.* **28**, 686 (1992).
5. J. Foss, A. Longtin, B. Mensour and J. Milton, *Phys. Rev. Lett.* **76**, 708 (1996).
6. B. Mensour and A. Longtin, *Phys. Lett. A* **205**, 18 (1995).
7. D.E. Sigeti, *Physica* **82D**, 136 (1995). *Phys. Rev. Lett.* **49**, 1467 (1982).
8. B. Mensour and A. Longtin (*Physica D*, in press).
9. K. Ikeda and K. Matsumoto, *J. Statist. Phys.* **44**, 955 (1986).
10. K. Ikeda, H. Daido, and O. Akimoto, *Phys. Rev. Lett.* **45**, 709 (1980).
11. M. Le Berre, E. Ressayre, A. Tallet, H.M. Gibbs, D.L. Kaplan, and M.H. Rose, *Phys. Rev. A* **35**, 4020 (1987).

NONLINEAR STOCHASTIC PHENOMENA

Adiabatically Rocked Quantum Ratchets

Peter Reimann, Milena Grifoni, and Peter Hänggi

Universität Augsburg, Memminger Str. 6, D-86135 Augsburg, Germany

Abstract. We investigate quantum Brownian motion in adiabatically rocked ratchet systems. Above a cross-over temperature T_c tunneling events are rare; yet they already substantially enhance the classical particle current. Below T_c, quantum tunneling prevails and the classical predictions grossly underestimate the transport. Upon approaching $T = 0$ the quantum current exhibits a tunneling induced reversal, and tends to a finite limit.

INTRODUCTION

Traditional heat engines are devices to extract useful work out of thermal fluctuations by way of transferring heat between equilibrium baths at different temperatures. More realistic set-ups, involving also non-thermal forces, have been addressed quantitatively only since a few years under the label of "Brownian motors", "molecular motors", or "ratchets" [1,2]. Besides their principal interest and the diverse astonishing effects they can produce, they also entail a variety of interesting technological applications [2,3], and may be of relevance for intracellular transport as well [4]. In this note we highlight the intriguing features of a Brownian motor when quantum effects start to play an important role [5]. At sufficiently low temperatures, our predictions should be observable in mesoscopic structures such as the superconducting quantum interference device (SQUID) proposed in [6]. Using recent technical developments [7], semiconductor superlattices could be designed which, too, exhibit a quantum ratchet effect. On top of that, our results are also of potential relevance for biological transport phenomena that involve transfer of light particles such as electron- or protons-reactions.

MODEL

Our starting point is the system-plus-bath Hamiltonian

$$\mathbf{H}(t) = \mathbf{p}^2/2m + V(\mathbf{x}) - \mathbf{x} f(t) + \mathbf{H}_B \,, \quad (1)$$

where $\mathbf{x}$, $\mathbf{p}$, and m are the coordinate operator, momentum operator, and mass of the quantum particle, respectively. Furthermore, the "ratchet"-potential $V(x)$ is assumed to be asymmetric and periodic, for instance (cf. Fig. 1)

$$V(x) = V_0 \left[\sin(2\pi x/L) - 0.22 \sin(4\pi x/L) \right] \,, \quad (2)$$

and $f(t)$ represents an unbiased non-thermal driving force. Finally, $\mathbf{H}_B$ describes the heat bath interacting with the particle and we adopt its usual modelization by an ensemble of harmonic oscillators at thermal equilibrium with a coupling bilinear in the bath and particle coordinates [8]

$$\mathbf{H}_B = \sum_{j=1}^{N} \left(\frac{\mathbf{p}_j^2}{2\,m_j} + \frac{m_j \omega_j^2}{2} \left(\mathbf{q}_j - \frac{c_j \mathbf{x}}{m_j \omega_j^2} \right)^2 \right) \,, \quad (3)$$

where $\mathbf{q}_j$ and $\mathbf{p}_j$ are the coordinate and momentum operators of the bath oscillators. The effect of the remaining model parameters m_j, ω_j, and c_j are completely fixed in the continuum limit $N \to \infty$ by the spectral density

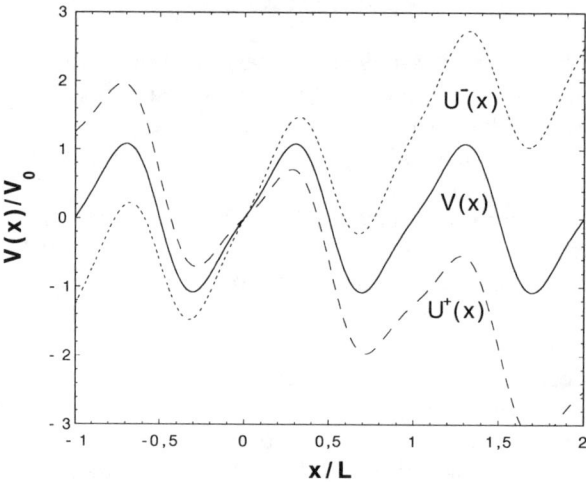

FIGURE 1. Solid: ratchet potential $V(x)$ in (2). Dashed and dotted: "tilted washboard potentials" $U^\pm(x)$ in (8) with $Fl = 0.2\,V_0$, $l = L/2\pi$.

$$J(\omega) = \frac{\pi}{2} \sum_{j=1}^{\infty} \frac{c_j^2}{m_j \omega_j} \delta(\omega - \omega_j) \ . \tag{4}$$

Focusing on so-called Ohmic friction [8], i.e.,

$$J(\omega) = \omega \eta \ , \tag{5}$$

the bath oscillators can be integrated out and the dynamics of the quantum particle in (1) can be rewritten as operator-valued quantum Langevin equation

$$m \ddot{\mathbf{x}}(t) = -\eta \dot{\mathbf{x}}(t) - V'(\mathbf{x}(t)) + f(t) + \boldsymbol{\xi}(t) \ . \tag{6}$$

Here, η is the viscous damping coefficient and $\boldsymbol{\xi}(t)$ a self-adjoint thermal noise operator with a Gaussian statistics of vanishing mean $\langle \boldsymbol{\xi}(t) \rangle$ and a symmetrized correlation $\frac{1}{2}\langle \boldsymbol{\xi}(t)\boldsymbol{\xi}(0) + \boldsymbol{\xi}(0)\boldsymbol{\xi}(t) \rangle = k_B T \eta \frac{d}{dt} \coth(\pi k_B T t/\hbar)$ (fluctuation dissipation theorem with T, k_B, and $\hbar$ representing temperature, Boltzmann's constant, and Planck's constant over 2π, respectively).

The quantity of foremost interest in our above defined ratchet dynamics is the particle current in the steady state

$$J = \lim_{t \to \infty} \langle \dot{\mathbf{x}}(t) \rangle \ , \tag{7}$$

where $\langle \ \rangle$ denotes the quantum statistical mechanical expectation value together with a time average over the driving force.

In general, this requires the solution of a highly non-trivial far from equilibrium problem. To simplify matters, we restrict ourselves to very slowly varying forces $f(t)$ such that the system can always adiabatically adjust to the instantaneous thermal equilibrium state (accompanying equilibrium). We furthermore assume that $f(t)$ is basically restricted to the values $\pm F$, i.e., the transitions between $\pm F$ occur on a time scale of negligible duration in comparison with the time the particles in (6) are exposed to either of the "tilted washboard" potentials

$$U^{\pm}(x) = V(x) \mp F x \tag{8}$$

see also Fig. 1. As a final assumption we require a positive but not too large F, such that $U^{\pm}(x)$ still display a local maximum and minimum within each period L. Apart from these premises, the driving $f(t)$ can be either of stochastic or of deterministic nature. In particular, our results presented in the next section are valid both for stochastic and deterministic choices of $f(t)$.

To completely fix the model, we still have to specify the 5 parameters m, η, V_0, F, and $l := L/2\pi$ in (2),(6),(8). We do this by prescribing 5 dimensionless numbers as follows: First, we fix V_0, F, l and thus $U^{\pm}(x)$ through $Fl/V_0 = 0.2$, $\Delta U^{min}/V_0 = 1.423$, and $|U''_+| l^2/V_0 = 1.330$, where U''_+ denotes the curvature of $U^+(x)$ at a local maximum and ΔU^{min} is the smallest of the 4 different potential barriers between adjacent local minima of $U^+(x)$ and $U^-(x)$. This choice of V_0, F, and $l = L/2\pi$ corresponds to the situation depicted in Fig. 1. Next we choose $\eta/m\Omega_0 = 1$ with $\Omega_0 := [V_0/l^2 m]^{1/2}$, meaning a moderate damping as compared to inertia effects. To see this, we notice that Ω_0 approximates rather well the true ground state frequency ω_0^+ in the potential $U^+(x)$, $\omega_0^+ = 1.153\,\Omega_0$, and similarly for $U^-(x)$. In particular, $\eta/m\Omega_0 = 1$ rules out the occurrence of "deterministically running classical solutions" both in $U^+(x)$ and $U^-(x)$. Before specifying our last dimensionless number we remark that the temperature T will not be fixed but rather used as control parameter. We, however, will restrict ourselves to thermal energies $k_B T$ much smaller than ΔU^{min} (so-called semiclassical condition) such that meaningful transition rates between adjacent minima of $U^\pm(x)$ can be defined and employed to determine the transport property (7) of our ratchet dynamics [5]. It then can be shown [8] that in the potential $U^+(x)$ genuine quantum tunneling events "through" the potential barrier are rare above a crossover temperature

$$T_c^+ = \frac{\hbar\,\mu^+}{2\pi k_B}\,, \qquad \mu^+ = \frac{\sqrt{\eta^2 + 4m|U_+''|} - \eta}{2m}\,, \qquad (9)$$

while for $T < T_c^+$ tunneling yields the dominant contribution to the transition rates. An analogous crossover temperature T_c^- arises for the potential $U^-(x)$ which is typically not identical but rather close to T_c^+. With the definitions

$$T_c^{max} = \max\{T_c^+, T_c^-\}\,, \qquad T_c^{min} = \min\{T_c^+, T_c^-\} \qquad (10)$$

we now fix our last dimensionless quantity through $\Delta U^{min}/k_B T_c^{max} = 10$. In this way, the weak noise condition is safely fulfilled for $T \leq 2T_c^{max}$, i.e., up to temperatures well above both T_c^+ and T_c^-. At the same time, the so-called semiclassical condition [8] can be taken for granted when evaluating the quantum mechanical transition rates for all $T \leq 2T_c^{max}$. Adopting a path integral treatment of the full system-plus-bath problem (1) this condition allows one to work within a saddle point approximation scheme [8]. For more details regarding the calculation of those rates and their relation to the current (7) we refer to [5].

RESULTS

We performed in our work [5] the first numerical dissipative, low-temperature calculations to tackle the involved saddle point problem arising in the determination of the exponentially leading contribution (bounce action) to the transition rates in a generic ratchet potential (2); moreover, we have evaluated the full prefactors (ratios of functional determinants) which dominate the non-exponential contributions to the incoherent, dissipative quantum tunneling rates in the semiclassical approximation. Our results for the quantum ratchet model as specified in the previous section are depicted in Fig.2. Shown are the current J_{qm} from the above sketched quantum mechanical treatment together with the result J_{cl} that one would obtain by means of a purely classical calculation. The small dashed part in J_{qm} in a close vicinity of the crossover temperatures T_c^{max} and T_c^{min} from (10) signifies an increased uncertainty of the semiclassical rate theory in this temperature domain.

Our first observation is that even above T_c^{max}, quantum effects may enhance the classical transport by more than a decade. They become negligible only beyond several T_c^{max}. In other words, significant quantum corrections of the classically predicted particle current set in already well above the cross-over temperature T_c, where tunneling processes are still rare. (They can be associated to quantum effects other than genuine tunneling "through" a potential barrier.) With decreasing temperature, $T < T_c^{min}$, quantum transport is even much more enhanced in comparison with the classical results [1b,1e]. A fur-

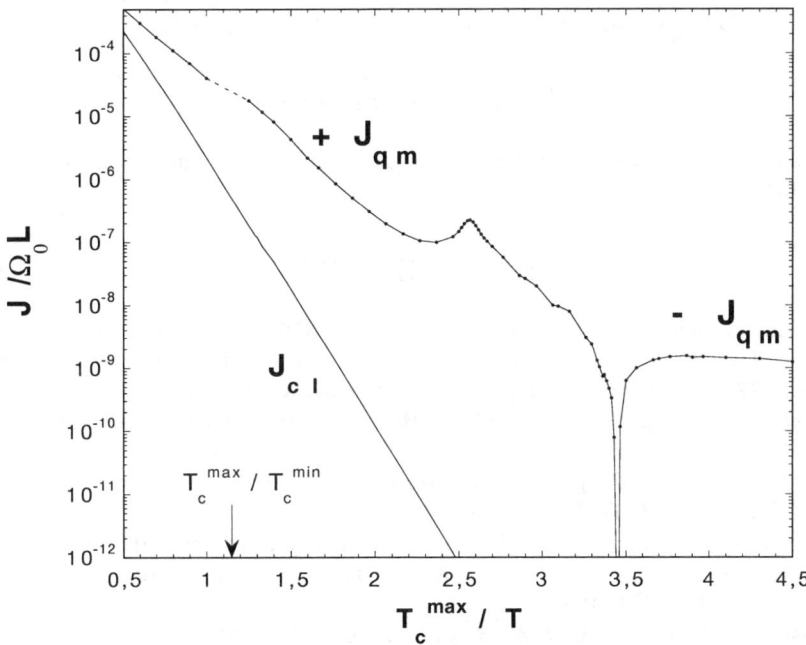

FIGURE 2. The quantum mechanical steady state current J_{qm} from (7) and its classical counterpart J_{cl} for the ratchet potential from Fig. 1 in dimensionless units $J/\Omega_0 L$. Note the change of sign, the finite $T \to 0$ limit, and the non-monotonicity of J_{qm}. For more details see main text.

ther remarkable feature caused by the intriguing interplay between thermal noise and quantum tunneling is the inversion of the quantum current direction at very low temperatures. In a classical description, such a reversal for adiabatically slow driving is ruled out. Finally, J_{qm} approaches a finite (negative) limit when $T \to 0$, implying a finite (positive) stopping force [2,6] also at $T = 0$. In contrast, the classical prediction J_{cl} remains positive but becomes arbitrarily small with decreasing T. A curious detail in Fig. 2 is the non-monotonicity of J_{qm} around $T_c^{max}/T \simeq 2.5$, caused by a similar resonance-like T-dependence in one of the underlying quantum mechanical transition rates. A better understanding of this issue is the subject of ongoing work.

We also studied other parameter values than those used in Fig. 2 as well as somewhat modified potentials (2). Basically, the same qualitative results are found except that the non-monotonous temperature dependence disappears for sufficiently large $\Delta U^{min}/k_B T_c^{max}$ values. Thus all the above described novel features appear to be typical for a large class of quantum ratchet systems. Such effects clearly become of paramount importance for applications in mesoscopic systems at low temperatures. Note that T_c can reach values larger than 100K in some physical and chemical systems, while it is in the mK region in Josephson systems [8].

Acknowledgements

M.G. and P.H. gratefully acknowledge the Deutsche Forschungsgemeinschaft (Ha 1517/14-2 and Ha 1517/13-1).

REFERENCES

1. Ajdari A., and Prost J., *C. R. Acad. Sci. Paris* **315**, 1635 (1992); Magnasco M. O., *Phys. Rev. Lett.* **71**, 1477 (1993); Astumian R. D., and Bier M., *Phys. Rev. Lett.* **72**, 1766 (1994); Doering C. R., Horsthemke W.,and Riordan J., *Phys. Rev. Lett.* **72**, 2984 (1994); Bartussek R., Hänggi P., and Kissner J. G., *Europhys. Lett.* **28**, 459 (1994).
2. For a review see: Hänggi P., and Bartussek R., in *Lecture Notes in Physics* ed. by J. Parisi et al., Berlin, Springer, 1996, pp. 294-308.
3. Rousselet J., Salome L., Ajdari A., and Prost J., *Nature* **370**, 446 (1994); Faucheux L. P., Bourelieu L. S., Kaplan P. D., and Libchaber A. J., *Phys. Rev. Lett.* **74**, 1504 (1995).
4. Astumian R. D., and Bier M., *Biophys. J.* **70**, 637 (1996).
5. Reimann P., Grifoni M., and Hänngi P., *Phys. Rev. Lett.* **79**, 10 (1997).
6. Zapata I., Bartussek R., Sols F., and Hänggi P., *Phys. Rev. Lett.* **77**, 2292 (1996).
7. Keay B. J. et al., *Phys. Rev. Lett.* **75**, 4098 (1995); *ibid.* **75**, 4102 (1995).
8. For a review see: Hänggi P., Talkner P., and Borkovec M., *Rev. Mod. Phys.* **62**, 251 (1990).

Quantum Steps in Hysteresis Loops

M. Thorwart*, P. Reimann*, and P. Jung+

*Institut für Physik, Universität Augsburg, Memminger Straße 6, 86135 Augsburg, Germany and +School of Physics, Georgia Institute of Technology, Atlanta, GA 30332, USA

Abstract. We are studying bistable quantum systems coupled to a thermal environment and driven by a strong external force. Thermal and quantum effects on the switching hysteresis are studied numerically by evaluating a time-dependent real-time double path-integral with the recently developed iterative tensor multiplication scheme for the *Quasi-adiabatic path-integral propagator method*. For temperatures below the quantum-classical crossover temperature, the response of the system exhibits steps for slow external driving. We show that these steps are due to dissipative resonant tunneling and appear in the vicinity of avoided level-crossings of the spectrum of the undamped Hamiltonian. We discuss possible relations to the recently discovered hysteresis steps in organic high-spin molecules.

INTRODUCTION

In recent experiments, macroscopic quantum tunneling has been observed in measurements of the magnetization of a macroscopic sample of a crystalline organic compound, Mn_{12} acetate, by Friedman et al. [1]. They found steps in the magnetization at regular intervals of the magnetic field in the hysteresis loop of oriented Mn_{12} molecule crystals. This macroscopic quantum effect has been confirmed by serveral groups [2–4]. The steps in the hysteresis loops are interpreted as evidence for thermally assisted, field-tuned resonant tunneling of the macroscopic magnetization of an ensemble of identical high–spin molecules. The Mn_{12} molecule has the spin quantum number $S = 10$. In the absence of an external magnetic field, the spin of the molecule has two degenerate ground states corresponding to spin parallel $(m = S)$ (m is the magnetic quantum number) and spin antiparallel $(m = -S)$ to the easy-axis. Experiments indicate a large magnetocrystalline anisotropy implying that the two degenerate ground states are separated by a large anisotropy barrier. Applying an external magnetic field breaks the degeneracy of the ground states and all other states. The energy levels with $m < 0$ and $m > 0$ are shifted against each other. At certain values of the external magnetic field, however, levels at $m > 0$ and $m < 0$ can become degenerate again, allowing for *resonant tunneling* of the spin between the degenerate states. In the ensemble of magnetic molecules, this can become observable as a *macroscopic effect* and is therefore called macroscopic tunneling of magnetization. At the values of the magnetic field where this degeneracy occurs, the macroscopic hysteresis curves exhibit steps.

While these experiments are carried out with spin-systems, we will study here the question whether the observed macroscopic behavior is a general feature of metastable quantum dynamics. To this end, we investigate quantum tunneling of an ensemble of massive particles in a periodically driven *bistable* potential. The potential barrier reflects the anisotropy-barrier of the molecules and the potential minima the ground states. The periodic forcing mimics the shift of the energy-levels due to the interaction of the molecules with the external magnetic field.

CLASSICAL DYNAMICAL HYSTERESIS

A model for classical hysteresis is the strongly damped motion of a particle with mass M in a double–well potential $V(q)$, tilted by an external periodic force $F(t)$. The double well potential is given by

$$V(q) = \frac{M^2 \omega_0^4}{64 \Delta U} q^4 - \frac{M \omega_0^2}{4} q^2, \qquad (1)$$

where $M\omega_0^2$ is the curvature of the potential at the minima and ΔU is the barrier height. Using the same dimensionless units as in the quantum-case below, i.e. $t \to \omega_0 t$, $q \to \sqrt{M\omega_0/\hbar}\, q$, $\gamma \to \gamma/\omega_0$, $\Delta U \to \Delta U/\hbar\omega_0$, $A \to A/\sqrt{\hbar M \omega_0^3}$, the equation of motion reads

$$\ddot{q} + \gamma \dot{q} + \frac{1}{16 \Delta U} q^3 - \frac{1}{2} q = -A \cos(\Omega t), \qquad (2)$$

with the damping constant γ, the amplitude A and the frequency Ω of the external force. For $\Omega \ll \gamma$ the system relaxes to its adiabatic steady states and the hysteresis loop is described by the parametric equation

$$\frac{dq(s)}{ds} = \frac{1}{2} q - \frac{1}{16 \Delta U} q^3 - A \cos(\gamma \Omega s). \qquad (3)$$

with the scaled time $s = t/\gamma$. In Figure 1a, we plot the classical hysteresis loop for $\gamma = 2$. For large damping, the hysteresis curve is a smooth curve with well defined switching points, given by the saddle-node bifurcation points $q_{1,2} = \pm(8\Delta U/3)^{1/2}$, $F_{1,2} = \mp(8\Delta U/27)^{1/2}$.

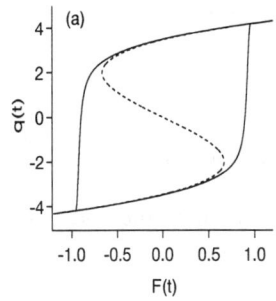 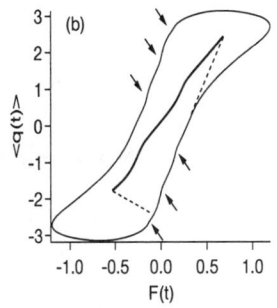

FIGURE 1. The classical hysteresis loop (a) is shown at $\Delta U = 1.5, A = 1.2, \Omega = 0.01$ for $\gamma = 2$ (solid line) with $F(t) = -A\cos(\Omega t)$. The dashed line corresponds to the stationary solution of Eq. (3). In (b) we show a quantum hysteresis loop for $\Delta U = 1.5, A = 1.2, \Omega = 5.0 \times 10^{-4}, \gamma = 0.005, T = 0.2$. Arrows point to the steps. The inset shows an enlarged part of the hysteresis loop where steps can be observed.

QUANTUM HYSTERESIS

To study the impact of tunneling through the barrier, we consider the quantum system described by the system Hamiltonian

$$\mathbf{H}_S = \frac{\mathbf{p}^2}{2M} + V(\mathbf{q}) + \mathbf{q}A \cos(\Omega t). \quad (4)$$

Here, $\mathbf{q}, \mathbf{p}$ are the coordinate and momentum operators and $V(\mathbf{q})$ is the double-well potential in Eq.(1). Dissipation is modeled according to the Caldeira-Leggett model [5] by coupling our system to an ensemble of harmonic oscillators yielding the total Hamiltonian

$$\mathbf{H} = \mathbf{H}_S + \mathbf{H}_B + \mathbf{H}_{SB}, \quad (5)$$

where $\mathbf{H}_B$ denotes the Hamiltonian of the heat bath and $\mathbf{H}_{SB}$ the Hamiltonian of the coupling. They are given by

$$\mathbf{H}_B = \sum_i \frac{\mathbf{p}_i^2}{2m_i} + \frac{1}{2}m_i\omega_i^2\mathbf{x}_i^2, \quad (6)$$

$$\mathbf{H}_{SB} = -\sum_i c_i\mathbf{x}_i\mathbf{q} + \sum_i \frac{c_i^2}{2m_i\omega_i^2}\mathbf{q}^2, \quad (7)$$

with the momentum and position operators of the bath oscillators $\mathbf{p}_i, \mathbf{x}_i$, its masses and angular frequencies m_i, ω_i and their coupling constants to the system c_i. The bath is fully characterized by its spectral density $J(\omega)$ (for details see e.g. [5]). Throughout this paper, we assume an Ohmic bath with exponential cut-off at $\omega_c \gg \omega_0$

$$J(\omega) = M\gamma\omega \exp\left(-\frac{\omega}{\omega_c}\right), \qquad (8)$$

where γ is the constant damping coefficient.

Numerical Solutions

The dynamics of the (one-dimensional) system of interest is described by the time evolution of the reduced density matrix

$$\rho(q'_f, q_f; t_i, t) = Tr_{bath}\langle q'_f | \mathcal{T} e^{-i/\hbar \int_{t_i}^{t} H dt'} W(t_i) e^{i/\hbar \int_{t_i}^{t} H dt'} | q_f \rangle. \qquad (9)$$

Here, q_f and q'_f are the spatial coordinates of the system at time t, $\mathcal{T}$ denotes the chronological operator, $W(t_i)$ the density operator of the total system at time t_i and Tr_{bath} the trace over the harmonic oscillator coordinates. At the initial time t_i, the system is decoupled from the bath and the bath is in thermal equilibrium. The density operator $W(t_i)$ is therefore a product of the density operator of the system and the canonical density operator of the bath at normalized temperature $T \to k_B T/(\hbar\omega_0)$. In general, the time evolution can only be performed numerically. We use the method of the *quasi-adiabatic path-integral propagator*, which was developed recently by Makri [6] in combination with a powerful tensor multiplication scheme [7] to evaluate the quasi-adiabatic path-integral iteratively (see also [8]). Fitted with the asymptotically time periodic density matrix $\rho(q'_f, q_f, t) = \lim_{t_i \to -\infty} \rho(q'_f, q_f; t_i, t)$, the hysteresis loop of the position expectation value

$$\langle q(t) \rangle = \int_{-\infty}^{\infty} \rho(q, q, t) \, q \, dq \qquad (10)$$

can be computed. A typical quantum hysteresis loop is shown in Figure 1b.

For temperatures above the quantum-classical crossover temperature [9], the hysteresis loop is a smooth curve. Below the crossover temperature, we observe steps if the driving frequency Ω is small. To compare with the behavior of the corresponding non-dissipative system (see below), we now turn to the transient time dependent response $\langle q(t) \rangle$. We start with a maximal tilt to the left and let the system evolve. We find steps occurring at those parameters of the tilt, where the adiabatic energy levels of $\mathbf{H}_S$ show avoided crossings (see Figure 2a). At the steps, tunneling out of the meta-stable minimum through the barrier is enhanced. Thermal occupation of the levels leads to thermally assisted resonant tunneling. An avoided crossing contributes to a step, if the driving frequency Ω is sufficiently small.

In Figure 2b, we show the response $\langle q(t) \rangle = \langle \psi(q,t) | q | \psi(q,t) \rangle$ of the system for zero damping and temperature ($\gamma = 0, T = 0$), with the initial preparation of the wave function $\psi(q, 0)$ as the ground state (with the potential fully tilted to the left). As the potential is ramped up, the response also shows distinct steps *at the same values of the external field than in the damped system*. These results for vanishing damping and temperature have been obtained by solving the time dependent Schrödinger equation numerically with the Crank-Nicholson method.

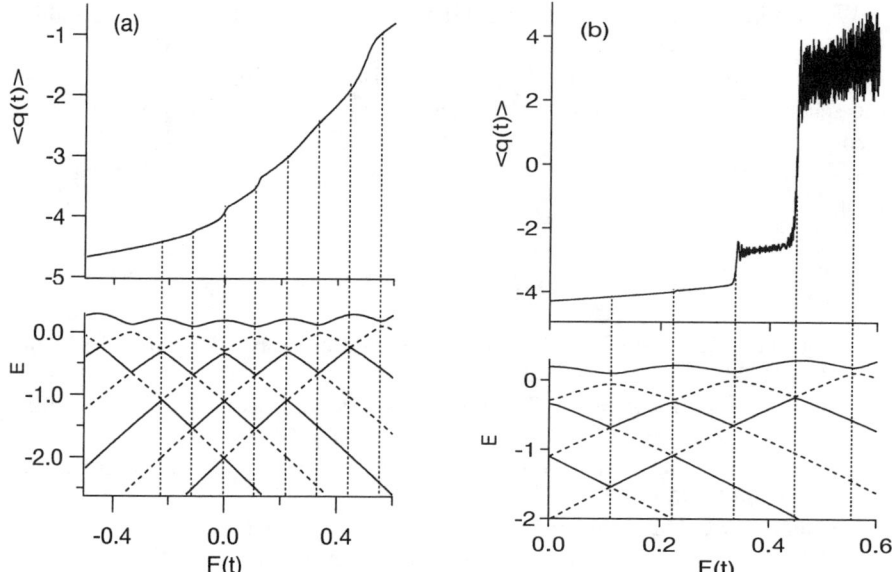

FIGURE 2. The time dependent response $\langle q(t) \rangle$ is shown for the damped system (a) as a function of the time dependent control force $F = -A\cos(\Omega t)$ where the system was prepared at $t = 0$ (i.e. $F = -A$) in the momentary ground state. The parameters are $\Delta U = 2.5, A = 1.8, \Omega = 10^{-4}, T = 0.1, \gamma = 0.001$. Below, the spectrum of the system Hamiltonian, computed parametrically as a function of the control force F, is shown. The steps in the response match the avoided level crossings in the parametric spectrum shown below. (b) Again, the time-dependant response $\langle q(t) \rangle$ is shown, now for the undamped system at zero temperature ($\gamma = 0, T = 0$). The parameters are $\Delta U = 2.5, A = 0.7, \Omega = 10^{-4}$.

CONCLUSIONS

We have shown that the response of a metastable quantum system to external forcing exhibits a surprisingly rich structure. Steps can occur at avoided crossings of the adiabatic energy levels when the driving frequency is sufficiently small. The role of the thermal environment is twofold. On the one hand, dissipation tends to smoothen out the steps, while one the other hand, thermal fluctuations can induce these steps by thermally assisted resonant tunneling. In a forthcoming paper [10], we will describe an intriguing multi-resonance structure of the relaxation rate connected with the observed steps in the response.

ACKNOWLEDGEMENTS

We would like to thank Milena Grifoni and Peter Hänggi for valuable discussions. We gratefully acknowledge financial support by the Deutsche Forschungsgemeinschaft and the State of Bavaria within the postgraduate scheme (Graduiertenkolleg GRK 283) "Nonlinear Problems in Analysis, Geometry and Physics" (M.T. and P.R.), the Friedrich–Naumann-Stiftung (M.T.) and the Deutsche Forschungsgemeinschaft within the Heisenberg-Program (P.J.).

REFERENCES

1. Friedman, J.R., Sarachik, M.P., Tejada, J., and Ziolo, R., *Phys. Rev. Lett.* **76**, 3830 (1996).
2. Hernandez, J.M., Zhang, X.X., Luis, F., Tejada, J., Friedman, J.R., Sarachik, M.P., and Ziolo, R., *Phys. Rev. B* **55**, 5858 (1997).
3. Thomas, L., Lionti, F., Ballou, R., Gatteschi, D., Sessoli, R., and Barbara, B., *Nature* **383**, 145 (1996).
4. Chudnovsky, E.M., *Science* **274**, 938 (1996).
5. Caldeira, A.O. and Leggett, A.J., *Ann. Phys. (N.Y.)*, **149**, 374 (1983).
6. Makri, N. and Makarov, D.E., *J. Chem. Phys.* **102**, 4600 (1995).
7. Makri, N. and Makarov, D.E., *J. Chem. Phys.* **102**, 4611 (1995).
8. Thorwart, M. and Jung, P., *Phys. Rev. Lett.* **78**, 2503 (1997).
9. Hänggi, P., Talkner, P., and Borkovec, M., *Rev. Mod. Phys.* **62**, 251 (1990).
10. Thorwart, M., Reimann, P., Jung, P., and Fox, R.F., submitted.

Synchronization in Ensembles of Stochastic Resonators

Alexander Neiman *,†, Frank Moss *,
Lutz Schimansky-Geier *,‡, and Werner Ebeling ‡

* Center for Neurodynamics, University of Missouri at St. Louis, St. Louis, MO63121
† Department of Physics, Saratov State University, Saratov 410073, Russia
‡ Institute for Physics, Humboldt University at Berlin, Berlin 10099, Germany

Abstract. We study the response of a parallel ensemble of stochastic resonators to external signals. For small signals linear response theory is used. We show that for a large number of elements this ensemble can be used to process broadband signals without frequency distortions. Nonlinear effects manifest themselves by the phenomenon of synchronization. Although synchronization is not observed at the output of single elements the mean frequency of the collective response is locked by the input signal. It takes place for a sufficiently large number of elements in a wide range of internal noise intensities. We also show that the synchronization is accompanied by a nonmonotonous Shannon entropy of the collective output indicating noise-induced order.

I INTRODUCTION

One of the major motivations in research on stochastic resonance (SR) [1] is the idea of improving the quality of small information signals passing through an optimally tuned bistable or threshold system [2,3]. Previous theoretical studies have shown that for extremely *weak* signals SR can be correctly described in terms of linear response theory (LRT) [4,5] using as measures the signal-to-noise ratio (SNR) or the spectral power amplification (SPA). However, SR can also be understood as a synchronization phenomenon [6,7]. Using alternative measures employing residence-time distributions [8], Gammaitoni et al. have termed SR as a bona-fide resonance in [6]. For a sufficiently *large* magnitude of the input signal (but still subthreshold) the mean frequency of the output can be locked in a wide range of noise intensities [7]. Moreover, it has been shown that at the parameter plane "noise intensity–amplitude of the signal" regions of synchronizations can be obtained, similar to Arnold's tongues. Inside these tongues the mean frequency of the output is locked by the input signal. This phenomenon is accompanied by an increase of order at the output: the dynamical entropies and the source Shannon entropy can be minimized for an optimal noise intensity [2]. Thus for a large amplitude of the input signal a noise-induced order of the output can be observed.

It has been shown also that SR can be significantly enhanced if instead of a single stochastic resonator an array of coupled [9] or uncoupled [10] resonators is taken. For an uncoupled parallel array of resonators convergent on a summing center, *stochastic resonance without tuning* has been observed [11]. The characteristic maximum in the SNR curve disappears as the shape becomes broadened with an increasing number of elements. In result the collective response of such a summing network is optimized for any arbitrary noise level larger than some small value. The model of parallel stochastic resonators seems to be generic for a number of applications and in particular for a simple network of sensory neurons [12].

In the present work we study linear (Sec.II) and nonlinear (Sec.III) responses of an ensemble of N stochastic resonators acting in parallel to a weak input signal. Each stochastic resonator $i = 1 \ldots N$ is subjected to the same input signal $s(t)$ and includes an internal Gaussian noise source $\xi_i(t)$ with intensity D. The noise is uncorrelated in the array $\langle \xi_i(t)\xi_j(t') \rangle = 0$ for $i \neq j$. The outputs of the elements $x_i(t)$ are converging on a summing center giving a collective variable $x_M(t) = (1/N) \sum x_i(t)$.

Synchronization in an ensemble of uncoupled chaotic oscillators by external noise has been studied also in [13]. This model differs from the system under investigation. We throughout assume statistical independence of the noise in different elements. Surprisingly, the system will be synchronized by a weak external signal for optimally tuned internal noise.

II LINEAR RESPONSE

We consider an array of identical stochastic resonators. A single resonator should have the susceptibility $\chi(\omega, D)$, where ω is the frequency and D is the intensity of the internal noise. Typically for SR, $\chi(\omega, D)$ shows a maximum in dependence on D. In the linear response theory (LRT) the susceptibility connects the Fourier forms of the input signal s and the output response δx as $\delta x(\omega, D) = \chi(\omega, D)s(\omega)$. It contains all informations of the response of the system to a weak input signal which can be also noisy [14–17].

The spectral density of the array $G_{MM}(\omega, D)$ is obtained in LRT as [17]

$$G_{MM}(\omega, D) = \frac{1}{N} G_{xx}^{(0)}(\omega, D) + |\chi(\omega, D)|^2 G_{ss}(\omega). \tag{1}$$

$G_{xx}^{(0)}(\omega, D)$ is the spectral density of single element in the absence of a signal resulting from the independent internal noise sources. The second part arises from the signal, where $G_{ss}(\omega)$ is the spectral density of the input signal. It follows immediately that in the limit of large N the weight of first item in

Eq.(1) becomes vanishingly small. The whole ensemble of nonlinear elements behaves as a linear system with the *transfer function* $\chi_{x,s}(\omega, D)$.

Let us consider the particular case of a noise contaminated weak periodic signal $A\sin(\Omega t)$. $s(t)$ is the sum of the signal and of a weak Gaussian noise $n(t)$ with spectral density $G_{nn}(\omega)$. The SNR at the input is $SNR_{in} = (1/4)A^2/G_{nn}(\Omega)$. The signal part at the output is $(1/4)A^2|\chi(\Omega, D)|^2$ whereas the noise part has two contributions, $(1/N)G_{xx}^{(0)}(\Omega, D)$ from the internal noise and $|\chi(\Omega, D)|^2 G_{nn}(\Omega)$ from the transferred input noise. The ratio of the SNR at the output to the SNR at the input reads

$$\eta = \frac{SNR_{out}}{SNR_{in}} = \frac{|\chi(\Omega, D)|^2 G_{nn}(\Omega)}{(1/N)G_{xx}^{(0)}(\Omega, D) + |\chi(\Omega, D)|^2 G_{nn}(\Omega)}. \qquad (2)$$

This ratio is less than 1 [15], unless N tends to infinity, as must be always true for an equivalent linear system. As N becomes large, the noise dependence of the numerator/denominator nearly cancels. It explains the broadening of the SNR and the appearance SR without tuning [11,17].

As is well known SR is often more pronounced for low-frequency signals [18]. In practice it means that the low-frequency domain of the signal will be processed better than the high-frequency one and as a result we will get frequency distortions at the output [19]. In this view Eq.(2) is of practical importance, because for large N the frequency dependence disappears in the same way as the dependence on the noise intensity. Therefore a parallel array of SR elements can be used for processing of broad-band signals.

III SYNCHRONIZATION

According to LRT, in the limit of infinite number of elements in the array the output signal is periodic signal with the amplified amplitude, $A|\chi(\Omega, D)|$. For a large, but finite N the periodic part of the output signal will be contaminated by a weak noise. The intensity of this noise scales as $(1/N)$ and also depends on the internal noise intensity D (see Eq.(1)). Thus it is natural to introduce the notion of instantaneous phase and the mean frequency as in theory of noisy self-sustained oscillators [20]. We also note that the concept of phase synchronization has been recently extended to chaotic systems in [13,21]. The case when the mean frequency of the output matches the frequency of periodic signal corresponds to the synchronization. This phenomenon occurs in a single stochastic resonator for a large enough amplitude of periodic signal [7].

For a numerical study of nonlinear effects we use the ideal Schmitt trigger as elements in the array which obey the equation

$$x_i(t + \Delta t) = \text{sgn}\left[\Lambda x_i(t) - s(t) - \xi_i(t)\right], \quad s(t) = A\sin(\Omega t) + n(t), \qquad (3)$$

where Λ is the threshold level of the triggers, $\xi_i(t)$ is Gaussian internal noise with cutoff frequency ω_c and intensity D, and $n(t)$ is external Gaussian noise with the same cutoff frequency and the intensity Q.

The mean frequency of the summed output has been calculated by counting the time intervals between successive zero-crossings. In the limit $N \to \infty$ this definition gives the frequency of the periodic force Ω, while, on the other hand, for the single element, ($N = 1$), the mean frequency refers to the Kramers rate of the element and equals to the mean switching frequency between metastable states of the system. The mean frequency is shown in Fig.1,(a) as a function of the internal noise intensity D for different numbers of elements in the array.

For a single element $N = 1$ the mean frequency displays the exponential Kramers-like behavior. However, with the increase of the number of elements in the array the dependence changes drastically. For a large enough N the mean frequency remains almost constant in a finite region of noise intensity. Thus for a large array the mean frequency of the summed output can be locked by weak input periodic forcing, therefore, demonstrating synchronization. In the limit $N \to \infty$ the ensemble of nonlinear elements behaves as an equivalent linear system and the mean frequencies of the input and of the output match for any value of noise intensity D (dashed line in Fig.1(a)).

This synchronization-like behavior can also be characterized by calculating the entropy or complexity measures of the output signal [2]. For the case of phase synchronization of chaotic systems the spectrum of the Lyapunov exponents is shown to be an appropriate measure of the phase synchronization [21]. For noisy bistable systems it is convenient to use complexity measures based on the stochastic symbolic dynamics [22]. In the present study we used

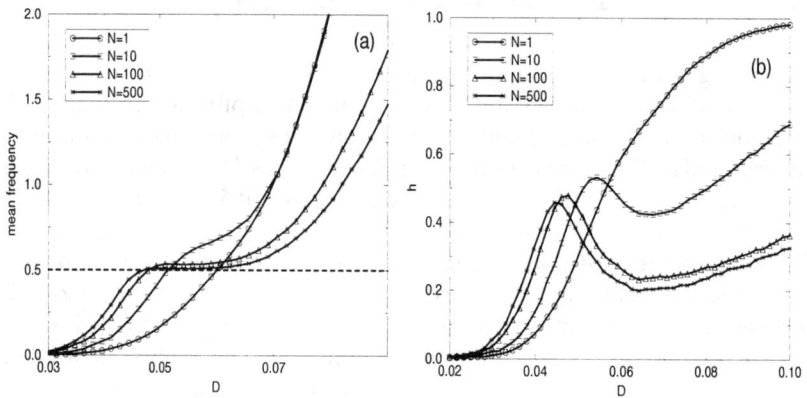

FIGURE 1. The mean frequency of the output (a) and the source entropy (b) versus D for different numbers of elements. The parameters values are: $\Lambda = 0.2$, $A = 0.03$, $\Omega = 0.5$, $\omega_c = 100.0$, $Q = 0.03$, $\Delta t = 0.01$.

a binary alphabet for the output sequence with respect to the sign of $x_M(t)$.

The standard measure to quantify the order in a symbolic sequence is the Shannon entropy of the source (or the source entropy) h [23]. For a binary sequence $h = 1bit$ corresponds to maximal randomness and $h = 0$ to an ordered (e.g. strictly periodic) structure. Although the Shannon entropy is a quantity which is not easy to calculate [24], several techniques for estimation of h exist. One method, which is widely used, was proposed by Ziv and Lempel [25]. The Lempel-Ziv complexity gives an upper estimate of the source entropy. Here we use this estimate employing the algorithm proposed by Kaspar and Schuster [26].

The results are presented in Fig.1(b). For $N = 1$ the entropy increases monotonically with D to $h = 1$ corresponding to complete randomness. In other words the increase of noise leads to a greater uncertainty in predicting the next symbol in the output sequence. However, with increasing the number of elements in the array the behavior of the entropy changes drastically. The entropy takes its maximum at the noise intensity corresponding to the onset of the synchronization region (see e.g. Fig.1,(a)) and then takes its minimum which corresponds to the maximum of the output SNR. Therefore the output sequence can be made more ordered by increasing the noise at least in a definite region.

IV CONCLUSION

We have studied a simple but generic model of a parallel array of stochastic resonators subjected to a weak input signal consisting of a periodic and a noisy component. The linear response theory applied to this model gave the explanation of the effect of SR without tuning. At the same time, numerical simulations have shown that the output of the array can be synchronized with the input signal in a finite region of noise intensities. Moreover, we have shown that the system exhibits the noise-induced order as the source Shannon entropy, estimated by means of the Lempel-Ziv complexity, possesses a minimum within the synchronization region.

This work was supported in part by the U.S. Office of Naval Research Physics Division and by common research project of DFG and RFRF [Grant 436 RUS 113/334/0(R)]. A.N. is supported by a U.S. National Academy of Science COBASE grant.

REFERENCES

1. A full bibliography can be found at WWW-cite *http://www.pg.infn.it/sr/*.
2. Neiman A., Shulgin B., Anishchenko V., Ebeling W., Schimansky-Geier L., and Freund J., *Phys. Rev. Lett.* **76**, 4299 (1996).

3. Levin J.E., and Miller J.P., *Nature* **380**, 165 (1996); Bulsara A.R., and Zador A., *Phys. Rev. E* **54**, R2185 (1996).
4. Dykman M.I., Mannella R., McClintock P.V.E., and Stocks N.G., *Phys. Rev. Lett.* **65**, 2606 (1990); *JETP Lett.* **52**, 144 (1990).
5. Jung P., and Hänggi P., *Phys. Rev. A* **44** (1991) 8032.
6. Gammaitoni L., Marchesoni F., and Santucci S., *Phys. Rev. Lett.* **74**, 1052 (1995).
7. Shulgin B., Neiman A., and Anishchenko V., *Phys. Rev.Lett.* **75**, 4157 (1995).
8. Gammaitoni L., Marchesoni F., Manichella-Saetta E., and Santucci S., *Phys. Rev. Lett.* **62**, 349 (1989); Zhou T., Moss F., and Jung P., *Phys. Rev. A* **42**, 3161 (1990).
9. Jung P., Behn U., Pantazelou E., and Moss F., *Phys. Rev. A* **46**, R1709 (1992) ; Neiman A. and Schimansky-Geier L., *Phys. Lett. A* **197**, 379 (1995); Linder J., Meadows B., Ditto W., Inchiosa M., and Bulsara A.R., *Phys. Rev. Lett.* **75**, 3 (1995); Schimansky-Geier L., and Siewert U., in: *Stochastic Dynamics*, ed. by L. Schimansky-Geier and T. Pöschel (Springer, Berlin, Heidelberg, New York, 1997)
10. Pantazelou E., Moss F., and Chialvo D., in: *Noise in Physical Systems and 1/f Fluctuations*, AIP Conference Proceedings 285, eds. P. Handel and A. Chung, (AIP Press, NY, 1993), p.549-552.
11. Collins J.J., Chow C.C., and Imhoff T.T., *Nature* **376**, 236 (1995).
12. Pei X., Wilkens L., and Moss F., *Phys. Rev. Lett.* **77**, 4679 (1996).
13. Pikovsky A.S., Rosenblum M.G., Osipov G.V., and Kurths J., *Physica D* **104**, 219 (1997).
14. Neiman, A. and Schimansky-Geier, L., *Phys. Rev. Lett.* **72**, 2988 (1994).
15. Dykman M.I., Luchinsky D.G., Mannella R., McClintock P.V.E., Stein N.D., and Stocks N.G., *Il Nuovo Cim. D* **17**, 661 (1995).
16. Collins J.J., Chow C.C., and Imhoff T.T., *Phys. Rev. E* **52**, R3321 (1995).
17. Neiman A., Schimansky-Geier L., and Moss F., *Phys. Rev. E* **56** R 12 (1997).
18. McNamara B., and Wiesenfeld K., *Phys. Rev. A* **39**, 4584 (1989).
19. Anishchenko V., Neiman A., Safonova M., and Khovanov I., in: *Chaos and Nonlinear Mechanics*, ed. by T.Kapitaniak and J. Brindley, p.41-53, (World Scientific, Singapore, 1995).
20. Stratonovich R.L., *Topics in the Theory of Random Noise* (Gordon and Breach, New York, 1963).
21. Rosenblum M.G., Pikobsky A.S., and Kurths J., *Phys. Rev. Lett.* **76**, 1804 (1996).
22. Witt A., Neiman A., and Kurths J., *Phys. Rev. E* **55**, 5050 (1997).
23. Shannon C.E., and Weaver W., *The Mathematical Theory of Communication* (University of Illinois Press, Urbana, 1949).
24. Ebeling W., Pöschel T., and Albrecht K.-F., *Int. J. Bifurcation and Chaos* **5**, 51 (1995).
25. Lempel A., and Ziv J., *IEEE Trans. Inf. Theory* **IT-22**, 75 (1976); Ziv J., and Lempel A., *IEEE Trans. Inf. Theory* **IT-23**, 337 (1977).
26. Kaspar F., and Schuster H.G., *Phys. Rev. A* **36**, 842 (1987).

Nonlinear Dynamics of Large Fluctuations

Mark I. Dykman*, Vadim N. Smelyanskiy*,
Colin J. Lambert†, Dmitrii G. Luchinsky†[1],
Peter V.E. McClintock†

*Department of Physics and Astronomy, Michigan State University,
East Lansing, MI 48823, USA
†School of Physics and Chemistry, Lancaster University
Lancaster, LA1 4YB, UK

Abstract. The distribution of paths for large fluctuations away from a stable state has been investigated, theoretically and by analog experiment. We have found *critical broadening* of the distribution of the paths coming to a cusp point representing the simplest generic singularity in the pattern of most probable (optimal) fluctuational paths in a non-equilibrium system. The critical behavior can be described by a Landau-type theory.

INTRODUCTION

A wide variety of physical phenomena, ranging from nucleation at phase transitions to failures of electronic devices, are driven by large rare fluctuations. In many cases the fluctuating systems of interest are far from thermal equilibrium. It was recognized by Onsager and Machlup [1] that an insight into the physics of large fluctuations can be gained from an analysis of the *distribution of fluctuational paths* along which the system moves to a given state. This distribution is a fundamental characteristic of the fluctuation *dynamics*, and its understanding paves the way to controlling the fluctuations. It peaks sharply at the *optimal* (i.e. most probable) path; we report below the first experimental observations of this pattern.

In thermal equilibrium, with white noise, the optimal paths are time-reversed deterministic paths [2]. But this is not true for nonequilibrium systems, because they lack time reversibility. The statistical distribution of

---

[1] Permanent address: Russian Institute for Metrological Service, Ozernaya 46, 119361 Moscow, Russia.

nonequilibrium systems may have singular features [3], and so may the pattern of optimal paths. Optimal paths represent the extrema of a variational problem. A generic property of the pattern of extreme paths is the occurrence of caustics [4], in general starting in pairs from a cusp (focal) point: see Fig. 1.

Caustics and cusps in the pattern of extreme paths have been found numerically [5] and investigated analytically [6]. Note that contributions to the probability distribution from different paths are all positive: interference effects do not occur and the singularities in the pattern of optimal paths are therefore different from those known for wave fields [6]. Although optimal fluctuational paths do not encounter caustics, but they can focus into cusp points. Use of the prehistory probability density [7] (see below) enables us to reveal and analyze the singular behavior, theoretically and experimentally [8].

SINGULAR BEHAVIOR OF THE DISTRIBUTION OF FLUCTUATIONAL PATHS

The simplest system that displays singularities of optimal paths is an overdamped Brownian particle driven by a periodic force $K(q;t)$,

$$\dot{q} = K(q;t) + \xi(t), \quad K(q;t) = K(q;t+T), \qquad (1)$$
$$\langle \xi(t)\xi(t')\rangle = D\delta(t-t').$$

For small noise intensities D, within a relaxation time τ_r the system (1) will approach the stable periodic state $q^{(0)}(t)$, $\dot{q}^{(0)} = K(q^{(0)};t)$, $q^{(0)}(t+T) = q^{(0)}(t)$, and a periodic stationary probability distribution will be formed in the basin of attraction to this state.

Analysis of the prehistory probability density, $p_h(q,t|q_f,t_f)$ [7] enables the distribution of paths in large fluctuations to be investigated and visualized. It is the conditional probability density for a system that had been fluctuating about $q^{(0)}(t)$ for a time greatly exceeding τ_r, and arrived to the point q_f at the instant t_f, to have passed through the point q at the instant t ($t < t_f$). It can be written as a path integral:

$$p_h(q,t|q_f,t_f) = \text{const} \times \int_{q(t_i) \approx q^{(0)}(t_i)}^{q(t_f)=q_f} \mathcal{D}q(t') \exp\left[-\frac{S[q(t)]}{D}\right]$$
$$\times \delta(q(t) - q), \quad t_i \to -\infty, \quad \int dq\, p_h(q,t|q_f,t_f) = 1 \qquad (2)$$

Here, $S[q]$ determines the probability distribution over the paths of a Markovian system. To lowest order in the noise intensity D it takes the form of the action functional for an auxiliary dynamical system with the Lagrangian $L(\dot{q},q;t)$ (cf. [9]):

$$S[q(t)] = \int_{t_i}^{t_f} dt L(\dot{q},q;t), \quad L(\dot{q},q;t) = \frac{1}{2}[\dot{q} - K(q;t)]^2. \qquad (3)$$

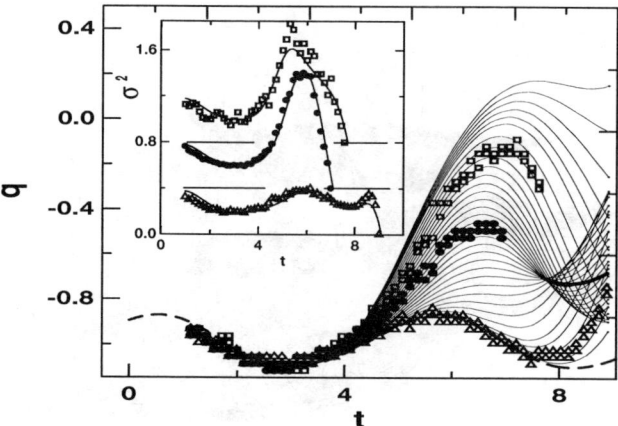

FIGURE 1. Extreme paths of (1) for $K = q - q^3 + 0.264\cos 1.2t$. The stable state $q^{(0)}(t)$ from the vicinity of which the paths start is shown by the dashed line. The bold line emanating from the cusp point is the switching line calculated for $D \to 0$. The data points show the maxima of the prehistory probability distribution measured for three final points away from the cusp. Reduced variances of the corresponding Gaussian distributions σ^2 (displaced along the ordinate axis for clarity) are compared with the theory in the inset.

The optimal (most probable) fluctuational path $q_{\text{opt}}(t|\, q_f, t_f)$ along which the system arrives at q_f at time t_f is found by solving the variational problem

$$\frac{\delta S[q]}{\delta q(t)} = 0, \quad q_{\text{opt}}(t_f|\, q_f, t_f) = q_f, \qquad (4)$$

$$q_{\text{opt}}(t_i|\, q_f, t_f) \to q^{(0)}(t_i) \quad \text{for} \quad t_i \to -\infty.$$

But Eq.(4) describes *extreme* fluctuational paths, which are not necessarily the *optimal* paths that provide the global minimum of the action $S[q]$ and are therefore of physical significance. Extreme paths $q(t)$ as given by (3), (4) can intersect, since a dynamical system with the Lagrangian (3) is nonintegrable. In contrast, generically only one optimal path can arrive at a given point.

The shape of the prehistory probability density p_h (2) can be found by expanding the coordinate $q(t)$ into the orthonormal functions $\psi_n(t)$ which diagonalize the second variation of the action:

$$q(t) = q_{\text{opt}}(t|q_f, t_f) + \sum_n a_n \psi_n(t). \qquad (5)$$

It follows from (3) that the functions $\psi_n(t)$ satisfy a Schrödinger-type equation

$$-\ddot{\psi}_n + V(t)\psi_n = \lambda_n \psi_n, \; V(t) = \left[\frac{\partial^2 K}{\partial q \partial t} + \frac{1}{2}\frac{\partial^2 K^2}{\partial q^2}\right]_{\text{opt}} \qquad (6)$$

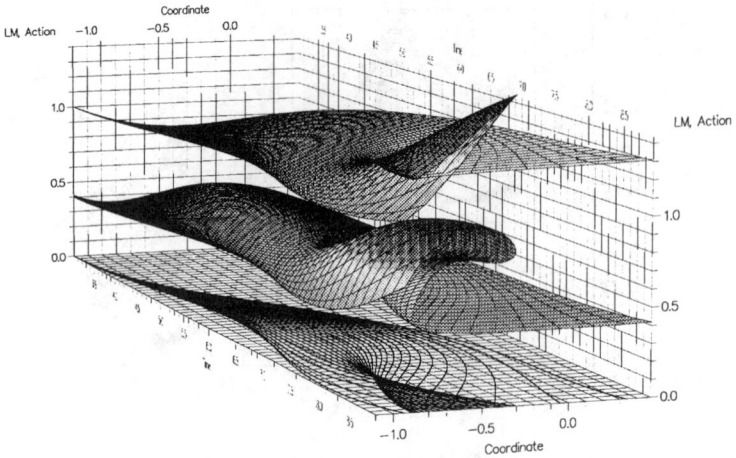

FIGURE 2. Generation of singularities: (middle plot) LM in the space (t, q, p) with two folds; (lower plot) projection of LM onto (t, q) plane; (upper plot) multivalued action surface $S(q, t)$.

with the boundary conditions $\psi_n(t_i) = \psi_n(t_f) = 0$ (in (6) the derivatives of $K \equiv K(q; t)$ are evaluated for $q = q_{\text{opt}}(t| q_f, t_f)$).

For trajectories $q(t)$ close to the optimal path, the a_n in (5) are small, and the action $S[q(t)]$ is quadratic in a_n (unless (q_f, t_f) is close to the cusp: see below),

$$S[q(t)] = S(q_f, t_f) + s(\{a_n\}), \ s(\{a_n\}) = \frac{1}{2} \sum_n \lambda_n a_n^2, \qquad (7)$$

$$S(q_f, t_f) \equiv S[q_{\text{opt}}(t| q_f, t_f)].$$

If one writes the path integral (2) as an integral over all a_n and substitutes Eqs. (5), (7) into (2), one obtains

$$p_h(q, t| q_f, t_f) = M \exp\left(-\frac{[q - q_{\text{opt}}(t|q_f, t_f)]^2}{2D\sigma^2(t| q_f, t_f)}\right),$$

$$\sigma^2(t| q_f, t_f) = \sum_n \lambda_n^{-1} \psi_n^2(t), \ M = (2\pi D\sigma^2)^{-1/2}. \qquad (8)$$

Near the maximum, the distribution p_h is evidently Gaussian in the distance of (q, t) from the optimal path $q_{\text{opt}}(t| q_f, t_f)$ (cf. [7]). Therefore by investigating p_h one can find directly the optimal path itself, and also analyze the shape of the tube of paths arriving at a given point (q_f, t_f). Away from the cusp point, the width of this tube is $\propto D^{1/2}$. The reduced width of the distribution (8) $\sigma(t| q_f, t_f)$ is independent of q_f, t_f for $t_f - t \gg \tau_r$: it gives the reduced

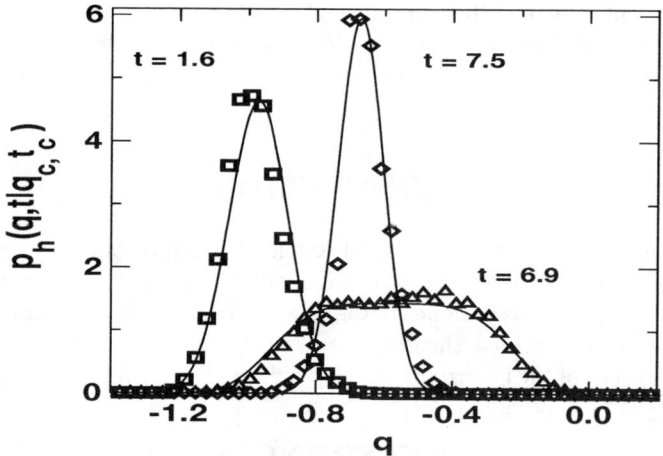

FIGURE 3. Cross-sections of the prehistory probability distribution calculated (curves) and measured (points) for fluctuations to the cusp point in Fig. 1 ($(q_c \approx -0.70, t_c \approx 7.69)$, for three values of t. The distribution is Gaussian very close to, and far from, the cusp; but it is critically broadened and strongly non-Gaussian at intermediate values of t.

width of the stationary Gaussian distribution about $q^{(0)}(t)$. Experiments on an analog electronic model of (1), based on a standard technique [10], have yielded results which are in a very good agreement with Eqs. (4), (8); see Fig. 1 (the explicit form of $\sigma(t|q_f, t_f)$ is given in [8]).

Eq.(8) does not apply if the final point (q_f, t_f) is close to a singularity of the pattern of extreme paths [8]. The origin of the singularities can be understood from topological arguments [6]: see Fig. 2. The trajectories of the auxiliary Hamiltonian system (3) form a Lagrangian manifold (LM) in the phase space of the system. Projections of these trajectories onto the (q, t) plane are optimal paths. The Lagrangian manifold in the space (t, q, p) is also shown: note that $p \equiv \partial L/\partial q = \dot{q} - K(q, t)$ is a momentum of the auxiliary system corresponding to the coordinate q. Generically the LM can have folds emerging in pairs from the cusp, and it is their projections onto the plane (q, t) that creates the caustics. Caustics and cusp points are the only generic structurally stable singularities of the pattern of extreme paths of $S[q]$ (3) [11]. Since caustics may not [6] be observed in the pattern of optimal paths, it is particularly interesting to investigate the distribution p_h near a cusp point (q_c, t_c). If the final point (q_f, t_f) coincides with (q_c, t_c), then the smallest eigenvalue $\lambda_0 = 0$ [4] and Eq. (8) does not apply; in particular, σ diverges for $\lambda_0 = 0$.

At the cusp point it is necessary [4] to keep in the expansion of the action s (7) the higher-order terms in the amplitude a_0 of the "soft mode" $\psi_0(t)$. Detailed analysis shows [8] that, when (q_f, t_f) approaches (q_c, t_c), fluctuations about the optimal path become strongly non-Gaussian. The characteristic width of the distribution, $\sim D^{1/4}$, is determined by the soft mode $\psi_0(t)$. The measured and calculated evolutions of the distribution with $t - t_c$ are in good agreement: see Fig. 3.

CONCLUSION

In conclusion, we have used general topological arguments, and the concept of the prehistory probability distribution, to analyze singular features of the pattern of optimal fluctuational paths in systems away from thermal equilibrium. We have described theoretically, and observed, critical broadening of the distribution of paths arriving in the vicinity of a cusp point.

REFERENCES

1. L. Onsager and S. Machlup, *Phys. Rev.* **91**, 1505, 1512 (1953).
2. See M. Marder, *Phys. Rev. Lett.* **74**, 4547 (1995) and references therein.
3. R. Graham, in *Noise in Nonlinear Dynamical Systems*, edited by F. Moss and P. V. E. McClintock (Cambridge University, Cambridge, 1989), vol. 1, p. 225.
4. M.V. Berry, *Adv. Phys.* **25**, 1 (1976); L.S. Schulman, *Techniques and applications of path integration* (Wiley, New York, 1981).
5. H.R. Jauslin, *J. Stat. Phys.* **42**, 573 (1986); *Physica* **144A**, 179 (1987); M.V. Day, *Stochastics* **20**, 121 (1987); *Ann. Prob.* **20**, 1385 (1992); V.A. Chinarov, M.I. Dykman and V.N. Smelyanskiy, *Phys. Rev. E* **47**, 2448 (1993); R.S. Maier and D.L. Stein, *Phys. Rev. Lett.* **71**, 1783 (1993); *Phys. Rev. E* **48**, 931 (1993); S.J.B. Einchcomb and A.J. McKane, *Phys. Rev. E* **51**, 2974 (1995).
6. M.I. Dykman, M.M. Millonas, and V.N. Smelyanskiy, *Phys. Lett. A* **195** 53 (1994); R.S. Maier and D.L. Stein, *J. Stat. Phys.* **83**, 291 (1996); V.N. Smelyanskiy, M.I. Dykman, and R.S. Maier, *Phys. Rev. E* **55**, 2369 (1997).
7. M.I. Dykman, P.V.E. McClintock, V.N. Smelyanskiy, N.D. Stein, and N.G. Stocks, *Phys. Rev. Lett.* **68**, 2718 (1992); J. Gómez-Ordóñez, J.M. Casado, and M. Morillo, *Phys. Rev. E* **54**, 2125 (1996); M. Morillo, J.M. Casado, and J. Gómez-Ordóñez, *Phys. Rev. E* **55**, 1521 (1997); B. Vugmeister, J. Botina, and H. Rabitz, *Phys. Rev. E* **55**, 5338 (1997).
8. M.I. Dykman, D.G. Luchinskiy, P.V.E. McClintock and V.N. Smelyanskiy, *Phys. Rev. Lett.* **77**, 5229 (1996).
9. M.I. Freidlin and A.D.Wentzell, *Random Perturbations in Dynamical Systems* (Springer Verlag, New York, 1984).
10. P.V.E. McClintock and F. Moss in ref. [3], vol. 3, p. 243.
11. H. Whitney, *Ann. Math.* **62**, 374 (1955); V. I. Arnold, *Catastrophe Theory* (Springer-Verlag, New-York, 1984).

Ratchet Effects for Simple and Compound Objects Driven by Correlated Noise

T. E. Dialynas*, Katja Lindenberg† and George Tsironis*

*Physics Department, University of Crete and
Research Center of Crete
P. O. Box 2208
71110 Heraklion-Crete, Greece
†Department of Chemistry and Biochemistry 0340 and
Institute for Nonlinear Science
University of California San Diego
La Jolla, CA 92093-0340

Abstract. We investigate the motion of a point particle and also of a rigid dimer in a periodic asymmetric ratchet potential driven by time-correlated forces. First we show that ratchet motion does not require noise at all - it can also occur in the extreme monochromatic deterministic case. We then consider a particle driven by colored noise, and derive a relatively simple analytical expression for the induced current at large noise correlation times τ. We also consider a rigid dimer diffusing on an asymmetric ratchet and show that there is current reversal as a function of the length of the dimer. These results may provide a model for differential protein motion on microtubules.

The stochastic motion of a particle in a periodic asymmetric potential has been a focus of attention in the last few years [1–5]. A particle may acquire net macroscopic motion in such a ratchet even in the absence of any deterministic directional forcing or any bias in the noise itself. The current is created by an interplay of the asymmetry of the potential and the characteristics of the external noise. The noise must be external, that is, the system must be *open*. This is one example of an array of problems in which noise actually plays a creative role.

The original motivation for the interest in the particle "ratchet effect" arose in a biological context: When a microtubular associated protein (MAP) executes motion on a microtubule, its diffusive dynamics has a specified direction. This directionality in the protein motion was associated with the non-symmetric form of the periodic potential of the microtubule and was thought to be induced by the correlated character of the ATP hydrolysis mechanism. Following this original motivation there have been improvements to the basic model that might explain why different but similar proteins may move in opposite directions on the same microtubule. One way to achieve this result, we have found, is to take into account the finite size of the moving particle. We consider a Brownian rigid dimer of length l, and show that the *direction* of the current is a function of the length of the dimer [6].

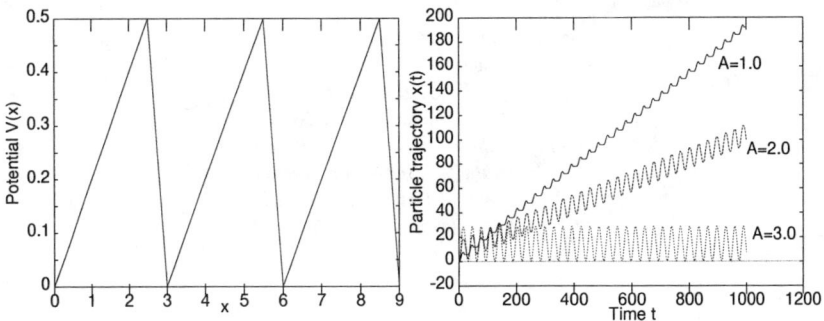

FIGURE 1. Left panel: Asymmetric piecewise linear ratchet potential with amplitude $Q = 0.5$, upward slope 0.4, downward slope -2. Right panel: Trajectories $x(t)$ for a sinusoidal driving force, with $\omega = 0.2$ and $A = 1.0, 2.0, 3.0$.

We first consider an overdamped point particle under the influence of two forces: a spatial asymmetrical periodic force $f(x)$ and a time-dependent force $\xi(t)$ whose time-average is zero to insure that it includes no systematic component. The motion of the particle is described by the equation for the trajectory $x(t)$ together with the initial position of the particle, $\dot{x} = f(x) + \xi(t)$ with $x(0) = x_0$. Our measure of the presence of a current is the *mean velocity* v, defined as $v = \lim_{t \to \infty} v(t)$, where $v \equiv [x(t; x_0) - x_0]/t$. If v is positive (negative), there is a current to the right (left). If v vanishes, there is no current.

The simplest ratchet force $f(x) \equiv -dV(x)/dx$ arises from the sawtooth potential shown in the left panel of Fig. 1. When $\xi(t)$ is zero-centered exponentially correlated Gaussian noise of any *non-zero* correlation time, this system leads to a current to the right, that is, to a positive value of v [7]. We will exhibit this result shortly. It is perhaps surprising to realize that even a perfectly periodic force may also lead to a net current to the right. Consider, for example, the sinusoidal force $\xi(t) = A\sin(\omega t)$. In the right panel of Fig. 1 we show three trajectories $x(t)$ obtained from numerical integration. Two of them clearly lead to a positive current, while the third does not. Indeed, the dependence of v on A and ω is quite complex [7].

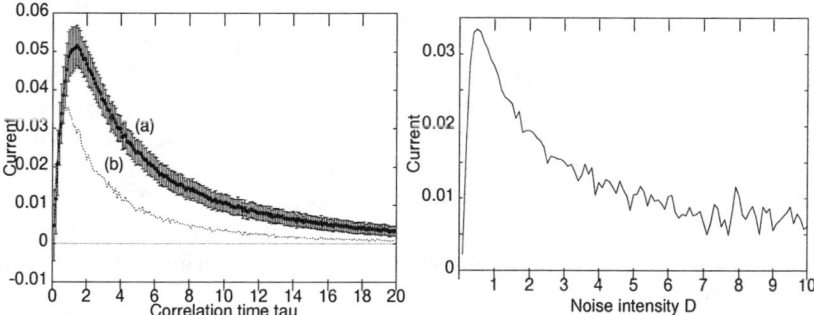

FIGURE 2. Left panel: Simulation results for the current on the sawtooth ratchet as a function of the correlation time of the noise for (a) $D = 0.5$, (b) $D = 0.3$. Simulation error bars are explicitly indicated for one of the simulations. Right panel: Simulation results for the current as a function of the noise parameter D with $\tau = 0.5$.

Consider now the noisy system in which the force $\xi(t)$ is zero-centered exponentially correlated Gaussian colored noise with correlation time τ and amplitude $D/2\tau$. When $\tau \gg 1$ we can derive an asymptotic expression for the current using the following argument [8]. Let us initially place the Brownian particle at the bottom of the potential. Extremely correlated noise is essentially quasistatic because the fluctuations of $\xi(t)$ are extremely slow. The particle escapes to the next well, left or right, when a fluctuation leads to a quasiconstant value of ξ of the appropriate size to cancel the force due to the potential. In this picture, the average time that the particle waits before passing from one well to another is given by the mean first passage time for the noise ξ to reach the appropriate critical value to cancel the effects of the potential and allow the particle to go over the barrier and reach the next minimum. For the particle to escape to the left (right), the noise must reach the critical value $\xi_i^\tau \equiv Q/d_i + d_i/\tau$, $i = 1$ (2), where d_1 (d_2) is the x-distance between a minimum and the immediate maximum on its left (right); the first term alone would be the critical value for the particle to reach the top of the barrier, and the second term is the correction that allows for the "rolling down" time. The mean first passage time for $\xi(t)$ to reach a value ξ_i^τ is [8,9]

$$T_i(\xi_i^\tau) \sim \frac{\sqrt{2\pi D \tau}}{|\xi_i^\tau|} \exp\left(\frac{{\xi_i^\tau}^2 \tau}{2D}\right). \tag{1}$$

The color-induced current is proportional to the net rate R of escape from a well,

$$R = \frac{1}{2}\left[\frac{1}{T_1(\xi_1^r)} - \frac{1}{T_2(\xi_2^r)}\right]. \qquad (2)$$

Equation (2) leads to a current that is always to the right on the ratchet as drawn in Fig. 1. The current increases with increasing correlation time τ, reaches a maximum, and then decreases back to zero at large values of the correlation time. The maximum current first increases with increasing noise intensity parameter D, and then decreases. This decrease is reasonable: if the noise is too intense it swamps out the ratchet effect entirely. Although there is no reason to expect our result to be accurate as $\tau \to 0$, the expression does remain analytic and continuous even in that limit. The formula captures the very large τ behavior quantitatively and the entire finite-τ dynamics qualitatively.

Our simulation results were obtained by calculating $v(t)$ N times for different realizations of the colored noise $\xi(t)$ and computing the average over N realizations. Ideally $N \to \infty$ and $t \to \infty$, but of course we have to be content with large but finite N and t. Simulation results for the current obtained with different values of the noise intensity parameter D are shown in the left panel of Fig. 2. For $D = 0.5$ we have explicitly indicated the experimental error to convey the limits of accuracy of our simulations. To confirm that strong noise does drown out all ratchet effects we show in the right panel of Fig. 2 simulation results for the current as a function of the noise parameter D. For small values of D the current rises, but it then decreases when D becomes comparable to the potential barrier height.

Our results lead directly to a simple description of the effect of the color of the noise on the Brownian particle in the ratchet. In the limit of large correlation time the noise acts like a constant force that at times (when it takes on the appropriate values) opposes the action of the ratchet potential. Since it is more likely for the noise to attain the value needed to counteract the smaller force exerted on the particle, the particle is moved by the noise more often to the right than to the left, thus leading to a net current. As the correlation time τ becomes smaller, two "opposite" effects begin to play a role. On the one hand, as the noise changes value more rapidly, the difference between the number of trajectories going to the right and going to the left diminishes. On the other hand, this same more rapid change leads to a reduction in the mean exit time from a well (in either direction) because larger values of the noise are more frequently attained (albeit retained for a shorter time) within a given time period. The maximum current occurs when these two effects are optimized together in the way typical of opposing tendencies. As τ becomes smaller the current again diminishes until, in the white noise limit $\tau \to 0$, there is no net current since the rates of escape to the left and to the right become equal.

The model just described leads to a net current, but the current is always in the same direction, that is, to the right on the ratchet of Fig. 1. It is known

[10], however, that structurally similar proteins such as kinesin and dynein move in *opposite* directions on the surface of the same microtubule and in the same environment. Such direction sensitivity may also be relevant for particle separation [11] and other biological applications [12], but can not be explained without the introduction of some scale that models protein differences. Such a scale might be the length of the protein. To model such a scale we introduce the length of the Brownian particle, l. We replace the point particle by a rigid dimer that experiences a net potential $\tilde{V}(x)$ that is a superposition of the periodic ratchet potential $V(x)$ evaluated at x and at $x + l$, i.e., $\tilde{V}(x) = V(x) + V(x + l)$.

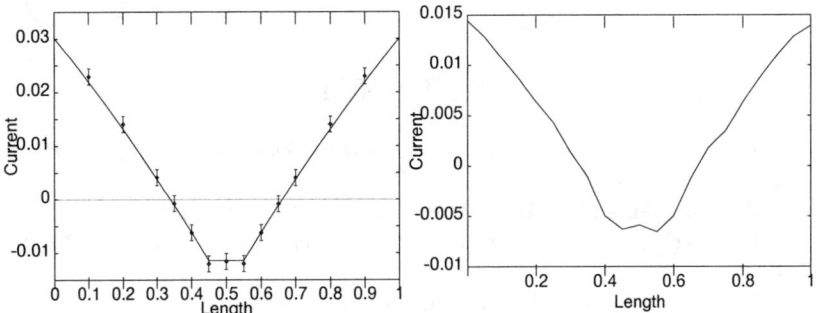

FIGURE 3. Left panel: Analytical predictions and numerical simulation results for the current on the sawtooth ratchet as a function of dimer length with dichotomous noise. Potential parameters are $Q = 1.0$, upward slope $1.818...$, downward slope $-2.2222...$. Noise parameters are $D = 0.5$, and $\tau = 0.1$. The error bars denote the uncertainties in the simulations. Right panel: Simulation results for the current as a function of dimer length with Gaussian noise and otherwise the same parameter values as in the left panel.

On the basis of this model, we have found current inversion analytically for a dimer on the sawtooth ratchet when the driving noise is dichotomous [6]. The results are shown in Fig. 3. To confirm these results we also show in the figure simulation results for the same problem but with Gaussian colored noise. For very short rods (not unlike point particles) and for long rods the current is to the right, but for rods of intermediate length there is a regime where the ratchet current is actually to the left. Greater accuracy would more clearly confirm that the current in the Gaussian noise case is also symmetric about a rod length of 0.5 for the potential parameters used in the figure.

We have presented a number of results for a particle in an asymmetric ratchet potential driven by a deterministic unbiased periodic force and also by zero-centered colored noise. In general, a Brownian point particle will acquire a finite velocity to the right in the potential as shown in Fig. 1. A Brownian rigid dimer, on the other hand, acquires a finite velocity to the right *or* to the left, depending on the length of the dimer. This bi-directional mechanism may offer an explanation for the observed opposite directionality of similar proteins in microtubules under otherwise apparently identical conditions. Our presentation has included a combination of analytic and numerical results.

ACKNOWLEDGMENTS

Two of us (K. L. and G. P. T) gratefully acknowledge NATO Travel Grant CRG 950399 that provided partial support for this work. One of us (K. L.) gratefully acknowledges the partial support of this research by the U. S. Department of Energy through Grant No. DE-FG03-86ER13606.

REFERENCES

1. M. O. Magnasco, Phys. Rev. Lett. **71**, 1477 (1993); **71** 1477 (1993); **72** 2656 (1994).
2. R. D. Astumian and M. Bier, Phys. Rev. Lett. **72**, 1766 (1994).
3. C. R. Doering, W. Horsthemke and J. Riordan, Phys. Rev. Lett. **72**, 2984 (1994).
4. M. M. Millonas and M. I. Dykman, Phys. Lett. A **185**, 65 (1994).
5. R. D. Astumian, Science **276**, 917 (1997),
6. T. E. Dialynas and G. P. Tsironis, Phys. Lett. A **218**, 292 (1996).
7. A more detailed version of this work together with more extensive results are presented elsewhere: T. Dialynas, Katja Lindenberg and G. P Tsironis, submitted for publication.
8. G. P. Tsironis and P. Grigolini, Phys. Rev. Lett. **61**,7 (1988); Phys. Rev. A **38**, 3749 (1988).
9. F. J. de la Rubia, E. Peacock-López, G. P. Tsironis, K. Lindenberg, L. Ramirez-Piscina and J. M. Sancho, Phys. Rev. A **38**, 3827 (1988).
10. B. Alberts, *Molecular Biology of the Cell* (Freeman, New York, 1988).
11. J. Rouselet, L. Salome, A. Adjari and J. Prost, Nature **370**, 446 (1994).
12. C. Doering, B. Ermentrout and G. Oster, Biophys. J. **69**, 2256 (1995).

Optimal control of large fluctuations by nonadiabatic fields

M.I. Dykman*, H. Rabitz†, V.N. Smelyanskiy*, and
B.E. Vugmeister†

*Department of Physics and Astronomy, Michigan State University, East Lansing, MI
48824, USA
† Department of Chemistry, Princeton University, Princeton, NJ 08544, USA

Abstract. We analyze the probabilities of large infrequent fluctuations in systems driven by external fields. In a broad range of the field magnitudes, the logarithm of the fluctuation probability is linear in the field magnitude, and the response can be characterized by a *logarithmic susceptibility*. This susceptibility is used to analyze optimal control of large fluctuations. For nonadiabatic driving, the activation energies for nucleation and for escape of a Brownian particle display singular behavior as a function of the field spectrum.

LOGARITHMIC SUSCEPTIBILITY FOR LARGE FLUCTUATIONS

It was pointed out by Debye that systems with coexisting metastable states may strongly respond to the driving field through the effect of the field on the probabilities of transitions between the states. For classical systems, the transition probability W is described by the activation law, $W \propto \exp(-R/kT)$. Even a relatively weak ac field h, for which the change of the activation energy $\Delta R \propto h$ is much less than R, can strongly affect W provided $|\Delta R|/kT$ is not small. This effect has been investigated for various systems and has attracted much attention recently in the context of stochastic resonance [1]. For $|\Delta R|/kT \gg 1$ the modulation of W becomes exponentially strong. So far this modulation has been analyzed for adiabatically slow driving, where the change of the field over the relaxation time of the system is small and fluctuational transitions occur "instantaneously", for a given value of the field (cf. [2]). The physical picture of transitions is different for *nonadiabatic driving*. In what follows we provide nonadiabatic theory of large fluctuations in spatially extended and lumped parameter systems. We show that the exponentially strong dependence of the fluctuation probabilities on the driving field can be described in terms of an *observable* characteristic, the *logarithmic susceptibility* (LS).

The notion of LS and the way to evaluate it are based on the idea of the optimal fluctuational path. This is the path along which the system moves, with overwhelming probability, when it fluctuates to a given state or escapes from a metastable state. The distribution of fluctuational paths to a given state peaks sharply at the optimal path, as first noticed by Onsager and Machlup.

Optimal paths in lumped parameter dynamical systems driven by Gaussian noise have attracted much theoretical interest [3] and were recently observed in experiments [4]. The notion of an optimal path applies also to continuous systems. Time-dependent fluctuations of the order parameter $\eta(\mathbf{x})$ were discussed in [5], and its optimal paths were considered in [6]. Optimal paths are "fluctuational counterparts" of dynamical trajectories: they map the problem of large noise-induced fluctuations onto the problem of noise-free dynamics of an auxiliary system (this dynamics depends on the properties of the noise driving the original system, see [3]).

We will first consider optimal paths and logarithmic susceptibility using as an example systems with a nonconserved order parameter [5](a). In these systems, fluctuations are described by the Langevin equation

$$\frac{\partial \eta(\mathbf{x},t)}{\partial t} = -\frac{\delta F}{\delta \eta(\mathbf{x},t)} + \xi(\mathbf{x},t), \tag{1}$$

with $\langle \xi(\mathbf{x},t)\xi(\mathbf{x}',t')\rangle = 2kT\delta(\mathbf{x}-\mathbf{x}')\delta(t-t')$ and with the free energy

$$F[\eta] = \int d\mathbf{x} \left[\frac{1}{2}(\boldsymbol{\nabla}\eta)^2 + V(\eta) - h(\mathbf{x},t)\eta\right], \tag{2}$$

where $V(\eta)$ is the biased Landau potential.

The probability density for the system to fluctuate to a state $\eta_f \equiv \eta_f(\mathbf{x})$ at a time t_f is described by the activation law, $W[\eta_f;t_f] \propto \exp(-R[\eta_f;t_f]/kT)$, with the activation energy given by the solution of the variational problem [3,6]

$$R[\eta_f;t_f] = \min \frac{1}{4}\int_{-\infty}^{t_f} dt' \int d\mathbf{x}' \left[\frac{\partial \eta}{\partial t'} + \frac{\delta F}{\delta \eta}\right]^2. \tag{3}$$

Here, the minimum is taken with respect to the paths $\eta(\mathbf{x}',t')$ that start from the stable state $\eta_{st}(\mathbf{x},t)$ at $t' \to -\infty$, and arrive at the final state $\eta_f(\mathbf{x})$ for $t' = t_f$. We consider pulsed or periodic driving fields h, in which cases the state η_{st} is stationary (it provides a minimum to $V(\eta)$) or periodic, respectively.

Eq. (3) defines the action for an auxiliary Hamiltonian system, with the Lagrangian given by the integrand in (3). Extreme paths of this system that minimize R are optimal fluctuational paths $\eta(\mathbf{x},t)$ of the original system (1).

In the absence of driving the system (1) is in thermal equilibrium, and the activation energy is $R \equiv R^{(0)} = F^{(0)}[\eta_f] - F^{(0)}[\eta_{st}^{(0)}]$ (the superscript 0 refers to the case $h = 0$). The optimal fluctuational path $\eta^{(0)}(\mathbf{x}, t|\eta_f, t_f)$ to the state η_f is $\dot{\eta}^{(0)} = \delta F^{(0)}/\delta \eta$.

To the first order in h, the field-induced change of R is

$$R^{(1)}[\eta_f; t_f] = \int_{-\infty}^{t_f} dt' \int d\mathbf{x}' \chi(\mathbf{x}', t_f - t'|\eta_f) h(\mathbf{x}', t'), \tag{4}$$

$$\chi(\mathbf{x}, -t|\eta_f) = -\dot{\eta}^{(0)}(\mathbf{x}, t|\eta_f, 0) \quad (t < 0). \tag{5}$$

The quantity χ describes the change $\propto h$ of the *logarithm* of the probability density to reach the state $\eta_f \equiv \eta_f(\mathbf{x})$, $\Delta \ln W \approx -R^{(1)}/kT$. This change may be *large*, and χ may be reasonably called the logarithmic susceptibility (LS). Like standard generalized susceptibility, LS has a causal form: We note that Eq. (4) suggests how to *measure* LS for various states $\eta_f(\mathbf{x})$.

Of special interest are effects of the field on the probability of escape from a metastable state of the system. For a system (1) escape occurs via nucleation. For $h = 0$, the critical nucleus $\eta_{cr}^{(0)}(\mathbf{x} - \mathbf{x}_c)$ is the unstable stationary solution of the equation $\dot{\eta} = -\delta F^{(0)}/\delta \eta$. In forming the critical nucleus the system is most likely to move along the optimal path $\eta_{nucl}^{(0)}(\mathbf{x} - \mathbf{x}_c, t - t_c) = \eta^{(0)}(\mathbf{x} - \mathbf{x}_c, t - t_c|\eta_{cr}^{(0)}, \infty)$.

The external field lifts the translational and time degeneracy of the optimal nucleation paths $\eta_{nucl}^{(0)}(\mathbf{x} - \mathbf{x}_c, t - t_c)$ with respect to the nucleus center $\mathbf{x}_c$ and the "instant" of nucleation t_c. Only one path per the field period goes from the stable to the unstable state and provides the minimum to R. The corresponding $\mathbf{x}_c$, t_c and the field-induced correction to the activation energy of nucleation $R_{nucl}^{(1)}$ are given by equations [7]

$$\frac{\partial \tilde{R}_{nucl}^{(1)}(\mathbf{x}_c, t_c)}{\partial \mathbf{x}_c} = 0, \quad \frac{\partial \tilde{R}_{nucl}^{(1)}(\mathbf{x}_c, t_c)}{\partial t_c} = 0, \tag{6}$$

$$\tilde{R}_{nucl}^{(1)} \equiv \int_{-\infty}^{\infty} dt \int d\mathbf{x} \, \chi_{nucl}(\mathbf{x} - \mathbf{x}_c, t - t_c) h(\mathbf{x}, t), \quad R_{nucl}^{(1)} = \min_{\mathbf{x}_c, t_c} \tilde{R}_{nucl}^{(1)}(\mathbf{x}_c, t_c),$$

$$\chi_{nucl}(\mathbf{x} - \mathbf{x}_c, t - t_c) = -\dot{\eta}_{nucl}^{(0)}(\mathbf{x} - \mathbf{x}_c, t - t_c) \tag{7}$$

Eqs. (6) - (7) provide the nonadiabatic theory of nucleation rate. They have a simple physical meaning: in the presence of a time- and coordinate-dependent field, the optimal fluctuation finds the "best" time t_c and place $\mathbf{x}_c$ to occur. For thermal equilibrium systems, the correction is given by the work done by the field along the optimal path.

The dependence of $R_{nucl}^{(1)}$ on the *shape* of the field may be singular. With the varying interrelation between the Fourier components of the field there occurs *switching* between different coexisting solutions of Eqs. (6), i.e. from one minimum of $\tilde{R}_{nucl}^{(1)}$ to another, with different $\mathbf{x}_c, t_c$ (cf. inset to Fig. 1).

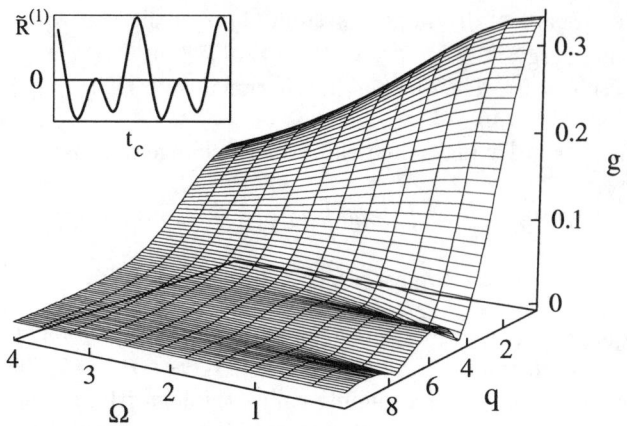

FIGURE 1. Reduced absolute value of the logarithmic susceptibility for nucleation $g(q, \Omega)$ in the case of a weakly asymmetric potential V. Inset: the correction $R^{(1)}(t_c)$ for nucleation in a uniform field $h = \cos \rho_c^2 t + 1.3 \cos(2\rho_c^2 t - 0.1)$.

Analytical results for the logarithmic susceptibility for nucleation χ_{nucl} (7) have been obtained for a weakly asymmetric double-well potential $V(\eta) = \frac{1}{4}u\eta^4 - \frac{1}{2}r\eta^2 - H\eta$, $|H| \ll r^{3/2}u^{-1/2}$. The critical nucleus in this case is a thin-wall droplet of a nucleating phase [8,10], and the Fourier transform $\tilde{\chi}_{\text{nucl}}(\mathbf{k}, \omega)$ of the logarithmic susceptibility $\chi_{\text{nucl}}(\mathbf{x}, t)$ is of the form:

$$\tilde{\chi}_{\text{nucl}}(\mathbf{k}, \omega) \equiv \frac{6R_{\text{nucl}}^{(0)}}{|H|} g\left(\rho_c k, \rho_c^2 \omega\right), \tag{8}$$

$$g(q, \Omega) = \int_0^1 dz \, \frac{\sin qz}{q} e^{-i\Omega z/2}(1-z)^{-i\Omega/2}.$$

(the free energy of the critical droplet $R_{\text{nucl}}^{(0)}$ and the critical radius of the droplet ρ_c are given in [8,10]).

One can see from (7) that, for a field of the form of a running or standing sinusoidal wave, $h = h_0 \cos(\mathbf{k}\mathbf{x} - \omega t)$ or $h = h_0 \cos \mathbf{k}\mathbf{x} \cos \omega t$, the correction to the activation energy is $R^{(1)} = -|\tilde{\chi}_{\text{nucl}}(\mathbf{k}, \omega)|h_0$. The susceptibility $|\tilde{\chi}_{\text{nucl}}|$ as given by (8) is shown in Fig. 1.

OPTIMAL CONTROL OF LARGE FLUCTUATIONS

In optimal control, one has to minimize or maximize the activation energy $R[\eta_f; t_f] = R[\eta_f; t_f | h]$ (3) of reaching a state η_f in the presence of the field h

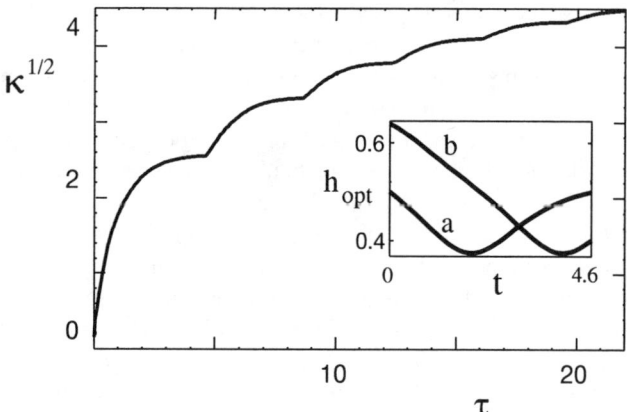

FIGURE 2. Reduced correction $\kappa^{1/2}(\tau)$ for optimal control of the escape rate of a Brownian particle in a cosine potential, $\ddot{q} + 2\Gamma\dot{q} - \sin q = \xi(t)$, for $\Gamma = 0.04$. Inset: the optimal fields for the pulse duration τ just below (a) and above (b) the switching value $\tau \approx 4.6$.

subject to a given constraint on the field, i.e. to a given value of the *penalty functional* $G[h] = \mathcal{G}$ (for example, the energy in the field pulse).

The optimal activation energy R_{opt} and the corresponding optimal control field $h_{\text{opt}}(t)$ can be found from the variational problem for the field

$$\delta\left[R[\eta_f; t_f | h] + \lambda\left(G[h] - \mathcal{G}\right)\right] = 0. \quad (9)$$
$$R_{\text{opt}} = R[\eta_f; t_f | h_{\text{opt}}]$$

where λ is the Lagrange multiplier.

For various optimal control problems in physics and chemistry the penalty functional G is quadratic in h [11]. Then, for comparatively weak fields where the field-dependent term in the activation energy R (4), (7) is linear in h, the problem (9) reduces to a linear equation for h. We give the solution of this equation for optimal control of the nucleation rate by a spatially uniform pulsed field $h(t)$ with a pulse duration τ, and the penalty functional is $G[h] = (1/2)\int_0^\tau dt\, h^2(t)$:

$$R_{\text{opt}} \approx R^{(0)} - [2\mathcal{G}\kappa(\tau)]^{1/2}, \quad \kappa(\tau) = \max_{t_c}\int_0^\tau dt\, \chi^2(t - t_c) \quad (10)$$

$$\chi(t) = \int d\mathbf{x}\, \chi_{\text{nucl}}(\mathbf{x}, t), \quad h_{\text{opt}}(t) = -\chi(t - t_{cm})\left[2\mathcal{G}/\kappa(\tau)\right]^{1/2} \text{ for } 0 < t < \tau,$$

where t_{cm} is the value of t_c which provides the maximum to the function $\kappa(\tau)$.

The above results can be easily reformulated for systems other than those described by the model (1). In particular, they apply to Brownian particles. For such systems, instead of the susceptibility for nucleation one should consider the susceptibility for escape from a metastable state.

In optimal control of escape of an underdamped Brownian particle by a pulsed field, the *shape* of the control field may change discontinuously with the varying pulse duration. This is related to the fact that the integral in the expression for $\kappa(\tau)$ may have several extrema for different t_c, and with the varying τ there may occur switchings between different extrema (cf. Fig. 1). Respectively, the activation energy is nondifferentiable at such τ (see Fig. 2).

In conclusion, we have provided the nonadiabatic theory of escape and nucleation rates in systems driven by time-dependent fields. The effect of the field on the probabilities of large fluctuations has been described in terms of the logarithmic susceptibility. This susceptibility can be measured experimentally even if the underlying dynamics of the system is not known. It has been used to formulate the problem of optimal control. We have demonstrated singular behavior of optimal modulation of the escape rate as a function of the parameters of the control field.

REFERENCES

1. See the special issue of *Nuovo Cim.* D **17**, Nos. 7-8 (1995); A.R. Bulsara and L. Gammaitoni, *Physics Today* **49**, no. 3, 39 (1996).
2. M.I. Dykman et al., *JETP Lett.* **52**, 144 (1990); T. Zhou, F. Moss, and P. Jung, *Phys. Rev. A* **42**, 3161 (1990).
3. See M.I. Freidlin and A.D. Wentzell, *Random Perturbations in Dynamical Systems* (Springer Verlag, New York, 1984); M.I. Dykman, *Phys. Rev. A* **42**, 2020 (1990); R.S. Maier and D.L. Stein, *J. Stat. Phys.* **83**, 291 (1996).
4. M.I. Dykman et al., *Phys. Rev. Lett.* **68**, 2718 (1992); M.I. Dykman et al., *Phys. Rev. Lett.* **77**, 5229 (1996).
5. (a) P.C. Hohenberg and B.I. Halperin, *Rev. Mod. Phys.* **49**, 435 (1977); (b) M.C. Cross and P.C. Hohenberg, *Rev. Mod. Phys.* **65**, 851 (1993).
6. R. Graham and T. Tél, *Phys. Rev. A* **42**, 4661 (1990); D.A. Kurtze, *Phys. Rev. Lett.* **77**, 63 (1996).
7. , V.N. Smelyanskiy, M.I. Dykman, H.Rabitz and B.E.Vugmeister, to be published.
8. J.S. Langer, *Ann. Phys.* **41**, 108 (1967); S. Coleman, *Phys. Rev. D* **15**, 2929 (1977).
9. J. Guckenheimer and P. Holmes, *Nonlinear Oscillators, Dynamical Systems and Bifurcations of Vector Fields* (Springer-Verlag, NY 1983);
10. J.D. Gunton, M. San Miguel, and P.S. Sahni, in *Phase Transitions and Critical Phenomena*, Vol.8, ed. by C. Domb and J.L. Lebowitz (Academic, NY 1983).
11. B. Kohler et al., *Phys. Rev. Lett.* **74**, 3360 (1995); J. Botina and H. Rabitz, *Phys. Rev. Lett.* **75**, 2948 (1995).

STOCHASTIC RESONANCE AT MOLECULAR LEVEL: THE POISSON WAVE MODEL

Sergey M. Bezrukov* and Igor Vodyanoy[†]

*National Institutes of Health, Bldg. 5, Rm. 405, Bethesda, MD 20892-0580, USA
and St.Petersburg Nuclear Physics Institute, Gatchina, 188350, Russia
[†]Office of Naval Research Europe, 233 Old Marylebone Road, London, NW1 5TH, UK

Abstract. Noise-facilitated signal transduction or the 'stochastic resonance' (SR) phenomenon, originally proposed as an explanation of the periodic recurrences of the Earth's ice ages [1], has been attracting rapidly growing interest from researchers working in different areas of science from physics to biology [2,3]. Only a few years ago it was generally accepted that SR can occur only in dynamical systems subjected to random forcing. Later it was hypothesized [4], and then shown both experimentally and theoretically [5,6], that the simplest 'stochastic resonator' consists only of signal, noise, and a threshold device. We introduce yet another class of systems where a noise-induced increase in the output signal-to-noise ratio can be observed [7]. These systems are both non-dynamical and threshold-free. We find SR in a very general model – a random pulse train where the probability of pulse generation is exponentially dependent on an input which is composed of a sine-wave signal plus random noise. We demonstrate that SR is a fundamental property of a wide variety of 'kT-driven' systems ranging from semiconductor p-n junctions [8] to modern mesoscopic electronic devices and voltage-dependent ion channels [9,10].

INTRODUCTION

Contrary to a threshold device, where subthreshold forcing does not generate any output signal, threshold-free systems are able to respond to input signals of however small-amplitude. In experiments with a parallel array of voltage-dependent ion channels reconstituted in a planar lipid bilayer [9] we have shown that at zero external noise this system transduces small signals with a finite, signal-amplitude-independent, coefficient. As the signal amplitude was decreased, the measured signal transduction coefficient approached a value that was in good agreement with a small-signal theoretical prediction. A very general random pulse train model with voltage-dependent pulse generation rate was used to obtain the expression:

$$\alpha = 1 + neV_h/kT \qquad (1)$$

relating the signal transduction coefficient, α, with ion channel gating charge, ne, and transmembrane potential, V_h. As a next step [7], we have analyzed this model in the presence of external noise and found that stochastic resonance is an inherent property of the voltage-dependent Poisson wave. Since the Poisson process describes elementary events in most physico-chemical (including biological) systems, we conclude that stochastic resonance is a fundamental feature of signal transduction in these systems, observable already at molecular level.

MODEL

We consider a system whose output can be described by a random train of identical pulses with the probability of pulse generation exponentially depending on the input signal. In particular, we assume that the pulse generation rate is given by:

$$r(V(t)) = r(0)\exp(neV(t)/kT) \qquad (2)$$

where $r(0)$ is the 'equilibrium rate' of pulse generation and ne is the parameter (effective charge) that defines system sensitivity to external voltage $V(t)$. As a mathematical object such a pulse train was introduced in 1955 by D.R. Cox who called it 'doubly stochastic Poisson process' [11].

We also assume that $V(t)$ is the sum of a slow zero mean Gaussian noise, $V_N(t)$, and a slow small-amplitude sine-wave signal, thus limiting ourselves to the adiabatic small-signal regime:

$$V(t) = V_N(t) + V_S\sin(2\pi f_S t) \qquad (3)$$

where $V_S \ll kT$ and f_S is much smaller than all other characteristic frequencies in the model.

By analyzing the statistical properties of such a pulse train [7] we describe the following features of signal transduction:

- threshold-free response – the ability to transfer small signals with a transduction coefficient independent of signal amplitude;

- noise-facilitated signal transduction – the property to increase the output signal amplitude by addition of noise to the system input;

- noise-induced improvement in the output signal-to-noise ratio – the existence of particular input noise levels that optimize the output signal quality.

For a random train of identical pulses the statistical properties of the process are entirely defined by the moments of pulse arrival. At low frequencies around the signal frequency f_S the power spectral density can be written in the form:

$$S_i(f) = 2Q^2 \langle r(V(t)) \rangle + 4(Qr(0))^2 \int_0^\infty \langle \exp(neV(t)/kT) \exp(neV(t+\tau)/kT) \rangle \cos(2\pi f \tau) d\tau \quad (4)$$

where Q is single pulse area. The first term of this expression describes the frequency-independent component of noise expected from an uncorrelated time-independent train of pulses with the average rate increased via the system nonlinearity, the second one accounts for the pulse rate modulation induced by the input signal and noise.

In the case of a small-amplitude signal and 'white' Gaussian noise that is band-limited by frequency f_c, the right-hand side of Eq.4 can be calculated straightforwardly. The output signal spectral density is given by

$$S_{iS}(f) = \frac{(Qr(0)neV_S)^2}{2(kT)^2} \exp\left(\left(\frac{ne\sigma}{kT}\right)^2\right) \delta(f - f_S) \quad (5)$$

where σ is the noise r.m.s. value. A threshold-free behavior and noise-facilitated signal trunsduction are clearly seen.

For the input noise with a Lorentzian spectrum (single pole filtering) we obtain the following expression describing the signal-to-noise ratio (SNR):

$$SNR = \frac{\left(\frac{neV_S}{kT}\right)^2 \frac{r(0)}{2\Delta f_a} \exp\left(\frac{1}{2}\left(\frac{ne\sigma}{kT}\right)^2\right)}{2 + \frac{2r(0)}{\pi f_c} \exp\left(\frac{1}{2}\left(\frac{ne\sigma}{kT}\right)^2\right) \sum_{m=1}^\infty \frac{1}{m!m}\left(\frac{ne\sigma}{kT}\right)^{2m}} \quad (6)$$

where Δf_a stands for the unit frequency window width of the spectrum analyzer that is to be used in measurements. It is seen that, except for the trivial dependence on the input signal amplitude (V_S) and the details of measurement technique (Δf_a), the SNR can be controlled by the noise intensity (σ), its frequency width (f_c), and 'equilibrium pulse rate' ($r(0)$).

Analysis of Eq.6 shows that external input noise can improve the output signal quality. Noise intensity corresponding to a maximum in the output SNR, increases with f_c but decreases with $r(0)$. For values of $f_c/r(0)$ smaller than 10^5 the maximum in the signal-to-noise ratio is achieved at the input noise close to

$$\sigma_{opt} \cong \frac{kT}{ne}\sqrt{\ln\frac{\pi f_c}{2r(0)}} \quad (7)$$

For the input noise with a sharp spectral cut-off (e.g. multipole Butterworth filtering), the factor ($\pi/2$) should be omitted. Interestingly, however good the initial statistics (large $r(0)$), the output signal quality can be further improved if the bandwidth of input noise is high enough for the condition $f_c > 2r(0)/\pi$ (or $f_c > r(0)$ for the noise with a sharp spectral cut-off) to hold.

CONCLUSIONS

We predict the existence of stochastic resonance in all systems that can be represented by a random pulse train with exponential statistics. We show that the two major SR attributes – noise-induced increase in signal transduction and improvement of the output SNR at some optimal noise value – are inherent properties of these systems. Since random pulse trains describe almost all physico-chemical processes at molecular level, stochastic resonance is probably much more universal phenomenon that it was believed previously.

ACKNOWLEDGMENT

We thank V.A.Parsegian for stimulating discussions and J.M.G.Vilar and G.Gomila for critical comments. This work was supported by a grant from the Office of Naval Research.

REFERENCES

1. Benzi, R., Sutera, S., and Vulpiani, A. *J.Phys. A* **14**, L453 (1981).
2. Wiesenfeld, K., and Moss, F. *Nature (London)* **373**, 33 (1995).
3. Bulsara, A.R., and Gammaitoni, L. *Physics Today* **49**, 39 (1996).
4. Wiesenfeld, K., Pierson, D., Pantazelou, E., Dames, C., and Moss, F. *Phys.Rev.Lett.* **72**, 2125 (1994).
5. Moss, F., Pierson, D., and O'Gorman, D. *Int.J.Bifurcation Chaos* **4**, 1383 (1994).
6. Gingl, Z., Kiss, L.B., and Moss, F. *Europhys.Lett.* **29**, 191 (1995).
7. Bezrukov, S.M., and Vodyanoy, I. *Nature (London)* **385**, 319 (1997).
8. Jung, P., and Wiesenfeld, K. *Nature (London)* **378**, 291 (1995).
9. Bezrukov, S.M., and Vodyanoy, I. *Nature (London)* **378**, 362 (1995).
10. Bezrukov, S.M., and Vodyanoy, I. *Biophys. J.*, in press.
11. Cox, D.R. *J.R.Statist.Soc.* **B17**, 129 (1955).

Theoretical Foundations for Nervous Networks

Brosl Hasslacher and Mark W. Tilden

Los Alamos National Laboratory, Los Alamos, New Mexico, 87545

Abstract. Following three years of study into experimental Nervous Net (Nv) control devices, various successes and several amusing failures have implied some general principles on the nature of capable control systems for autonomous machines and perhaps, we conjecture, even biological organisms. These systems are minimal, elegant, and, depending upon their implementation in a "creature" structure, astonishingly robust. Their only problem seems to be that as they are collections of non-linear asynchronous elements, only complex analysis can adequately extract and explain the emergent competency of their operation. The implications are that so long as Nv non-linear topologies can retain some measure of sub-critically coupled planar stability, the Piexito theorem will guarantee a form of plastic mode-locking necessary for broad-behavior competency. Further experimental evidence also suggests that if Nv topologies are kept in sub-chaotically stable regimes, they can be implemented at any scale and still automatically fall into effective survival strategies in unstructured environments. An explanation for how this is be possible in such minimal structures is presented.

INTRODUCTION

A Biomorphic robot (from the Latin for "of a living form") is a self-contained mechanical device fashioned on the assumption that chaotic reaction, not predictive forward modeling, is appropriate and sufficient for sustained "survival" in unspecified and unstructured environments. On the further assumption that minimal, elegant survival devices can be "evolved" from lesser to greater capabilities using silicon instead of carbon (using the roboticist as the evolutionary force of change), over two hundred different "Biomech" robots have been built and studied using solar power, motors, and minimal Nervous-Net control technology.

Nervous Networks (Nv) are a non-linear analog control technology that has been "evolved" to automatically solve real time control problems normally difficult to handle with conventional digital methods. Using Nv nets many sinuous robot mechanisms have been demonstrated that can negotiate terrains of inordinate difficulty for wheeled or tracked machines, as well as exhibiting very competent strategies for resolving immediate survival conundrums. The scale of devices developed so far has ranged from single "neuron" rovers to sixty neuron distributed controllers with broad terrain abilities, and from machines under one-inch long to several meters in length. They have recognizable behaviors that, if not efficient, are at least sufficient to resolve otherwise intractable sensory integration problems.

This work has concentrated on the development of Nv based robot mechanisms by electronic approximations of biologic autonomic and somatic systems. It has been demonstrated that these systems, when fed back onto themselves rather than through computer-based control generators, can realistically mimic many of the abilities normally attributed to lower survival-biased biological organisms. That minimal non-linear systems can provide this degree of control is not so surprising as the part counts for successful Nv designs. A fully adept insect-walker, for example, can be fully controlled and operated with as little as twelve standard transistor elements.

The initial focus of Nv technology was to derive the simplest control systems possible for robotic "cradle" devices. The reason for this is threefold. First, such systems would feature robustness characteristics allowing inexpensive machines reliable enough to be trusted with performing unsupervised work in unstructured environments. Second, using Nv technology we hoped to resolve one of the most enviable things about biological designs, namely how nature can stick large numbers of lightweight, efficient actuators almost anywhere and still have them operate effectively. Third, and most important, exploration of minimal control systems may explain the biological paradox of why biological mechanisms can get by on so few active control elements. A common garden ant has roughly twenty-thousand control amplifiers distributed throughout its entire body, whereas a digital watch may have as many as half a million amplifiers and still be unable to walk. How does nature do so much with so little? The question is, what are the fundamental properties of living control systems, and what relationship do they have to the implicit abilities of Nv control topologies? Does Nv technology use some approximation of natural living things, is it the other way around, or is it neither?

BACKGROUND THEORY BEHIND SCALABLE, ADAPTIVE, AUTONOMOUS MACHINES

Simple electronics and mechanics makes an entire Biomech device a plastic analog computer, but where is the complexity for walking? Answer: In a phase space of quasi-periodic, mode locking analog oscillators, capable of chaos and spun out dynamically. This implies that it is difficult (if not impossible) to write out equations of motion for the systems because of the variable chaotic dimensions involved (both in the robot and from the environment), but fortunately we can bound them because the dynamic parameters are globally organized. We can thus study domains in parameter space for various maps and characterize typical behaviors (the Principle of Genericity). The control systems discussed have a dynamic phase space that couples to the world fractality and uses it to compute its dynamics.

That is, we can assume the basic, unperturbed Nv microcore system as a double pendulum in a reactive fluid. The oscillation characteristics of both pendula are not independent but linked by indirect, subcritical, but highly variable mode-locking forces. These oscillations are typically rendered as the classic double Dynamicists Torus (2 Torus), and in real-world systems with random noise and therefore random initial conditions, this is the general case. The windings from two completely uncoupled oscillators eventually cover the entire torus surface. In marginally to critical coupled cases, the oscillators fall into recognizable harmonics and are very easy to analyze using classical Laplacian techniques. However, how do we classify systems where the coupling characteristics are continuously changing? This is best done using Arnold tongues from circle-map models.

ARNOLD TONGUES AND CRITICAL REGIONS

If ω is the winding number, the number of times the combined orbits wind or cover the torus, then $\omega = f_1/f_2$, where f_1 and f_2 represents the unperturbed frequency of the individual oscillator systems. k represents the degree of driving of a non-linearity, represented as a scalar. For example, in the standard Arnold's Circle Map, for the range of non-linearity of k up to 1 there are almost no irrationals locked. Mode locking is multiple and rational, and at all rationals under external perturbation we get periodic and quasi-periodic orbits. Above $k = 1$, the critical line, the situation is very complex. The Arnold tongues overlap, resulting in a hysteresis that could account for the evidence of multi-stability regions in Biomech systems.

The point is that at any moment, the stress of the world environment on the two-torus robot micro-model results in bifurcation forces on a "qualia" point. This points' motion around the Arnold space is based upon perturbation forces that either push it down a funnel into one of several quasi-periodic behaviors (avoid, fight, push, struggle), or upwards onto a noise regime that allows it to find, should the perturbations allow, another behavior path. In this way, Biomechs have the capacity of performing a complex search for behavioral solutions without the need of algorithmic structures. But above the $k = 1$ critical case, how is it possible that a rational behavior can be derived from what must be an irrational, chaotic regime?

If we theorize a single Arnold funnel, it can be imagined that much more structure exists than would be thought, especially when bounded by resolution complexities imparted by a quasi-fractal world on the torus structure. There is always a path to rationality, provided the world does not get too complex that the region loses coherence.

So the machine "adapts" by hopping from tongue to tongue depending on the coupling strength k, the feedback of the world onto the non-linear system. This way the machine can alter its global behavior into "basins", the Biomech classic example being the emergent walking gaits or search modes that combine to form the capable general problem solving ability seen in most robot systems to date. The disadvantage comes in topologically locked Nv structures that are too complex to escape from tightly coupled behaviors, then problems show up in the forms of "epileptic" responses where the robot falls into hyper-periodic behavior that it cannot get out of. The trick is therefore to find that class of Nv control architectures that can support the broadest k regions without getting trapped at the $k = 0$ line. This is a topic for another paper, but the big question here is what keeps a capable Biomech nervous system from becoming completely chaotic above critical k values?

PEIXOTO'S AMAZING THEOREM

The Peixoto (Pu-she-to) theorem on structural stability for general motion of coupled oscillators on a 2-torus (circa 1960) appears to be the complete theoretical foundation for the adaptive behavior characteristics of Biomorphic machines. For example, in the generic quasi-periodic case under Peixoto, the irregular phase winding will foliate into approximations of a leaf structure given all classes of tenuous coupling between oscillators.

Peixoto's Theorem states that a finite, even number of closed trajectories will always phase lock between alternating attractors and repellors. That is, the attractors are bisected by a giant repellor orbit basin. This leads to unpredictable but bounded braid structures on the torus surface (still structurally unstable under external perturbations), but any random walk in this extended phase space must eventually convert to motion on such an attractor basin.

The bottom line is that nervous nets under Peixoto is both robust and flexible, provided the world that provides the perturbation is interpreted by the Nv structure as not completely chaotic. Such dependence on bounded structure rather than predictive certainty is what constrains the behaviors into function rather than chaos. This implies that to scale the systems into higher degrees of function, the coupling between tori cannot be higher than the criticality threshold, but also not lower than the purely chaotic threshold. Biasing considerations and architecture are thus crucial for effective nervous net designs, but physical scaling is not. These devices can be any physical size and still expect the same homogeneity of Peixoto's control.

Indeed, so long as the Peixoto parameters for coupling of tori are maintained, experiment has shown that the stability of a system to match the complexity of concurrent inputs, be they from limbs, sensors, or even other networks, will always converge towards one of several behavior basins proportional to the complexity of the external image. Using Peixoto's theorem as the "bounder" for control attributes, and tori as the basic elements of nervous net structure, the implications are that there exist minimal structures that can form a survivability competence in the real world almost completely automatically. It has been seen in nature that limited, repetitive behavior is a leading cause of death, but the added complexity of a "brain" is not generally required. Ninety-eight percent of all species have no brain to speak of, so where do their most primitive adaptive traits evolve from? If natural objects survive by having the broadest behavioral structure with the smallest number of components, then it is interesting to speculate that perhaps some approximation of Nv systems and Peixoto might be a universal basis for the evolution of mobile life.

It must be noted that the Biomech robots built so far were not built to mimic any class of living organism. They evolved to their state through repetitive attempts by the designer to optimize behavior characteristics given the limitations of artificial materials, yet these robots do appear to manifest a distinctly biological behavior pattern. The snakebot, for example, gets an immediate rise from domestic animals and people with snake phobias. This sort of response must only come from the richness of behaviors seen in the complex process flow. Periodic behaviors, as one would expect from a toy, are lacking here. Complex, sinuous behaviors were not designed but emerged as a feature of the systems. If Peixoto's theorem is valid for these cases, then we conclude that such behaviors are necessary, and not just sufficient.

From an engineering point of view, it is extremely advantageous that design of these machines is directly related to the limit complexity of the environment to be traversed. Though not a calculus right now, eventually matching a robot to a task could be as simple as getting detailed environment specifications about the afflicted area. Such has been the case in the research of Biomech machines for automatic explosive ordnance removal in hostile desert regions. The idea is to make a survival-based machine cheap and smart enough to thoroughly explore an unspecified minefield, but not so smart as to realize that it is in a minefield wishing it were somewhere else.

Since the start of research in the spring of 1994, development of this technology has advanced to solving difficult sensory and cognitive problems. The goal is the reduction of currently complex systems down to an inexpensive but robust minimum, and Nv architecture stability to a simulateable calculus. Further efforts are also being made to apply this control strategy to the expanding nanotechnology field. At the nanometer scale Nv's may prove more feasible than nano-computers for control of self-assembling micro structures, and if Peixoto remains stable at this scale, then such must be the case.

In conclusion, we speculate that nervous nets may become to electrical neural nets the same way peripheral spinal systems are to the brain, a medium-independent version of biological actuation methods. But the biggest result here is that there really are minimal, elegant solutions to building systems capable of automatically negotiating real world complexity. Something we have always suspected but have been perhaps too lost in our computers to develop clearly. Further development of Nv technology is necessary, but it is encouraging to know that the future of machine life might not have to depend on ever more advanced compilers and operating systems, but instead on the advances and application of complexity science.

NONLINEAR MODELS OF SOCIETAL SYSTEMS

Complex Issues of Military Capability: Measurement, Assessment, Simulation

L. D. Miller, M. F. Sulcoski, and B. A. Farmer

National Ground Intelligence Center
220 7th St. N. E.
Charlottesville, Va. 22902

Abstract. The nonlinear science methodology behind the Military Capability Spectrum Project (PRISM) is presented. This project approaches assessments of military power of nations (worldwide) from the perspective that their military organizations are complex adaptive systems (CAS) locked in a threat/alliance coupled group that collectively evolves toward a self-organized critical state. Dissipative behavior can take the form of war and other forms of geopolitical instability. Measurement tools for assessing the characteristics of military organizations are presented and methodologies for displaying the resulting spectrum of worldwide military power are discussed. Tools for regional stability/instability analysis are suggested that rely upon time history data.

INTRODUCTION

The ending of the cold war has presented the US military with new challenges in planning and decision making. The weapon technologies, force levels, and command structures and philosophies that were developed during the cold war may not be appropriate for an age when our military is more likely to be called into action to support small peace keeping operations, humanitarian assistance, and regional conflicts rather than large symmetrical conflicts. Furthermore, the recent globalization of communications technology demands a temporal urgency for information processing and dissemination. Our future military operations are more likely to involve international coalitions than was the case in the recent past. Our experience in Somalia demonstrated that entering an operation for humanitarian reasons does not obviate the need for accurate and timely military capability assessments (MCA).

Fortunately, the geopolitical evolution that has presented these new challenges has been accompanied by a scientific evolution that promises to assist in the search for solutions. A welcome new framework for considering the

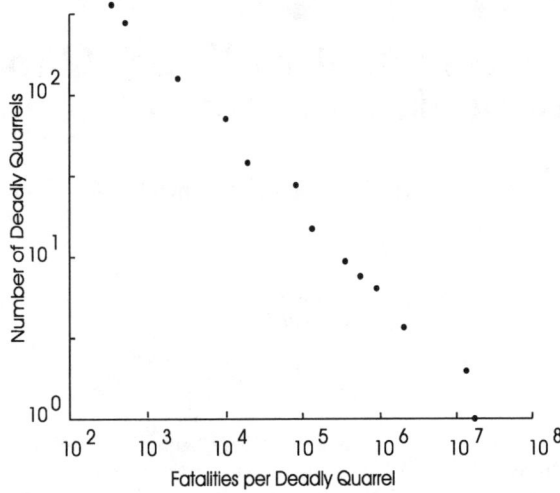

FIGURE 1. Richardson's data on the frequency of "deadly quarrels."

problems of military science now exists. Chaos Theory suggests that no degree of accuracy in MCA will permit absolute predictions of the outcome of combat operations; nevertheless, Dockery and Woodcock [1] have shown that a nonlinear approach to combat simulation merits careful study. Likewise, Mann [2] has considered the role that nonlinear science can play in strategic thought and planning. It remains an open question whether nonlinear science is more than just a good metaphor for military problems. Chaos Theory, in particular, seems too limited in scope to transcend metaphor.

Complexity Theory conditions us to look for patterns in spatial and temporal behavior as indicators of the underlying dynamics of complex adaptive systems (CAS). The tools of fractal geometry provide a means for pattern analysis. Computer simulation methods reveal that the behaviors of CAS are generic in ways suggestive of the universality classes of Chaos Theory. The phenomenon of self-organized criticality [3] (SOC) appears to be a generic class of behavior that holds promise for improving our understanding of military organizations as CAS. The slow and steady buildup of military capability of the nations of the world corresponds to the elastic storage mechanism of an open CAS, while wars and other forms of destructive behavior represent the dissipative avalanches. The work of Richardson [4] provided evidence in support of this conjecture prior to the discovery of SOC. Richardson analyzed the data on war casualties incurred over the period from 1820 through 1945 (see Figure 1). Since no better metric for the destructiveness of war exists than casualties, the approximate linear (power law) form of these data supports (though does not prove) the existence of SOC behavior in international relations (at least during this historical period).

ISSUES OF MEASUREMENT

Before applying complexity theory to MCA, we must confront issues of measurement. The first (for CAS) is one of scale. Because CAS are usually ordered hierarchical (approximate fractal) structures, no single scale suffices for their description. Three scales are customarily identified in military thought: 1) the technical/tactical scale (TTS) where the interest is in the individual soldier and his equipment, 2) the operational scale (OS) where the interest is in the mid-level commanders, the groups they command, and their areas of specialty, and 3) the strategic scale (SS) where the interest is in the high-level commanders, the networks they command, their overall objectives, and their plans for achieving them. No military capability assessment is complete without information at all three scales. Our measurement approach is to use subject-area experts (SAE) to aggregate information at the TTS into a number of different qualitative categories at the OS. These categories are mostly limited to land force capabilities. Decisions concerning category definitions, numbers of categories, and weightings for different categories were all made by committees of SAE. The twelve categories that are currently included in the PRISM database are Air Defense; Combat Engineer Support; Combined Arms Operations; Command, Control, Communications, Computers, and Intelligence; Economics; Fire Support; Joint Operations; Logistics and Sustainability; Maneuver Forces; Military Culture; Technology Base; and Training.

Once these categories were defined, committees of SAE (including experts in technologies, systems, and forces) identified ten levels of performance in each category ranging from the most primitive (1) to the most advanced(10). Each committee of SAE identified an ensemble of qualitative TTS characteristics associated with each level. The resulting Terms of Reference (TOR) document became the measurement tool for the PRISM project. Measurement then consists in knowledgeable SAE assigning TOR-derived category levels in each of the twelve categories for the armed forces of a given country.

ISSUES OF ASSESSMENT

Assessment is an aggregation process. Each OS category level assignment by an SAE team is an assessment aggregating information at the TTS. Unfortunately, this aggregation process is invisible to us because it is performed in the brains of our SAE. Understanding this process will provide guidance in aggregating information at the OS to achieve assessments at the SS. Few human experts are sufficiently knowledgeable to handle SS MCA via purely mental processes. Thus, we are led to the study of the science of perception.

The science of perception was initiated by Weber and Fechner [5] during the last century. They introduced the concept of the psychophysical function: the relationship between the human response to a stimulus and the physical magnitude of the stimulus. The Weber-Fechner studies suggested a logarithmic relationship. More recently, Stevens [6] clarified the nature of the psychophysical function by considering the differences between human responses evoked by magnitude estimation (ME) based upon a single reference standard and those evoked by category estimation (CE) based upon two reference standards (a minimum and a maximum stimulus). Stevens demonstrated that when responses are evoked by ME the psychophysical function is frequently a power law with the exponent dependent on the nature of the physical stimulus. He further showed that when human response is evoked by CE, a different functional form pertains - one which actually depends on the frequency distribution of the stimuli. Still more recently, Lefebvre [7] has suggested a simple human decision model that leads to a hyperbolic function for the ratio of Stevens' psychophysical functions of ME and CE type. Our own investigation has shown that for groups of stimuli skewed toward the minimum end of the scale, the Lefebvre form of the psychophysical function for CE is qualitatively similar to the original Weber-Fechner law when Stevens' power-law exponent for ME of that stimulus is near one. In all of these investigations the CE level assignments resulting from human judgment are assumed to be ordinal, whereas the application of the inverse psychophysical function yields ratio scale data.

The issue before us is how to aggregate the CE level assessment data at the OS to form SS assessments for a nation's military capability. Roberts [8] has shown that ordinal data are best aggregated (or averaged) with the geometric mean. We thus define the strategic complexity (C) of a nation's armed forces as the generalized geometric mean of the twelve CE level assignments:

$$C = \sqrt[m]{\prod_{k=1}^{12} F_k^{W_k}}, \tag{1}$$

where F_k is the CE level assignment for the kth category, W_k is the weight of the kth category, and $m = \sum_{k=1}^{12} W_k$. The above expression reduces to the geometric mean when all category weights are unity. Since all CE level assignments are integers (1-10), C is a number between 1 and 10. C is also best considered an ordinal number since the Fs are ordinal numbers. We expect that some generalized inverse psychophysical function is required to transform C into a ratio scale qualitative SS MCA of a nation's armed forces.

Before postulating a psychophysical function, however, we must address another issue. Two nations might have the same value of C: one through a uniform set of level assignments closely clustered around C while the second has a wild distribution of level assignments (some above and some below C). Our expectation is that the second nation does not deserve as high an SS assessment as the first due to the miss-match of capabilities and the synergistic couplings between them. No doubt, our experts mentally included such disparities in TTS characteristics before making their OS CE level assignments. To address this issue we introduce a category level distribution function $n_i = N_i/12$, where N_i is the number of categories receiving a level assignment of i (1-10), and define a category level entropy:

$$S = -\sum_{i=1}^{10} n_i \ln n_i. \qquad (2)$$

We now introduce an inverse psychophysical function that depends on both C and S and define a qualitative military capability potential (MCP):

$$MCP = A^{(C-1)/(1+S)}, \qquad (3)$$

where A is an undetermined universal constant (probably slowly varying with time). We consider $1 \leq (C+S)/(1+S) \leq 10$ to be the ordinal SS assessment variable. We choose an exponential form (the inverse of the Weber-Fechner law). Since MCP ($MCP \geq 1$) is now assumed to be a ratio scale number, it can be multiplied by total land force manpower (M, taken from the CIA World Fact Book) to define a total military capability potential ($TMCP$):

$$TMCP = M \times MCP. \qquad (4)$$

We view $TMCP$ as the ratio scale strategic assessment of a nation's total military strength, including both qualitative and quantitative factors.

The only remaining issue of assessment is the value of A. The choice of A determines the value of force multiplication that pertains between the most primitively trained, equipped, and led soldier (with $MCP = 1$) and the most advanced. We could have turned to our SAE for an estimate of this number; instead, we decided to attempt to estimate it from the results of actual combat. Predicting the results of combat is an extremely complex problem and we do not mean to imply that our SS assessments can be used for this purpose. Nevertheless, we reason that under certain circumstances (a saturated battlefield: a situation where forces are so evenly matched that the outcome of combat is a stalemate) the casualty ratio should be the inverse of the MCP ratio. This condition was met in the mid-1980s war between Iran and Iraq. Employing SAE assessment data (for the mid-1980s), we require:

$$\frac{IRAN\ CASUALTIES}{IRAQ\ CASUALTIES} = \frac{A^{(C_{IRAQ}-1)/(1+S_{IRAQ})}}{A^{(C_{IRAN}-1)/(1+S_{IRAN})}}, \qquad (5)$$

and find that $A = 2.30$. This implies a maximum force multiplier ratio of $MCP_{max} = 1800$ or A^9. These numbers are crude estimates (the casualty figures themselves are only crudely known). Nevertheless, they permit us to make relative comparisons of the SS strengths of ground forces and provide a tool for collecting ratio scale time history data on national military strengths.

In Figure 2 we display the frequency distribution of worldwide ground force TMCP that results from our SAE assessments. We display these data on a

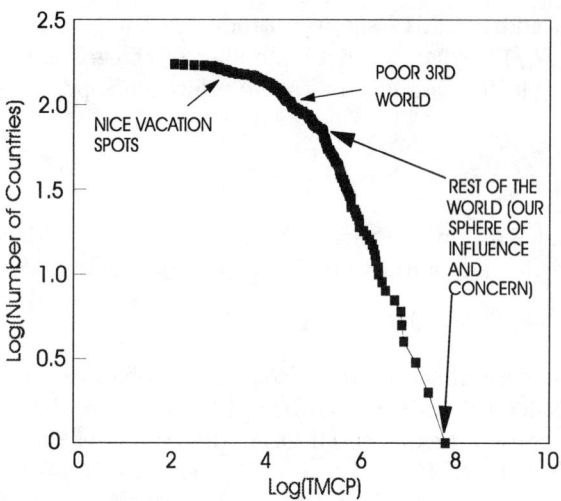

FIGURE 2. Current frequency distribution (rank ordered plot) of the world's ground force military power (TMCP).

rank ordered log-log plot (standard for SOC discussions). Each point on this plot is associated with a country (the point where it falls off the cumulative list by virtue of its failure to exceed the associated value of $TMCP$). We will not discuss the positions of individual countries on this plot; however, we call attention to the striking linearity of the data for values of $Log(TMCP) > 5$. Does this imply SOC behavior in the international relations of countries associated with this power law region? At first this appears to be an erroneous conclusion because SOC is generally associated with dissipative behavior and what we are displaying in Figure 2 is total strength (an elastic quantity).

Suppose (for the moment) that CAS are fractal networks with components capable of elastic deformation under slow energy input from external sources. Suppose further SOC avalanches are triggered by inelastic failure of a network component. The size of an SOC avalanche would be correlated with the stored elastic energy of the failed component, the nearby components, and the strengths of their couplings prior to their failure. Then we would expect that the rank ordered frequency distribution of stored elastic energy of network components would also display the SOC power law form. That is what we see for the linear region of Figure 2. The countries associated with the $Log(TMCP) < 5$ region of the figure are mostly of two types: 1) remote tropical islands (nice places to go on vacation) disinterested in defense for geographical reasons, and 2) very poor third world countries whose economies are incapable of supporting a military with more than police functions. The remainder we associate with our sphere of influence and concern: a group whose military evolution is governed by alliances and perceived threats.

TABLE 1. Richardson's Data on Military Spending just prior to WWI: Units are Millions of Pounds Sterling (M£).

Alliance	1909	1910	1911	1912	1913
fr	115.3	119.4	127.8	145.0	166.7
gah	83.9	85.4	87.1	93.7	122.3

ISSUES OF SIMULATION

The TMCP distribution of the threat/alliance coupled group identified in Figure 2 is a static snapshot. A goal of the PRISM project (not yet realized) is to collect time history data that can be analyzed for emergent patterns of behavior. The generalized Discrete Richardson Model [9] (DRM), a coupled map lattice of dynamic SS MCA, captures the nonlinearity of reflexive action, response, counter action, counter response... expected to dominate the group dynamics. We will illustrate the concept using data on military spending just prior to the outbreak of World War I (WWI) collected by Richardson [10]. The metaphor of WWI as an SOC avalanche event has occurred to many researchers. Richardson's data for currency normalized (£) military spending of the two major pre-war alliances (France/Russia (fr) and Germany/Austria-Hungary (gah)) for the period 1909-1913 are shown in Table 1.

We designate 1908 alliance strengths t_{1908}^{fr} and t_{1908}^{gah} (t is shorthand for $TMCP$). Their levels of t in succeeding years can be found as follows:

$$t_{19xx+1}^{gah} = (1-d)t_{19xx}^{gah} + e_{19xx+1}^{gah}$$
$$t_{19xx+1}^{fr} = (1-d)t_{19xx}^{fr} + k_e e_{19xx+1}^{fr}, \qquad (6)$$

where d is identified with a yearly military depletion factor, k_e is an efficiency factor (only applied to the fr alliance by unit definition), and $e_{19xx+1}^{fr/gah}$ are expenditures taken from Table 1. Repeated application of Equation 6 for the period 1909-1913 will yield ten polynomials for $t_{19xx}^{fr/gah}$ in a set of four unknowns: $t_{1908}^{fr}, t_{1908}^{gah}, d,$ and k_e. The DRM equations:

$$t_{19xx+1}^{gah} = t_{19xx}^{gah} + (t_{max}^{gah} - t_{19xx}^{gah})[i_{gah}(t_{min}^{gah} - t_{19xx}^{gah}) + k_{gah}(t_{19xx}^{fr} - t_{min}^{fr})]$$
$$t_{19xx+1}^{fr} = t_{19xx}^{fr} + (t_{max}^{fr} - t_{19xx}^{fr})[i_{fr}(t_{min}^{fr} - t_{19xx}^{fr}) + k_{fr}(t_{19xx}^{gah} - t_{min}^{gah})], \qquad (7)$$

can be iterated to yield the same quantities in a different set of ten unknowns: $t_{1908}^{fr}, t_{1908}^{gah}, t_{max}^{fr}, t_{max}^{gah}, t_{min}^{fr}, t_{min}^{gah}, i_{gah}, i_{fr}, k_{gah}, k_{fr}$. The t_{max} and t_{min} variables are interpreted as maximum sustainable and minimum level (police function) $TMCP$, the $i_{fr/gah}$ are internal reactivity parameters, and the $k_{fr/gah}$ are threat coupling parameters. Two unknowns (the $t_{1908}^{fr/gah}$ initial conditions) are common between these two sets. Equating the rhs of the iterates of Equations 6 and 7 yields ten (nonlinear) equations in twelve unknowns. An eleventh

TABLE 2. DRM Parameters for WWI (Units of t Parameters are M£).

Alliance	t_{1908}	t_{max}	t_{min}	i	k
FR	266.0	699.9	414.2	−0.0014	0.00104
GAH	437.5	560.7	207.3	−0.0037	0.00655

equation relating these unknowns was obtained by equating the ratio of t_{max} to the ratio of alliance gross national products (GNP) for the year 1913 [11].

Space does not permit a discussion of the procedure used to solve these equations. Although an unconstrained variable remains, our military historical knowledge of this period permits us to eliminate spurious solutions. A set of realistic DRM parameters (and initial conditions) is given in Table 2. The corresponding values for d and k_e are 0.23 and 0.65 respectively. The time variation of $TMCP$ for these DRM parameters is shown in Figure 3. The mapping (Equation 7) becomes unstable beyond 1918, shows a "power crossing" in the 1915 timeframe, and is characterized by extreme sensitivity to initial conditions at the point of the assassination (1914). No combination of initial conditions (t_{1908}) leads to stable behavior for this set of DRM parameters. Inspection of the mapping behavior suggests that one way to achieve stability is to reduce the threat to the gah alliance. An instability threshold exists for this variable when k_{gah} is near 90% of the value shown in Table 2. The area of stable behavior in the space of initial conditions (t_{1908}) just below this threshold is shown in Figure 4a. Note that the area of stability dissolves

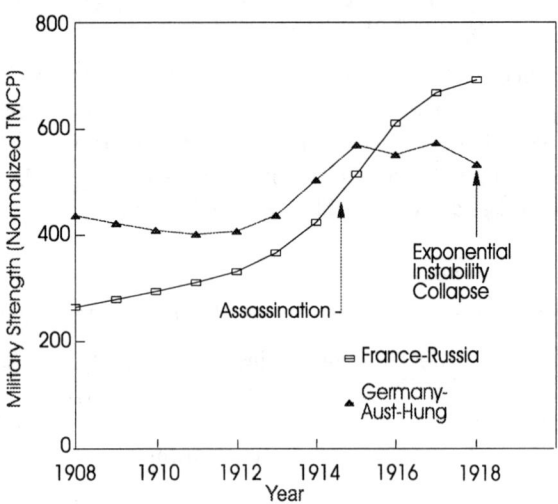

FIGURE 3. TMCP variation (M£) for WWI alliances for 1908-1918.

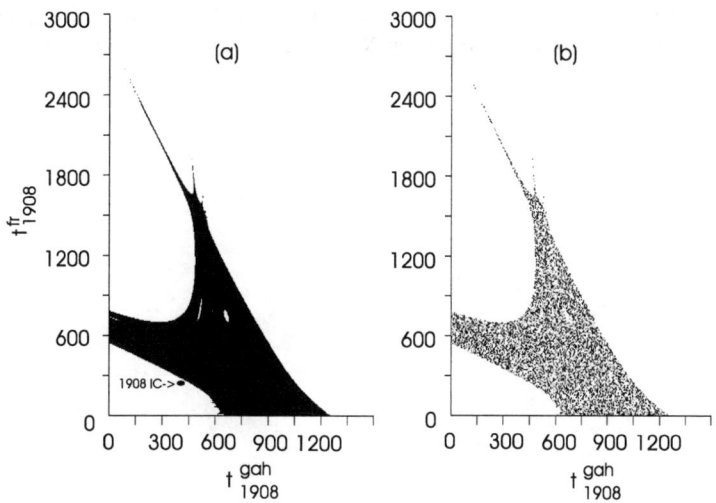

FIGURE 4. Region of stability for initial conditions (t_{1908}) for k_{gah} equal 91% (a) and 91.1% (b) of value in Table 2.

into a fractal dust (and disappears) as the threshold is exceeded (Figure 4b). In conclusion the DRM methodology offers promise for strategic stability analysis once the PRISM military capability assessment tools have been refined to the point where accurate (and sufficient) time history data become available.

REFERENCES

1. Dockery, J. T., and A. E. R. Woodcock, *The Military Landscape: Mathematical Models of Combat*, Woodhead Publishing Limited, Cambridge, England, 1993.
2. Mann, S. R., in *Essays on Strategy IX*, National Defense University Press, Washington, DC, 1993, p. 33, Edited by T. C. Gill.
3. Bak, P., and K. Chen, *Scien. Amer.* **46**, 46(1991).
4. Richardson, L. F., *Statistics of Deadly Quarrels*, Boxwood Press, Pittsburgh, Pa., 1960.
5. Schroeder, M., *Fractals, Chaos, Power Laws: Minutes from an Infinite Paradise*, W. H. Freeman and Company, New York, 1991, p. 70.
6. Stevens, S. S., *Percep. Psychophys.* **6**, 251(1969).
7. Lefebvre, V., *A Cosmic Subject*, Russian Academy of Sciences Institute of Psychology Press, Moscow, 1997, and personal communication.
8. Roberts, F. S., in *Handbooks in OR and MS, Vol 6*, Elsevier Science B. V., 1994, p. 621, Edited by S. M. Pollock et al.
9. Grossmann, S., and G. Mayer-Kress, *Nature* **337**, 701(1989).
10. Richardson, L. F., *Arms and Insecurity*, Boxwood Press, Pittsburgh, Pa. 1960.

11. Varynen, R., in *Arms Races: Technological and Political Dynamics*, Sage Publications, 1990, p. 325, Edited by N. P. Gleditsch and O. Njolstad.

Degree of Correlation Inside a Financial Market

Rosario Nunzio Mantegna

Istituto Nazionale per la Fisica della Materia, Unità di Palermo and Dipartimento di Energetica ed Applicazioni di Fisica, Università di Palermo, Viale delle Scienze, I-90128 Palermo, ITALIA

Abstract. I present an empirical study of the correlations observed between pairs of time series of stock prices in the New York Stock Exchange. I verify that various degrees of correlations or anti-correlations are present inside a financial market and I study the time evolution of these correlations. I briefly discuss how these empirical observations might be consistent with the well accepted hypothesis of absence of arbitrage in an efficient financial market (i.e. that there is no way of extracting money from the market in a continuous way without risk).

INTRODUCTION

Since the fifties several empirical studies have been performed to quantify the correlation observed in the time evolution of a stock price traded in a financial market (see, for example [1]) . The most accepted conclusion about this important problem is that time correlations are rather weak or absent in a time interval ranging from a trading day to several years. Consistent with this empirical observation the expected value of the discounted price $P(t_k)$ of an asset at time t_k is assumed to be equal to the last known price (observed at time t_{k-1}) provided the set of information consisting of the past price changes $\Delta P(t_i)$ [2]. In formal terms

$$E[P(t_k)|P(t_0), \Delta P(t_1), \Delta P(t_2), ..., \Delta P(t_{k-1})] = P(t_{k-1}) \qquad (1)$$

The rigorous mathematical definition of this stochastic process is the definition of martingale [3]. A random walk with zero drift is a special case of martingale. The converse is not true. Roughly speaking, in a random walk, increments are independent and identically distributed, while in a martingale increments are independent of previous information but not necessarily identically distributed.

When one attempts to model a stock exchange as a whole system taking into account the simultaneous presence of several stocks traded in the same market, the simplest hypothesis is of course to consider stock prices in a stock exchange behaving as an ensemble of martingales with no correlations between them. However this naive approach is not consistent with the common expectation that correlated and anti-correlated economic activities exist. The hypothesis that correlations between stock prices are present in financial markets is the one which provides the theoretical foundation to the theory of selecting the most efficient portfolio of financial goods [4]. The portfolio selection theory relies on the property, observed in empirical data, that the covariance between different stock price changes might be positive, negligible or negative in financial markets.

In this seminar I present empirical results on the degree of correlation between stocks traded in the New York Stock Exchange and on the temporal evolution of these correlations. Specifically I investigate the correlation coefficient [5] between all of the pairs of stocks used to compute the Dow Jones Industrial Average (DJIA) index and the Standard and Poor's 500 (S&P500) index.

I confirm the presence of correlations and anti-correlations between stocks. The correlations and anti-correlations are observed on a time scale of years and their values are only slowly time dependent with characteristic times longer than a year.

DEGREE OF CORRELATION

I analyze two sets of stocks of the New York Stock Exchange (NYSE): (i) the 30 stocks used to compute the DJIA; (ii) the 500 stocks used to compute the S&P500 index (in this last case some of the stocks used are traded in the Over the Counter (OTC) market). The investigated period ranges from July 3rd, 1989 to October 27th, 1995 . In addition to the investigation of the overall time interval, analyses are also performed with a time interval of one year.

I investigate the correlation coefficient between stocks i and j, where i and j are the numerical labels of the stocks. Starting from the daily logarithmic changes of the price

$$Y_i = \ln P_i(t) - \ln P_i(t-1) \qquad (2)$$

where $P_i(t)$ is the price of the stock i at the day t. I compute the correlation coefficient by following the mathematical definition

$$\rho_{ij} = \frac{<Y_i Y_j> - <Y_i><Y_j>}{\sqrt{(<Y_i^2> - <Y_i>^2)(<Y_j^2> - <Y_j>^2)}} \qquad (3)$$

TABLE 1. Average value and standard deviation of ρ determined year by year and in the overall time interval for the set of stocks used to compute the DJIA.

time period	$<\rho>$	σ_ρ
1990	0.370	0.135
1991	0.300	0.099
1992	0.154	0.099
1993	0.110	0.091
1994	0.187	0.091
7/89-10/95	0.224	0.076

ρ_{ij} quantifies the correlation observed between the price of the stock i with the price of the stock j. ρ is varying from -1 to 1. $\rho = 1$ means full correlation between Y_i and Y_j variables, $\rho = -1$ means fully anti-correlated variables while $\rho = 0$ is observed for uncorrelated variables.

DJIA

The number of ρ_{ij} correlation coefficients for this set of stocks is $435 = 30 \cdot 29/2$. The probability density function (PDF) $P(\rho)$ empirically measured, namely the probability to find values of ρ within the interval $(\rho - d\rho, \rho + d\rho)$ is a bell shaped curve approximately symmetrical with pronounced tails (especially when $\rho > 0$). More detailed conclusions about the PDF cannot be drawn from the empirical results because the measured PDFs are rather noisy due to the relatively limited number of correlation coefficients present in this set of stocks. The average values and standard deviation of ρ in the time intervals investigated are given in Table 1. The highest correlation coefficient on a time scale of 1 year is observed between Coca Cola and Procter & Gamble in 1990 ($\rho = 0.726$) while the highest anti-correlation is observed between Philip Morris and Exxon Corp ($\rho = -0.163$) in 1993.

The values of ρ are slowly time dependent. For example I measure in the case of Coca Cola and Procter & Gamble the values summarized in Table 2. The correlation coefficient between CocaCola and Procter & Gamble is always well above the average values of the DJIA set. The distance in units of standard deviations from the average value ranges from 1.25 to 2.62, varying year by year.

Standard & Poor's 500

The same analysis is also performed in the set of stocks used to compute the S&P500 index. In this case the number of correlation coefficients is of

TABLE 2. Correlation coefficient between CocaCola and Procter & Gamble measured from 1990 to 1994 on yearly time periods. $\delta = (\rho - <\rho>)/\sigma_\rho$ namely the distance from the average value given in units of standard deviation.

time period	ρ	$<\rho>$	δ
1990	0.726	0.370	2.62
1991	0.472	0.300	1.73
1992	0.279	0.154	1.25
1993	0.332	0.110	2.44
1994	0.395	0.187	2.27
7/89-10/95	0.459	0.224	3.09

the order of 10^5. This implies that the empirical determination of the PDF $P(\rho)$ is rather accurate. $P(\rho)$ measured in the overall time interval is a double peaked PDF with the main peak observed for $\rho \approx 0.15$ and the second peak observed for $\rho \approx -0.05$. The ratio between the height of the two peaks is approximately 30 (the highest peak is observed for $\rho \approx 0.15$). The PDF of the correlation coefficient slightly changes its shape depending on the investigated time period. The PDF always shows a bell shaped curve around the average value of ρ but also shows prominent tails in the intervals $-0.3 < \rho < 0$ and $0.4 < \rho < 0.9$. The tails are more evident during the years 1990, 1991 and 1994. The average value and standard deviation of ρ measured in the time intervals investigated are given in Table 3. The time dependence of the average value of ρ for the S&P500 set of data is qualitatively the same as that observed for the stocks of the DJIA. Specifically, a minimum is observed in 1993 and a maximum is observed in 1990 for both sets of stocks.

In this set of stocks, the highest correlation on a time scale of 1 year is observed between Homestake Mining and Placer Dome Inc. in 1994 ($\rho = 0.820$) while the highest anti-correlation is observed between Barrick Gold

TABLE 3. Average value and standard deviation of ρ determined year by year and in the overall investigated time interval for the set of stocks used to compute the S&P500.

time period	$<\rho>$	σ_ρ
1990	0.258	0.115
1991	0.207	0.097
1992	0.130	0.087
1993	0.095	0.091
1994	0.138	0.086
7/89-10/95	0.168	0.068

TABLE 4. Correlation coefficient between Barrick Gold and Nynex Corp measured from 1990 to 1994 on yearly time periods. $\delta =$ is defined as in Table 2.

time period	ρ	$<\rho>$	δ
1990	-0.298	0.258	-4.83
1991	-0.104	0.207	-3.21
1992	-0.052	0.130	-2.09
1993	-0.132	0.095	-2.49
1994	+0.005	0.138	-1.55
7/89-10/95	-0.123	0.168	-4.28

and Nynex Corp ($\rho = -0.298$) in 1990. As for the DJIA stock sets the values of ρ are slowly time dependent. As an example I shows in Table 4 the correlation coefficient of Barrick Gold and Nynex Corp. The correlation coefficient between Barrick Gold and Nynex Corp is always well below the average values of the S&P500 set. The distance in units of standard deviations from the average value ranges from -1.55 to -4.83. This behavior is not specific to this pair of stocks. It is indeed observed for other pairs of stocks. This means that several pairs of stocks maintain a certain degree of correlation on a long time scale.

DISCUSSION

Stock prices are correlated, uncorrelated or weakly anti-correlated in the New York Stock Exchange. The correlation coefficient between two stocks is slowly time dependent. The probability density function of the correlation coefficient $P(\rho)$ also evolves in time. The time evolution on a yearly time scale of the average value of ρ, $<\rho(t)>$ shows approximately the same behavior in the DJIA and in the S&P 500 stock sets.

To interpret the above summarized results from the point of view of a physicist we can describe our empirical investigation by saying that we are investigating the averages of ensembles of a "many body" system. The different realizations of the ensemble investigated being the set of values Y_i observed at trading day t_k. The system defined in this way presents long range interactions of "ferromagnetic" and "anti-ferromagnetic" type in the abstract "space" of stocks.

Are these findings of moderately strong and relatively stable correlations and anti-correlations consistent with the paradigm of efficient market?

To answer the question let's notice that even if one can extract from the market time series some information about correlations between pairs of stocks, this information of course cannot be used to predict the future value of stock prices because time series are essentially uncorrelated in time. Hence, the empirical findings of correlations and anti-correlations between stock prices inside a financial market on a time scale of months or years aren't inconsistent with the hypothesis of *efficient market*. On the other hand, the result of the analyses clearly show that a picture based on the assumption that all stock prices are uncorrelated isn't a realistic one. The stock exchange is by far more "complex" than a collection of several independent martingales.

ACKNOWLEDGMENTS

Financial support from MURST and INFM is gratefully acknowledged.

REFERENCES

1. Lo A. W., *Econometrica* **59**, 1279 (1991).
2. Samuelson, P. A., *Industrial Management Review* Vol. 6, No. 2, 41-49 (1965); reproduced as Chapter 198 in Samuelson, *Collected Scientific Papers, Volume III*, Cambridge, M.I.T. Press, 1972.
3. Doob, J. L., *Stochastic Processes* (John Wiley, New York) 1953.
4. Markowitz, H., *Portfolio Selection: Efficient Diversification of Investment* (John Wiley, New York) 1959.
5. Feller W., *An Introduction to Probability Theory and Its Applications* (Wiley, New York) 1971.

Structuring the IW Diagnosis Problem[1]

M. J. Coombs, A. Taha, and D. Birx

Physical Science Laboratory
New Mexico State University
Las Cruces, NM 8800

Abstract. The effect of command and control C^2 systems in 20th-century warfighting doctrine has been to distance commanders from the battlefield. Relying on a complex array of electronic sensing and communication systems, commanders now operate in a virtual battlespace. This is essentially an *information* battlespace consisting of intelligence assessments, operational plans, and logistic options, all integrated through a communication infrastructure. This infrastructure is a natural conduit for an attack on command and control functions through attrition of the information space. Such attacks are said to constitute Information Warfare (IW). This paper surveys some of the hard issues that will need to be addresses in the "diagnosis" of IW and proposes a research approach.

THE IW DIAGNOSIS PROBLEM

The digitization of the battlefield opens up new vulnerabilities, as well as advantages, to fighting forces. On the positive side, increased availability of accurate and timely information is a powerful force multiplier, since it provides better situation awareness and more flexibility of command at all echelons. On the negative side, however, the information networks that link command and control (C^2) functions to fighting forces are inherently vulnerable. Vulnerabilities include openness to external exploitation by an enemy as an instrument of disruption (e.g., through the jamming of essential information) or deception (e.g., through the corruption of information).

---

[1] This work was conducted with joint funding from the Defense Information Systems Agency (DISA), Contract PO No. DCA100-96-M-083 and from the National Ground Intelligence Center (NGIC) under ARL Contract No. DAAD07-91-C-0139, WAO No. 93-3.1-1, Rev 21. The authors wish to thank Dr. Mark Sulcoski of NGIC for discussions on the role of complex systems concepts and metrics in the management of military decision systems, and to Dr. John Dockery of DISA for discussions on the assessment of IW vulnerability.

Attacks on a commander's information support system are part of Information Warfare (IW). More generally, IW is the battlespace component of Information Operations (IO), the currently fast evolving branch of doctrine concerned with any action that "disrupts or destroys the cognitive processes of the command function or its physical support" (Dockery, 1996). Thus, offensive IW seeks to exploit an opponent's information vulnerabilities in order to adversely affect its decision processes, while defensive IW seeks to protect information essential for the direction and co-ordination of friendly forces from such attacks. Defense requires an ability to:

(i) *baseline* the information carrying properties of the communications networks it designs in the absence of IW;

(ii) *detect* network anomalies that indicate the presence of IW;

(iii) *diagnose* the type of the IW attack, including IW goals (e.g., modification of the temporal properties of the network versus the modification of information content) and methods (e.g., jamming of a communication channel or information insertion via some form of malicious computer code;

(iv) *diagnose* the seriousness of the attack relative to a network's ability to support operational decisions;

(v) *remediate* the IW attack through the design/selection and implementation of counter-measures.

These tasks have become increasingly hard as battlefield information networks have grown massively in complexity and sophistication. Exploiting commercial developments in networking and intelligent computing, for example, battlefield networks now:

(i) have *variable topologies*, connectivity fluctuating widely between sparse and dense configurations as the system adjusts to the demands of an engagement;

(ii) are *multi-layered*, integrating a large number of different protocols and transmission modalities (e.g., satellites versus fiber optic cables);

(iii) are *multi-functional*, integrating a large, fluctuating number of node of different type (e.g., databases, expert systems, human decision cells), importance, and observability;

(iv) are *intelligent*, integrating a dynamic mix of human and computer autonomous, adaptive agents; and

(v) incorporate *complex INFOSEC (security) measures* such as automated channel hopping and spread-spectrum encryption.

With such systems, detecting, diagnosing, and remediating anomalous conditions originating from IW attacks is exceptionally difficult. As has been amply demonstrated in the commercial world, intelligent communication networks function more like network ecosystems (Huberman and Hogg, 1988) than passive carriers. Connections are made or broken, utilized or ignored, as each agent seeks to improve its information environment relative to its decision priorities. Thus, different network configurations emerge dynamically as the system adapts to meet the varying collective decision needs of the agents that constitute its nodes.

In order to detect anomalous behavior, it is first necessary to define normal behavior. Conventional approaches to the analysis of battlefield information networks either define normal behavior directly in terms of system specifications, involving *a priori* principles of design, or indirectly in terms of symptoms of abnormal operations, involving deviations from doctrinal operational thresholds. They also make the simplifying assumptions that the physical configuration is fixed, that transmission is relatively free of noise, that interactions are linear, and that performance is stationary. Modern military networks, however, break all these assumptions, possessing the opposite set of properties. As indicated above, battlefield networks are:

(i) *dynamic*, in that human and automated agents may create or destroy links in order to improve some operational "fitness" condition through the manipulation of their information environment;

(ii) *noisy*, at least in the sense that information packets may be destroyed or reordered as they compete for transmission resources;

(iii) *nonlinear*, in that small difference in the availability of information to different nodes, or in the effectiveness of decisions, may become amplified to produce large differences in information availability of effectiveness as the system iterates;

(iv) *nonstationary*, in that envelopes characterizing acceptable system performance may vary with the mix of decision functions required to be supported;

(v) *distributed*, in that decision functions may be arrayed over multiple components, so making it difficult to implement an integrated monitoring or remediation plan.

A COMPLEX ADAPATIVE SYSTEMS (CAS) VIEW

The complexity of envisioned battlefield C^2 networks precludes reliance on the *a priori* definition of normal system behavior in the detection and diagnosis of IW attacks. Moreover, since we are no longer able to treat information networks as simple systems with predictable, "designed in" behavior, we need to find some alternative analytic foundation for IW defense. The approach proposed in this report is to view such networks as *complex adaptive systems (CAS)* and to have the diagnostic function arise as an *emergent system property*.

The *emergent diagnosis* approach is being explored for the protection of networks in other critical areas such as banking, medical monitoring, and nuclear plant control (e.g., Maxion, 1990). John Dockery (Dockery, 1996) is a proponent for its application to the defense of battlefield networks. The foundations for implementation arise naturally from the four basic properties of CAS which are discussed below. Out of these properties, we seek to provide a mechanism by which emergent network constraints may become manifest within an abstract representation and then attributed to the collective operation of specific functions on nodes within the network. We then seek to embed this mechanism within a simple diagnostic (abductive) reasoning architecture based on the construction, and evaluation, of set covers (e.g., Coombs et al., 1992)[2]. This architecture would be designed to automate the constructive search for network simulations to structure (explain) available real-world transaction data. The search would be driven by two consideration: (i) the degree to which simulations integrate fragmentary transaction data, and (ii) a minimality condition on the complexity of the simulations required to explain observed transactions.

There are four defining properties of CAS as they apply to C^2 networks.

(i) **Networks are made up from many independent agents.**

It is assumed that each component of the network acts independently to access the information necessary for it to accomplish its decision goals. While this independent action may involve either cooperative of competitive interaction with other agents, the decision to cooperate or compete is not imposed by some global control mechanism. Rather, the decision is taken by the agent itself after assessing the properties of its operational environment. This does not, however, preclude the possibility that some agents may encourage some degree of dependence in other agents by requiring them to follow given *negotiation protocols*. For example, agents specializing in monitoring network activity may require that agents queried follow a protocol that encourages honest accounts of their state. A typical protocol might employ a shared utility function that rewards agents passing evaluably correct information with an extension to the neighborhood function that defining their range of awareness (see Rosenschein and Zlotkin, 1994). The agent supplying the information may, of course, disregard the protocol.

---

[2] This architecture was the basis for a more elaborate conceptualization of an open-systems abductive reasoning architecture discussed in Coombs and Srikant, 1995.

Although networks are composed of agents, agents should not be regarded as atomic units but as *wholes*, or complex systems, in their own right. A given network model may choose to ground itself at a particular descriptive level. However, the level selected should be regarded as a matter of analytic effectiveness rather than necessity. For example, a command unit agent in a battlefield network may be decomposed into a number of human, computer, and software agents. A further decomposition would generate "wetware", silicon, or logic function sets respectively. It is anticipated that, because of this multi-resolutional property, it will not be possible to ground diagnostic inference for IW in a pre-defined level of network description. Rather, it will be necessary to arrive empirically at the appropriate level of system decomposition required to achieve some given diagnostic goal.

(ii) **Agents interact locally.**

Agents interact with other agents within locally restricted *neighborhoods*. A given agent may, of course, employ different sized neighborhoods for different interactive functions. The conceptual foundation for defining distance may also vary with function. For example, an IW monitoring agent under routine operation may define its attention neighborhood in terms of designated functions in the population of operations agents, while it may employ neighborhoods defined over information flows when tracking a located information anomaly.

While a given CAS agents may only interact with a small proportion of the agents in a network, they may be influenced by indirect connections that span a large proportion of them. Importantly, it is the dynamics of this indirect *global connective structure* that give complex systems their unique character. This includes such properties as emergence and phase transitions between frozen, fluid (edge-of-chaos - EOC) and chaotic states. It is, therefore, the influence of the global connective structure that makes it so hard to define normal behavior in a complex C^2 system. The global structure constitutes the evolving set of influence chains and feedback loops that constrain the inputs to locally interacting agents, and which, in turn, is modified by the local interactions. Thus, while CAS agents interact locally, the global connective structure sets bounds on their behavior and on the outcomes of their behavior. It is this structure that provides the main source of information on the factors determining the properties of an observed CAS.

A further important consideration is that the constraining structure might be expected to change with time since C^2 networks are *open, dissipative systems* in many different ways. In addition to the obvious need for electrical energy physically to sustain data flows, the involvement of human agents introduces an informationally dissipative element. We can expect novel patterns of constraints in C^2 networks that reflect the strategic creativity of human agents stimulated by the competitive interactions that the network sustains (see Leven, 1997 on "Information and Optimal Ignorance").

(iii) **Overall behavior is independent of the internal agent structure.**

In complex networks, global behavior is often multi-determined. In other words, there is unlikely to be a unique relationship between a given global behavior and a unique collection of agents or agent mechanisms. From a diagnostic perspective, this make it hard to pre-define relationships between symptoms of abnormal behavior and specific causes, even assuming that one can define abnormal behavior in the first place. The attribution of causal relationships must proceed indirectly by (i) modeling the rules governing the emergence of global constraints, (ii) identifying the clusters of agents affected by each rule, and (iii) identifying the candidate mechanisms implemented by each agent in the cluster as candidates for investigating as generators of each rule.

Inferences concerning the semiotics employed by specific agents are often an important source of information for guiding the search for mechanisms. Semiotic information might include, for example, knowledge of the semantics used by an agent in mapping observed time delays between events and ascription of causal dependence between the events. Given such knowledge, we might be better able to trace an change in the structure of constraint, to the rule governing that change, to the participants in that rule, to the reactions of individual participating agents, to the attribution of semantic drivers for agent reactions. A typical semantic driver might be the perception of a joint cause behind two separate, but temporally close, events that reduced some measure of the agent's fitness related to its designated role in the network. The potential importance of semiotic variables in the analysis of network behavior suggests that representations of global connective constraints should provide scaffolding for semiotic interpretation. Scheme 1 provides a hierarchy of logical domains relevant to C^2 systems over which semiotic types (syntax, semantics, pragmatics, etc.) may be expected to range.

(iv) **Overall behavior of the system of agents is well defined.**

In CAS we often find that, despite the complexity of interactions, the global system behavior is often highly structured. Emergent feedback structures in C^2 networks, for example, may be both stable and simple, even though they arise from highly dynamic and complex local interactions generated by the flux of battle. For example, the emergence of feedback loops may informationally segment the decision system into

GOAL SATISFACTION STRUCTURE

⇑

CANALIZATION STRUCTURE

⇑

NEGOTIATION STRUCTURE

⇑

DECISION STRUCTURE

⇑

ACTOR PROCESS STRUCTURE
(e.g., belief, affect, motor)

FIGURE 1. *

Scheme 1: Hierarchy of domains of interpretation for C^2 networks.

an number of loosely coupled, oscillating units. While these units may change in actual composition with time, as composites each unit may maintain the same size and relative oscillatory phase relation.

AN APPROACH

In brief, the problem with the analysis of CAS C^2 networks is that there is no pre-defined network to analyze, and, due to system openness to both structural and functional change, no well-defined set of network behaviors to determine normal from abnormal operation. These are both fundamental precondition for the application of existing network analysis tools, irrespective of whether they model systems close to the physical communications system (e.g., COMNET III - CACI, 1995) or as an abstract graph (e.g., GERT - Whitehouse, 1973). We are therefore seeking an analysis methodology that:

(i) enables us to induce hypothetical active networks for investigation from fragmentary and uncertain qualitative data;

(ii) enables us, through simulation, to locate the observed active network in a space of possible configurations (and behaviors);

(iii) enables us to identify controlling variables in the network;

(iv) enables us to attribute the manipulation of controlling variables either to benign or to malicious forces.

We conjecture (Coombs and Srikant, 1995) that, given an open system assumption, the first two requirements are fundamental to the achievement of the latter two. This paper is restricted to these and thus to the problem of structuring a particular IW diagnostic problem. In particular, we are concerned with (i) use of combinatorial algebra, in particular Q-analysis (Atkin, 1972; 1977), for constructing *virtual networks models* from transaction data derived from an *object network*, and for representing the structural properties of models, and (ii) the application of a diagnostic algorithm over these models for identifying changes in network parameters for use in explaining locally observed anomalies. Such an algorithm, implemented within an autonomous agent framework, might proceed as follows.

(i) A C^2 *Processing Agent* (CPA) notices that one, or more, of its variables is displaying values that are outside of the range to which the agent is currently adapted - termed its *zero-reaction level* by Maxion (Maxion, 1990).

(ii) The CPA classifies the change in its values in terms some hypothetical change in system properties and calls the attention of a specialized *Investigative Agent* (IA) to the change.

(iii) The IA builds *virtual network* models of an observed *object network* based on the results of a Q-analysis (or equivalent) of available *transaction data*.

(iv) The IA experiments with the *virtual network* models in order to identify an equivalence class of model which both map to extended data sets extracted from the anomalous *object network* and which display properties capable of generating the anomaly.

(v) The IA identifies those parameters and parameter values in the *virtual network* that control the generation of explanatory trajectories, and reports this information to the triggering CPA.

(vi) The CPA instantiates the parameters and parameter values from the *virtual network* with associated entities and values in the *object network*, focusing on those that may have changed prior to the observation of the anomaly, and then seeks evidence for their action as causes of the anomaly.

(vii) If the CPA establishes causation, it will continue by calling another IA to investigate possible malicious intent and to propose remediation; if the CPA fails to establish causation, it will cycle back to step (ii) and pursue other property classifications of the anomaly.

Ongoing work at the Physical Science Laboratory at New Mexico State University has had some success at using a generalization of Q-analysis to extract virtual networks of emergent regulators from transaction data collected on object networks. This is the first step implementing a diagnostic process, since it is likely that regulatory objects will be leading targets for mutation in an IW attack.

REFERENCES

1. Atkin, R. (1972). From cohomology in physics to q-connectivity in social science. *International Journal of Man-Machine Studies*, 4, 139-167.
2. Atkin, R. (1977). *Combinatorial Connectivity in Social Systems*. Birkhauser: Basel.
3. CACI (1995). COMNET III Network Planning Tools, CACI Products Company, La Jolla, CA.
4. Coombs, M.J., Pfeiffer, H.D., and Hartley, R.T. (1992) e-MGR: an architecture for symbolic plasticity. *International Journal of Man-Machine Studies*, 36, 247-263.
5. Coombs, M.J. and Srikant, M. (1995). *Dynamic Control in Problem Solving Tasks*, Final Report - Year 2, NSF Grant No. IRI-9122401, National Science Foundation.
6. Dockery, J. (1984). Mathematics of commend and control analysis. *European Journal of Operational Research*, 21, 172-188.
7. Dockery, J. (1996). *IW Workbook*. Internal Advisory Document, Defense Information Systems Agency.
8. Huberman, B.A. and Hogg, T. (1988). The behavior of computer ecologies. In B.A. Huberman (ed), *The Ecology of Computation*. North-Holland: Amsterdam, Netherlands.
9. Maxion, R.A. (1990). Towards diagnosis as an emergent behavior in a network ecosystem. *Physica D*, 42, 66-84.
10. Rosenschein, J.S. and Zlotkin, G. (1994). *Rules of Encounter: Designing Conventions for Automated Negotiation among Computers*. MIT Press: Boston, MA.
11. Whitehouse, G.E. (1973). *Systems Analysis and Design Using Network Analysis Techniques*. Prentice-Hall, Inc.: New Jersey.

Tests for robustness of dynamical models of arms races.

G. Mayer-Kress

College of Health and Human Development
The Pennsylvania State University
University Park, PA 16802

Abstract. The theory of dynamical systems -especially in the form of discrete maps and automata- makes it very easy to convert conceptual ideas of arms races or other forms of social interactions into formal, dynamical models. Modern interfaces and interpreters also allow a quick, interactive parameter search and display of attractors or other long term behavior. This situation makes it also easy to forgetthat the qualitative behavior of non-linear dynamical systems with a high- dimensional parameter (control) space can change dramatically with a change of parameters that often is well within the resolution accuracy of the model. In order to obtain trustworthy interpretations of numerical simulations, it is essential that multiple tests of robustness are performed. We discuss stochastic methods based on Chapman-Kolmogorov equations, search methods based on neural nets and genetic algorithms, geometrical methods based on multi-dimensional tableau- representations bifurcation diagrams and other methods whose application often depends on the context.

INTRODUCTION

Since the early work of L.F. Richardson [1] it was a challenge to the scientific community to describe the complex dynamics of international relations and especially arms races and military conflict with the help of dynamical, mathematical models (see e.g. [2], [3], [4], [5], [6], [7], [8]). Whereas the computational resources have increased by many orders of magnitude since the days of L.F. Richardson - who did the calculations by hand- one basic obstacle persists: The observables of those models are few in number and difficult to estimate with any degree of accuracy. This fact makes it very difficult to generalize methods from physical systems (with typically large numbers of data points observed at high accuracy) to applications of any type of societal dynamics.

In this paper we discuss examples of discrete dynamical models of arms-races and opinion formations and different methods of evaluating how robust the qualitative outcome of those models are with respect to uncertainties and perturbations in the dynamics and state of the system. The models studied are discrete in time which excludes the complication of numerical integration artefacts that are especially relevant in singular situations such as discontinuous transitions and bifurcations.

The examples we want to discuss here are:

A six-dimensional strategic arms-race model that estimated general implications of the introduction of new defense systems [3]

A three-nation generalization of a non-linear, discrete-time Richardson-type arms race model [[4], [5]]

A both continuous and discrete versions of a non-linear opinion dynamics model of two groups in two nations [6]

The test methods that we describe can be grouped into three different categories: stochastic methods, interactive, computer-graphical methods, and evolutionary search methods.

STOCHASTIC METHODS

The basic assumption for this method is that of limited knowledge of the state and the dynamics of the system under investigation. For example in typical arms-race models a nation **A** responds to a perceived threat by a nation **B**. There are two basic types of uncertainties involved in translating this conceptual statement into a quantitative, mathematical model: How well does nation **A** estimate the threat from nation **B**? To what degree will nation **A** respond to the threat of nation **B**? The first uncertainty is an uncertainty about the **state** of the system i.e. the variables of the system. Typically it is the threat-levels that are evolving and changing as the relevant observables in arms-race models. The second uncertainty involves the level of response of nation **A** to the perceived threat from a nation **B**. A common assumption is that this level of response is expressing a nations government and their international relations. It is typically assumed not to change in time and is therefore interpreted as a control parameter of the system. It is evident that both types of uncertainties can be amplified rapidly especially in crisis situations.

In the model that we discussed in [3] the observables (threat levels) were given as numbers of intercontinental ballistic missiles. The current numbers of those missiles are known accurately or at least there is an exact number of those missiles announced to the public. For a unclassified simulation we therefore could assume that there is no uncertainty in the initial state of the model. (The other weapons systems, namely strategic defense systems and anti-satellite weapons were not deployed at the time and therefore the initial estimates for those variables is exact, namely zero.)

Monte Carlo Simulations

The uncertainties therefore are exclusively in the parametric estimates. The assumption in our model was that all parameters have an uncertainty whose magnitude σ does not change in time. For the simulation we therefore add a stochastic perturbation ξ_n of standard deviation magnitude σ to each parameter p_i at each time-step n. Depending on the location in parameter- or state-space this perturbation either leads to a finite uncertainty Δx_m in state-space outcome (actual number of each weapon system after m time steps, in our case $m = 35$ decision periods (e.g. years)). We would call such an outcome *robust*. In cases where the uncertainty is still increasing even after the relevant time window we would call the outcome *sensitive* to uncertainties. The disadvantage of this method is that it is quite compute intensive as a variant of what is known as *Monte Carlo* simulation. The advantage of this method is that it provides information about the probabilities $P(x_n)$ of the system being close to state x_n at time n.

Measure Space Dynamics

Instead of iterating the dynamical model itself many million times in order to get a probability distribution of the outcomes we can study the time-evolution of probabilities (e.g. *Bowen-Ruelle-Sinai (BRS)- measures*) instead. When we view a probability distribution as a point in an infinite dimensional measure space then we can study the dynamics induced to this space by the original model. This can be done very naturally in the case of discrete-time dynamical systems with the help of Chapman-Kolmogorov equations ([12], [13]).

For many applications quantitative knowledge of probabilities is not required and it is sufficient to know in where it is different from zero (i.e. where we *have a chance* to see the system.) This question does not make much sense in the context of models formulated in the language of stochastic differential equations with Gaussian white noise:There is a finite probability for the system to be anywhere at any point in time $t_n > t_0$. For discrete time dynamical systems with bounded noise ((ϵ, δ)- diffusions, see [14]) we can compute the support of the transformed measure for any point in time. For example if we start with an initial point that has an uncertainty of size δ and use it as input for a model with uncertainties in the parameters then we will have an output with a finite uncertainty ϵ whose size can be estimated accurately. (Similar concepts have been formulated in the context of fuzzy logic and quasi-axiomatic theories (see [9])).

Interactive Computer-Graphics

This method is especially useful in situations with a moderate number of parameters and state-space dimensions. The generalized Richardson model discussed in [5] describes the arms-race among three nations under consideration of economical constraints. The two weaker nations will form an alliance against the strongest one until the balance of power shifts (see e.g. [2]). The alliance formation factor and economical constraints in the model introduce non- linearities which cause multiple stable solutions, bifurcations between fixed points solutions and time dependent attractors. We have identified parameter domains for which the attractors become chaotic. This fact can be interpreted as an irregular sequence of changes of alliance configurations.

This model has a three-dimensional state space (threat potential for each of the three nations) and a twelve dimensional parameter space [5]. If we are interested in conditions under which we will expect sensitivities we can take an ensemble approach instead of the stochastic approach discussed above: We now treat the space of initial configurations as well as the parameter space as sub-spaces of a larger space of uncertainties for which we can estimate upper bounds of feasibility.

With the help of computer-graphics we can visually explore sensitivities with respect to three-dimensional sub-spaces: We can select a combination of three parameters and initial conditions which define a three-dimensional neighborhood in the fifteen dimensional space the size of which is given by the estimated bounds of uncertainties.

We now can partition this block of parameters/ initial conditions with a resolution that is determined by computational limitations (on a typical workstation for quasi-realtime interactivity this would be of the order of 100000 scenarios. Each of these scenarios is simulated for a given target time and the result is represented in a three- dimensional, color-coded rendering. In [10] an application is discussed where the outcomes are categorized into four different classes: unbounded solutions and three different bounded solutions with one of the three nations dominating at the end of the simulation period.

The program then searches the 3-D cube of parameters and initial conditions for connected domains namely those with the same category. The boundaries of those categories are then displayed and color-coded. These boundaries correspond to critical surfaces where a switch in the alliance configuration takes place and instabilities are to be expected. In this context we would then consider those solutions to be robust that are not close to one of the transition surfaces. (This includes diverging solutions discussed above as a special case.)

This method allows a very fast efficient search through a large number of scenarios. Critical domains can be identified instantly through a fractionated (fractal ?) boundary surface. We have done some preliminary explorations to also include auditory cues to detect critical domains in high-dimensional control spaces.

Four-Dimensional Opinion Formation Models

In democratic societies the decision of the government on international relations is reflecting the public opinion. In [6] we discuss a simple model that describes the formation of public opinion in two competing nations: The population is divided into *hawks* and *doves* and the influence of each of the groups on their own government will determine the strength of the corresponding configuration in the competing nation. The structure of this model correspond to two coupled cusp systems. We have used computational models to unfold the bifurcations in this model [6]. We have studied both continuous time and discrete time models and found large regions in parameter space, for which the solutions and the dynamics of the two classes of models are equivalent. Furthermore we were able to extend the range of the discrete model and identify parameter domains for which multiple periodic solutions and also chaotic solutions become stable.

The visualization method that we used for analyzing this four-dimensional model is similar to the method described for the three-nation generalized Richardson model. Instead of using surfaces in three-dimensional cubes we represented the outcome in the form of tableaux. That means the four-dimensional space is decomposed into a cartesian product of two two-dimensional spaces or in other words: we attach to each point in the plane a two-dimensional plane orthogonal to the first plane.

This is possible because we are restricting ourselves to a finite lattice of points (determined by a finite resolution). By creating gaps between the lattice points we now can rotate the two-dimensional interval that was orthogonal to the original plane in a way that it fits into the gaps of the previously described lattice. With the help of this trick we can create tableaux of outcomes that can be visually examined for boundary points that indicate crises.

Evolutionary Search Methods

In many models of conflict (e.g. chess) it is possible to evaluate a configuration (state of the system) and rank it. This means we can introduce a *fitness function* into the state/parameter space of the model. For example we can search for solutions that have a maximal distance to the nearest crisis indicating surface.

In [11] the fitness value was determined by the degree to which the solution satisfies a *Balance of Power* conditions. We used genetic algorithms, a machine learning tool that simulates the process of biological evolution for finding not optimal solutions (which are very brittle as the four different attempts to naturally evolve sabre-tooth tigers shows) but solutions that are robust and still close to the optimum.

For the analysis we rescaled the parameters to be in the range [0, 1], so it is straightforward to discretize them into bins. This results in one binary-coded integer for each parameter designating its bin. The size of the bin is determined by the number of bits used in the discretisation (for example 8 bits corresponding to $2^8 = 256$ different bins for each parameter were used in [11]). The genetic algorithm finds balance-of-power solutions which were not known to exist previously. Manually setting the parameters had not suggested that such points existed. These results are non-trivial in that the genetic algorithm found many different solutions and most of the solutions were not degenerate (that is, parameters were not set to 0.0 or 1.0). This is in contrast to a similar search using neural networks, which has produced results which often suggest trivial solutions. Studying hyper-planes of parameter space shows that it contains regular sub-regions with close to perfect fitness. That suggests for our model that once we have found one solution with the genetic algorithm, it is easy to find other solutions in a neighborhood due to the smoothness of the system.

Acknowledgement: I would like to acknowledge partial support for this project by the Army Research Office under Contract Number DAAE30-97-C-0006.

REFERENCES

1. Richardson L.F., *Arms and insecurity*, Boxwood, Pittsburgh 1960
2. K. Waltz, *Theory of International Politics*, Addison-Wesley Publishing Company, Reading Mass. 1979
3. Saperstein A. and Mayer-Kress G., *J. Conflict Resolution*, **32**, 636-670 (1988)
4. Grossmann S., and Mayer-Kress G., *Nature*, **337**, 701-704 (1989)
5. G. Mayer-Kress, in: *The Ubiquity of Chaos*, S. Krasner (Ed.), Proceedings AAAS conference, San Francisco, January 1989, Los Alamos preprint LA-UR-89-1355
6. Abraham R., Keith A., Koebbe M., and Mayer-Kress G., *Intl. J. Bifurcations and Chaos* **1(2)**, 417-430 (1991)
7. Mayer-Kress G., in *Proc. Third Woodward Conference: "Modelling Complex Systems"*, (Lui Lam, Ed.), San Jose, 4/12-13/91
8. Kadtke J., Kremliovsky M., and Pentek A., in [9], pp. 129-142
9. Coombs M., Sulcoski M., (Eds.), *Proc. of the 1996 Intl. Workshop Control Mechanisms for Complex Systems*, Las Cruces, NM 1996
10. Challinger J., *Int. J. Bif. Chaos*, **2**, No.2, 251-261 (1992)
11. Forrest S., and Mayer-Kress G., in: *The Genetic Algorithms Handbook*, L. Davis, (Ed.),Van Nostrand Reinhold, New York 1991
12. Haken H., and Mayer-Kress G., *Z.Phys.B-Cond. Matter*, **43**, 183 (1981)
13. Krause J., and Mayer-Kress G., unpublished results
14. Mayer-Kress G. and Haken H., *Physica* **10D**, 329-339 1984

DEVICES/EXPERIMENTS

Stochastic Resonance and Resonant Trapping in a Schmidt Trigger

L. Gammaitoni*, F. Marchesoni† and S. Santucci‡

*Dipartimento di Fisica, Università di Perugia and Istituto Nazionale di Fisica Nucleare,
VIRGO Project, I-06100; gammaitoni@perugia.infn.it Perugia (Italy)
†Department of Physics, University of Illinois, 1110 W. Green St., Urbana, IL 61801 and
Istituto Nazionale di Fisica della Materia, Università di Camerino,
I-62032 Camerino (Italy)
‡Dipartimento di Fisica, Università di Perugia and Istituto Nazionale di Fisica della
Materia, Sezione di Perugia, I-06100 Perugia (Italy)

Abstract. An hysteretic symmetric two-thresholds system (Schmidt trigger) driven with colored noise and periodic signal allows a direct observation of *Stochastic Resonance*, when the amplitude of the periodic signal is smaller than the trigger threshold, and *Resonant Trapping*, when the trigger operates in the supra-threshold regime. Evidences of the two phenomena are presented and discussed within a common theoretical approach.

INTRODUCTION

The occurrence of the *Stochastic Resonance* (SR) [1] phenomenon in a Schmidt Trigger (ST) device [2] was first shown in 1982 [3]. A ST is an hysteretic two state system. If we call $+y_m$ and $-y_m$ its output y states, y rests in $-y_m$ as long as the input signal $x(t)$ is smaller than a threshold value b. As $x(t) = b$ the trigger switches (almost) instantaneously into state $+y_m$ and sits there as long as $x(t) > -b$.

We consider the input signal $x(t) = \xi(t) + f(t)$, where $f(t) = A_0 \cos(\Omega t + \phi)$ and $\xi(t)$ is a Gaussian stationary noise with average $\langle \xi(t) \rangle = 0$ and autocorrelation function

$$\langle \xi(t)\xi(0) \rangle = \sigma^2 \, e^{-|t|/\tau}. \tag{1}$$

We discriminate between two operating ST configurations:
(a) the *sub-threshold* regime $A_0 < b$. Under such a condition *Stochastic Resonance* (SR) is clearly observable [3]: the amplitude $\bar{y}(\sigma)$ of the output harmonic

component with angular frequency Ω shoots up with increasing the noise intensity σ until it reaches a maximum at $\sigma = \sigma_{SR}$ and then it dies away for larger σ values [6].

(b) the *supra-threshold* regime $A_0 > b$. In this configuration the trigger switches are tightly driven by the input modulation, random failure events occurring sparsely in time due to the noise input component $\xi(t)$. A similar occurrence was predicted recently for a wide class of continuous bistable systems: since such a phenomenon attains its best evidence for a certain value of the noise strength, the term *Resonant Trapping* (RT) was coined [8,13].

THE SUB-THRESHOLD REGIME: STOCHASTIC RESONANCE

In the limit of *weak* forcing $A_0 \ll b$ the switch statistics is governed by one characteristic rate μ_0 independent of the forcing amplitude A_0.

Moreover, the rate μ_0 can be determined by computing the mean first-passage time (MFPT) for $\xi(t)$ to diffuse from the lower $-b$ to the upper threshold $+b$ (or vice-versa). Such a MFPT coincides with the mean residence time of the ST in one stable state [9].

In view of Eq.(1), it is clear that the noise source $\xi(t)$ is an Ornstein-Uhlenbeck process [10]. The corresponding MFPT from $-b$ to $+b$ with reflecting barrier at $\xi = -\infty$, reads

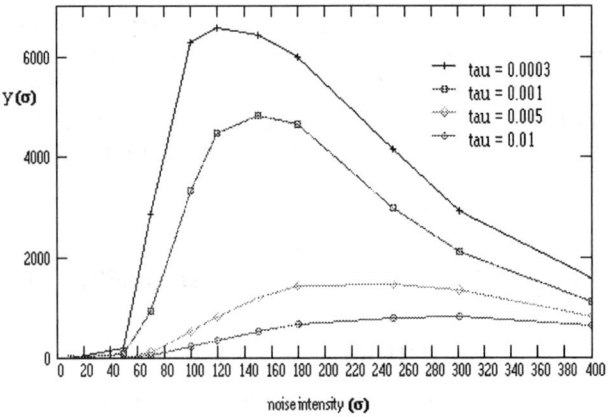

FIGURE 1. $\langle y(\sigma) \rangle$ versus σ for different values of τ. The parameters and σ are expressed in dimensionless units, i.e. $A_0 = 200$ and $b = 300$, $\Omega = 200$.

$$T_0(b) = \frac{\tau^2}{D} \int_{-b}^{b} \frac{dy}{p(y)} \int_{-\infty}^{y} p(x)dx, \qquad (2)$$

The switch rate μ_0 is related to $T_0(b)$ by the identity $\mu_0(b) = [2T_0(b)]^{-1}$ [9], [10], [13].
In the adiabatic limit $\Omega\tau \lesssim 1$, the spectral amplitude $\bar{y}(\sigma)$ at the forcing frequency of the ST output can be easily computed in the two-state model approximation of Ref. [4]. A straightforward calculation yields [13]

$$\bar{y}^2(\sigma) = y_m^2 \left(\frac{A_0 b}{\sigma^2}\right)^2 \frac{\mu_0^2(b)}{\Omega^2 + \mu_0^2(b)} \qquad (3)$$

and $\mu_0(b) \simeq \sqrt{2/\pi}(b/\sigma\tau)\exp(-b^2/2\sigma^2)$, in agreement with Melnikov's earlier analysis [7]. The results of a numerical simulation of the sub-threshold ST are displayed in Fig. 2. The comparison with the theory, Eq. (3), proves particularly close at low noise intensities [13].

THE SUPRA-THRESHOLD REGIME: RESONANT TRAPPING

In this configuration the forcing amplitude A_0 is taken larger than the threshold b, so that the switching dynamics is (at low noise intensity) mostly driven by the forcing signal itself. On increasing σ it might happen that, when $f(t)$ crosses the upper (lower) threshold, the noise signal $\xi(t)$ is smaller than $-(A_0 - b)$ [larger than $(A_0 - b)$]. In such a case the switch event gets frustrated, unless $\xi(t)$ re-crosses the boundary $-(A_0 - b)$ [or $(A_0 - b)$] prior to the subsequent sign reverse of $f(t)$.

A similar failure mechanism was observed first [8] in a dynamical system, namely, a quartic double-well potential: there, a frustrated switch is attributable to the fact that the finite escape time out of the unstable well increases with noise at low intensities and is maximum for an optimal value of σ, whence the term *resonant trapping*.
To investigate such a mechanism in the ST we consider here the simple case of an input square-wave signal (for a more general discussion see Ref. [13]). The conditions for a failure event can be stated as follows:

(a) at the time $t = 0$, when the forcing signal changes sign from $-A_0$ to A_0 (or from A_0 to $-A_0$), we require that $\xi(0) < -(A_0 - b)$ [or $\xi(0) > (A_0 - b)$]. This condition occurs with probability

$$P[\xi < -(A_0 - b)] = \int_{-\infty}^{-(A_0-b)} p(\xi)d\xi = (1/2)\,P[|\xi| < (A_0 - b)]; \qquad (4)$$

(b) the noise $\xi(t)$ is not allowed to re-cross the boundary $-(A_0 - b)$ [or $(A_0 - b)$] for a whole half forcing period, i.e. for $t \leq T_\Omega/2$, lest a switch event occurs anyway, though delayed in time. The failure probability decays exponentially with T_Ω, the relevant time constant $T_1(b)$ being the average re-crossing time from $\xi(0)$, with $\xi(0) < -(A_0 - b)$, to $-(A_0 - b)$ [or from $\xi(0)$, with $\xi(0) > (A_0 - b)$, to $(A_0 - b)$].

To calculate the re-crossing time constant $T_1(b)$, let us assume that at $t = 0$ the noise $\xi(t)$ satisfies condition (a), $\xi(0) \equiv \xi_0 < -(A_0 - b)$. The MFPT for $\xi(t)$ to diffuse from ξ_0 up to the absorbing boundary $-(A_0 - b)$ (with $\xi = -\infty$ a reflecting barrier) is [9]:

$$T_{1-}(b, \xi_0) = \tau \sqrt{\pi} \int_{\bar{\xi}_0}^{-(\bar{A}_0 - \bar{b})} e^{y^2} [1 + \Phi(y)] dy, \quad (5)$$

with $\bar{b} = b/\sqrt{2\sigma^2}$, $\bar{\xi}_0 = \xi_0/\sqrt{2\sigma^2}$ and $\bar{A}_0 = A_0/\sqrt{2\sigma^2}$, $\Phi(x)$ being the error function. Finally, we take the average of $T_{1-}(b, \xi_0)$ over ξ_0 in the allowed range $(-\infty, -(A_0 - b)]$, i.e.

$$T_1(b) = \int_{-\infty}^{-(A_0-b)} p(\xi_0) T_{1-}(b, \xi_0) d\xi_0 / \int_{-\infty}^{-(A_0-b)} p(\xi_0) d\xi_0. \quad (6)$$

Combining conditions (a) and (b) gives the probability that a failure takes place, namely

$$(1/2) P[|\xi| > (A_0 - b)] \exp[-(1/2) T_\Omega / T_1(b)]. \quad (7)$$

The probability (7) vanishes for $\sigma \to 0+$ and jumps to an horizontal asymptote $(1/2) \exp(-T_\Omega/2\tau)$ in the neighborhood of $\sigma \sim A_0 - b$. Most importantly, no

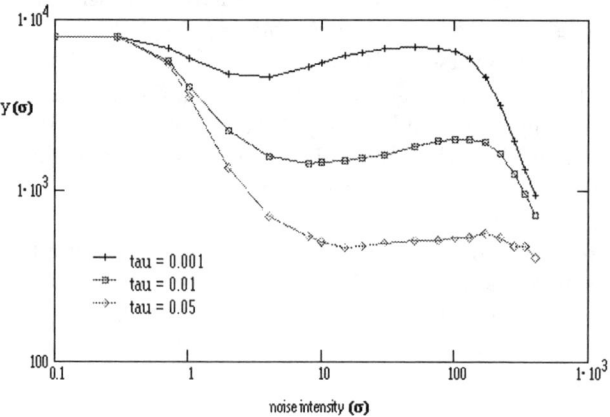

FIGURE 2. $\langle y(\sigma) \rangle$ versus σ for the sinusoidal forcing. The parameters and σ are expressed in dimensionless units, i.e. $A_0 = 200$ and $b = 199$, $\Omega = 200$.

resonant behavior is predicted.

The trigger input/output synchronization requires that one more condition holds true:

(c) at time $t = 0$ when $f(t)$ flips sign, say from $-A_0$ to A_0, a trigger switch follows suit only if the further restriction $\xi(0) < A_0 + b$ applies, that is with probability $P[-(A_0-b) < \xi(0) < (A_0+b)]$. Correspondingly, we must exclude situations where in the following half forcing period $\xi(t)$ takes on values more negative than $-(A_0+b)$, lest a spurious (noise induced) switch occurs opposite in phase to $f(t)$. The time constant of such a mechanism coincides with the MFPT for $\xi(t)$ to diffuse from ξ_0 in the interval $[-(A_0-b), +\infty)$ up to $-(A_0+b)$, namely

$$T_{2-}(b, \xi_0) = \frac{\tau^2}{D} \int_{-(A_0+b)}^{\xi_0} \frac{dy}{p(y)} \int_y^\infty p(x)dx. \qquad (8)$$

Following the procedure developed for calculating $T_1(b)$, we take the average of $T_{2\pm}(b, \mp\xi_0)$ over the starting point $\xi_0 = \xi(0)$.

Also in this case the spectral amplitude $\bar{y}(\sigma)$ can be computed explicitly within the framework of the two-state model [4], [13]:

$$\bar{y}^2(\sigma) = \left(\frac{4}{\pi}y_m\right)^2 \frac{T_1^{-1} - T_2^{-1}}{\sqrt{(T_1^{-1} - T_2^{-1})^2 + \Omega^2}}. \qquad (9)$$

Eq. (9) predicts that the curves $\langle y^2(\sigma) \rangle$ versus σ for large noise correlation times drop sharply at $\sigma \sim A_0 - b$ and $\sigma \sim A_0 + b$ and flatten out in the between, thus forming a plateau.

When the ST is pumped by the supra-threshold sinusoidal signal (instead of a square wave signal), the relevant curves $\langle y^2(\sigma) \rangle$ versus σ show a shallow minimum instead of the plateau just described. This is clearly illustrated [13] in fig 3.

The local maximum of $\langle y^2(\sigma) \rangle$ at $\sigma \sim A_0 + b$ is a signature of the RT phenomenon introduced in Ref. [8]. Indeed, the overall failure mechanism just described for the square wave case can be extended, at least in principle, to the present case, too. The reentrant synchronization mechanism can be explained qualitatively as follows. For $\sigma \ll A_0 - b$ a trigger switch occurs only after $f(t)$ has crossed the levels $\pm b$. On increasing the noise intensity σ, it may happen that that a switch event gets anticipated in time as $f(t)$ approaches $\pm b$ from sub-threshold values. Analogously, the failure condition (a) must be retained a little longer, since a re-crossing could occur also at times when $|f(t)|$ is smaller than, but close to b. The noise-induced reentrant synchronization is thus due to the noise-induced prolongation of the time interval when $f(t)$ acts effectively as a supra-threshold signal.

REFERENCES

1. R. Benzi, A. Sutera, A. Vulpiani; J. Phys. **A14**, L453 (1981). For a recent review on SR see A. Bulsara, L. Gammaitoni; Phys. Today **49** 39 (1996)
2. J. Millman, *Microelectronics* (McGraw-Hills, New York, 1983)
3. S. Fauve and F. Heslot, Phys. Lett. **97A** 5 (1983)
4. B. McNamara and K. Wiesenfeld, Phys. Rev. **A39** 4854 (1989)
5. L. Gammaitoni, F. Marchesoni, E. Menichella-Saetta and S. Santucci, Phys. Rev. Lett. **62** 349 (1989)
6. P. Jung and P. Hänggi, Europhys. Lett. **8** 505 (1989)
7. V. I. Melnikov, Phys. Rev. **E48** 2481 (1993)
8. F. Apostolico, L. Gammaitoni, F. Marchesoni and S. Santucci, Phys. Rev. **E55** 36 (1997)
9. H. Risken, *The Fokker-Planck Equation* (Springer, Berlin, 1984)
10. P. Hänggi, P. Jung and F. Marchesoni, J. Stat. Phys. **54** 1367 (1989)
11. L. Gammaitoni, E. Menichella-Saetta, S. Santucci, F. Marchesoni and C. Presilla, Phys. Rev. **A40** 2114 (1989)
12. P. Hänggi, P. Jung, Ch. Zerbe and F. Moss, J. Stat. Phys. **70** 25 (1993)
13. L. Gammaitoni, S. Santucci, F. Marchesoni, preprint (1997)

New Regime in the Stochastic Resonance Dynamics of SQUIDs

Andrew D. Hibbs, Brian R. Whitecotton

Quantum Magnetics Inc., 7740 Kenamar Court, San Diego CA 92121

Abstract. We have measured the response of DC SQUIDs to various levels of internal and external noise. By varying external experimental parameters, the internal potential can be modified to allow the SQUID to be operated in modes in which it switches states due to internal thermal noise alone. With appropriate selection of bias conditions, a highly non-linear transfer function between output and input flux can be established. Five SQUIDs with different junction and loop parameters have been studied. When biased into the thermally switching regime, the output signal and signal to noise ratio are found to go through maxima as a function of both applied magnetic flux and bias current. Typical enhancements of order 30 dB and 10 dB respectively are observed for all devices. When driven by external noise, we find that it is possible to fully recover the input SNR while producing amplification.

Owing to the very small energies involved there have been few studies of the role of internal thermal noise in stochastic resonance (SR) [1]. This is unfortunate because thermal noise is always present, and in many practical applications it is internal thermal noise that limits the ultimate sensitivity of high performance devices. One such device is the superconducting quantum interference device (SQUID). SQUIDs have the highest energy resolution of any detector at low frequencies (< 1MHz) and are widely used for measuring electromagnetic fields when the ultimate sensitivity is desired. However, apart from issues of improved sensitivity, SQUIDs also offer a unique opportunity to study a well-defined system with a complex, but controllable internal potential.

The electrons in a superconducting material may be described by a single, macroscopic wavefunction and SQUIDs are fundamentally quantum devices. In a closed superconducting loop, requirements for phase continuity force the magnetic flux within the ring and the stored energy, to be quantised. If the superconducting loop is interrupted by one or more Josephson junctions of appropriate design, continuous or hysteretic transitions may be made between distinct quantised states. Systems comprised of a loop and one or two junc-

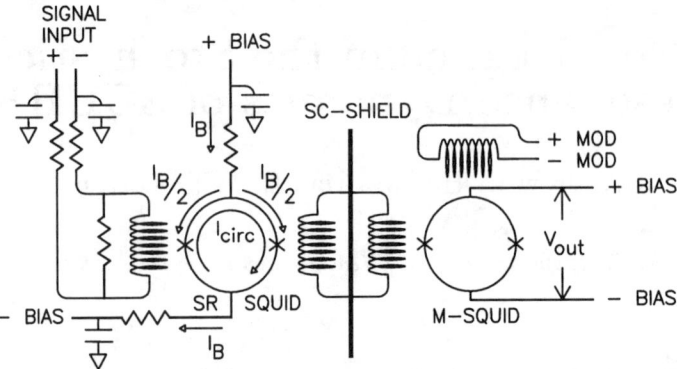

FIGURE 1. Circuit Diagram for the DC SQUID Showing Variables Used in the Analysis.

tions are usually called RF or DC SQUIDs respectively [2].

By varying the applied magnetic flux and, in the case of the two junction DC SQUID, biasing the junctions with an electric current, the potential of the superconducting system can be adjusted to have a wide range of barrier heights and asymmetry. In a controllable manner it is possible to reduce the potential barrier to the point at which the SQUID switches between states corresponding to n and $n \pm 1$ flux quanta (Φ_0) under the effect of thermal noise alone. The measured noise power due to this switching has a random telegraph noise spectrum, which is white up to a knee and then decreases as the frequency squared. The roll-off knee can be adjusted from zero frequency to perhaps as high as 1 GHz, although we are limited to an upper measurement frequency of 50 kHz by our SQUID readout electronics. As the roll-off frequency is increased the white noise level of the switching is reduced, allowing us to vary the noise power at the signal frequency by approximately 60 dB, at which point it is below the noise level of our measurement apparatus. In general, in the high switching regime there is a voltage across at least one of the SQUID junctions although the device is still superconducting. This resistive regime was not accessible in prior SR studies of SQUIDs [3–5] but appears to correspond to the regime of enhanced SNR. For example, we find that the peak in the signal amplification and SNR always occurs in the resistive regime.

A simplified diagram of a DC SQUID as used in our experiments is shown in Figure 1. For simplicity we assume the Josephson junction and SQUID inductance in each arm to be identical. A bias current I_B is applied across the two junctions in parallel and an external magnetic flux Φ_x (in units of Φ_0) is applied to the SQUID loop.

The current through each junction is the sum of the net circulating shielding current I_{CIRC}, which comprises the entire asymmetric component of the current distribution, and half the bias current, $I_B/2$. If the junctions are in the resistive state an additional current equal to $\frac{1}{R}\frac{d\theta}{dt}$ passes through each junction, where R is the normal state resistance of the junction and θ the phase difference across the junction. The potential energy of the SQUID, $U(\theta_1 \theta_2)$, may be written in terms of the phase differences across each Josephson junction as follows [6]:

$$U(\theta_1 \theta_2) = \frac{1}{2\beta}(\theta_1 - \theta_2 - 2\pi\Phi_x)^2 - \cos\theta_1 - \cos\theta_2 - I(\theta_1 + \theta_2) \quad (1)$$

in which $I = \frac{I_B}{2I_J}$ where I_J is the critical current of each junction. The units of energy are $I_J\Phi_0/2\pi$ and β is a parameter that characterizes the inductive screening properties of the SQUID [7]. The potential is the sum of three terms. A magnetic energy associated with the shielding current flowing in the SQUID the potential energy of each individual junction, and a term corresponding to the free energy of the current source which biases the SQUID.

In our experiment, and in most arrangements involving a DC SQUID, I_B and Φ_x are imposed parameters while the phase differences θ_1 and θ_2 are free variables which adjust in such a way that U is minimized. The stable operating states correspond to the minima of the potential. In the zero voltage state the condition $\frac{dU}{d\theta_1} = \frac{dU}{d\theta_2} = 0$ holds, which allows us to simplify the potential to a function of one phase variable. In the resistive state a more complex analysis leads to the same expression [8].

$$U(\theta_2) = \frac{1}{2\beta}(\beta\sin\theta_2 - \beta I)^2 - I(2\theta_2 + \beta\sin\theta_2 - \beta I + 2\pi\Phi_x) -$$
$$\cos\theta_2 - \cos(\theta_2 + \beta\sin\theta_2 - \beta I + 2\pi\Phi_x) \quad (2)$$

The variation of $U(\theta_2)$ as a function of phase for $\beta = 1.5$ and 8 is shown in Figure 2 for zero applied I_B and Φ_x. We see that the potential contains local minima. For $I_B > 0$, the potential is tilted but is still periodic in phase as shown in Figure 3. If Φ_x is varied the relative height of the minima is altered. In general, as Φ_x is increased from just below $\Phi_0/2$ to just above it the phase corresponding to the potential minimum jumps discontinuously. The result is a "flux jump", sudden change in internal state of the SQUID in which the internal flux changes by an amount of order Φ_0. This process is illustrated in Figure 4 for $\beta = 1.5$. We also see that for $I_B = 0$ and $\Phi_x = 0.5$ the potential contains two equal minima centered around $\theta = 2n\pi$. Thus under these conditions the general form of the potential is the same as the RF SQUID biased at $\Phi_x = 0.5$ [5].

For some combinations of β and I_B, such as shown in Figure 4, there is a barrier between the two minima at $\Phi_x = 0.5$ that prevents the SQUID from

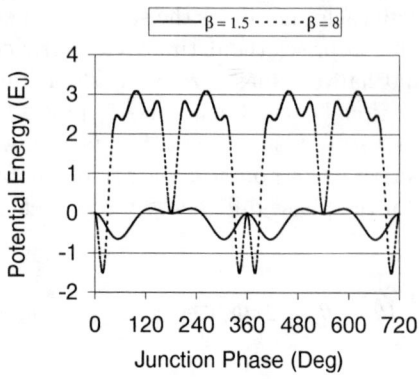

FIGURE 2. Potential for $\beta = 1.5$, and $\beta = 8$. I_B and $\Phi_x = 0$.

jumping to the lowest energy minimum. This leads to hysteresis in the transition, which does not occur until some higher value of Φ_x (for increasing Φ_x). The effect of hysteresis is to strongly reduce the switching rate. Hysteresis can be removed by applying a bias current. For example for the SQUID shown in Figure 4, a current of 0.7 I_J is necessary.

Our experiment comprises two coupled commercial Nb SQUIDs [9] in a shielded liquid helium Dewar connected as shown in Figure 1. The measurement (M-) SQUID is operated in the conventional manner and measures the flux state of the switching (S-) SQUID by inductive coupling to the circulating current I_{CIRC}. The variation of I_{CIRC} as a function of Φ_x produces the input to output transfer function (TF) that characterizes the two SQUID

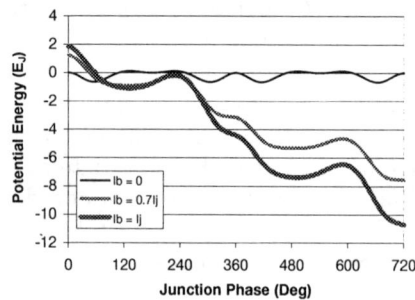

FIGURE 3. Potential for $I_B \neq 0, \beta = 1.5, \Phi_x = 0$.

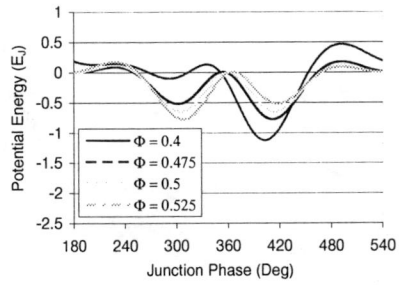

FIGURE 4. Potentials for Φ_x below, at, and above 0.5, $I_B = 0, \beta = 1.5$

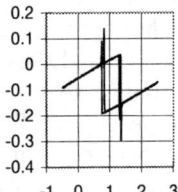 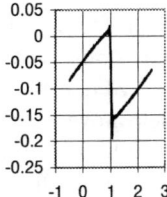

FIGURE 5. TFs for S-3, (a) $I_B = 0$, (b) $I_B = I_{ZH}$.

system. The same M-SQUID was used throughout. Five S- SQUIDs, denoted S-1 through S-5, with different values of β were tested. All SQUIDs have integral thin film superconducting input and modulation coils. The S- and M- SQUIDs are coupled by connecting their input coils with superconducting wire, resulting in a coupling efficiency of around 20%. DC flux bias and the signal is applied to the S-SQUID via its modulation coil. Current bias is applied to the junctions in parallel as in conventional SQUID bias techniques.

We now consider the experimental results of varying I_B and Φ_x. As the current bias, I_B, to the S-SQUID is increased from zero, characteristic variations in the transfer function between I_{CIRC} and Φ_x occur. Figure 5 shows TFs for device S-3 for $I_B = 0$ and 4.5 μA. All devices tested exhibit this basic form of TF. In the hysteretic regime ($I_B \lesssim 2I_J$); the TF is comprised of linear regions connected by quasi- linear discontinuous jumps in I_B, which occur when $I_{CIRC} = \pm I_J$. The linear regions correspond to the mutual inductance between I_{CIRC} and Φ_x, which is present from the geometry of the coils and the phase to current coupling of the SQUID junctions. Owing to hysteresis, the transitions corresponding to $+I_J$ do not occur at the same Φ_x as those for $-I_J$. The hysteresis width decreases approximately linearly with I_B. At

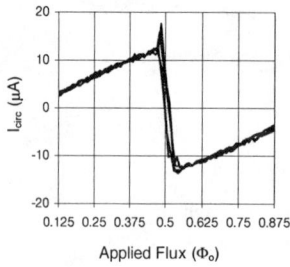 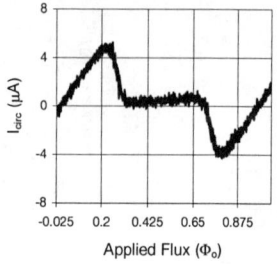

FIGURE 6. Circulating current vs. applied flux for S-1: a) $I_B = 1.1 I_J$, b) $I_B = 1.5 I_B$.

of two linear regions connected by a region of essentially infinite gradient. As I_B is further increased the infinite slope of the flux transition is reduced, as shown in Figure 6a for S-1. However, once I_B reaches approximately 1.2 I_{ZH} different behavior is observed for the four low β SQUIDs and the one with high β. For the low β SQUIDs the transfer function splits into two symmetrical transitions, which move apart in proportion to I_B. A typical TF is shown in Figure 6b. For the highest β SQUID (S-4), the TF does not split, instead the gradient of all regions of the TF gradually reduce to zero. The response of the S-SQUID to a steady sine wave input signal of frequency 97 Hz is shown in Figure 7. The signal amplitude is chosen to produce an output SNR of 20 dB (a factor of 10) from the linear, non-switching, region of the TF. In practice, the input signal amplitude is of order $100\mu\Phi_0$ (2 x 10^{-19} Wb). The vertical axis in Figure 7 is the difference between the SNR at the output and the effective SNR applied to input. For low applied currents there is a deep trough centered on the hysteretic part of the TF. This allows us to calibrate Φ_x to be 0.5 at the trough center. As I_B is increased the SNR first decreases and then increases rapidly, reaching a peak of around 8 dB higher than the value when no switching is occurring. For higher bias current there is a bifurcation in the SNR peak corresponding to the split in the TF shown in Figure 6b. The SNR remains at the peak value but no longer occurs at $\Phi_x = 0.5$. The separation in flux value between the SNR peaks corresponds to the separation of the regions of maximum gradient in the TF. For the highest β SQUID(S-4) there is no bifurcation in the SNR peak, corresponding to the absence of a split in the TF. The additional SNR peaks at the top and bottom of Figure 7 are the ends of bifurcations in the SNR peaks originating at $1.5\Phi_0$ and $-0.5\Phi_0$. The maximum value of the SNR for the five SQUIDs tested is shown in Table 1. Despite the wide range of β, the variation in SNR is relatively constant.

We can probe the SNR gain effects by plotting the signal and noise level separately as shown in Figure 8 for S-4. For low I_B (e.g. $< 2I_J$ for S-4) the TF is of the form shown in Figure 5(a). The potential barrier between

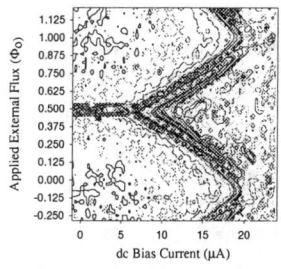 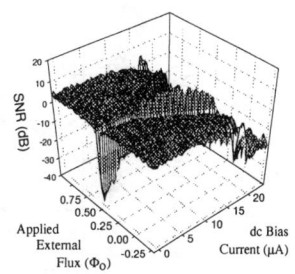

FIGURE 7. Increase in SNR vs. applied current and flux for S-1. a) contour plot, b) 3-dimensional view of the same data

states and the hysteresis in Φ_x prevents switching and the noise of the S-SQUID is correspondingly low ($\approx 5\mu\Phi_0/\sqrt{Hz}$). As I_B is increased the S-SQUID begins to switch state under internal thermal noise and the flux noise level of the S-SQUID increases dramatically. The output signal amplitude increases also by approximately the same amount. For further increases in I_B the noise decreases monotonically to its original level. However, the output signal reaches a roughly stable level typically a factor of ten higher than at the input. As a result, once this point is reached the signal exceeds the noise producing a net gain in SNR.

The simplest interpretation is that the SNR gain is due to a multiplication effect of the steep part of the TF, the steep region being created as a result of the switching between the quantized states and of the intrinsic properties of the SQUID. However, for all devices measured, there is a characteristic region within which the output signal increases while the steepest gradient of the TF appears to be decreasing. This is the very region in which the signal begins to exceed the noise, producing the net increase in SNR. The behavior of the S-SQUID and its TF in this region is still under investigation. We also note that at the SNR peak, the signal frequency is at least 10^4 times smaller than the upper frequency of the flux noise in the S-SQUID. Thus there appears to be no correlation between the switching times of the signal and noise, contrary

TABLE 1. SNR Increase and β for all SQUIDs Tested.

SQUID sample	β	SNR Gain (dB)
S-1	1 to 1.4	8
S-2	2	10
S-3	2.5	12
S-4	10	11
S-5	2.7	10

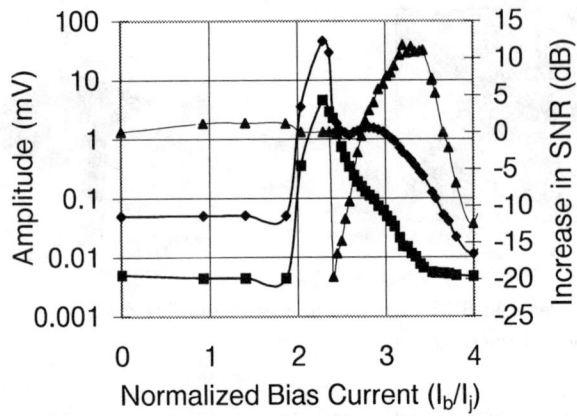

FIGURE 8. Output of S-4 vs. Bias Current for $\Phi_x = 0.5$. ◇=signal (left axis) □=noise (left axis), and △=increase in SNR (right axis).

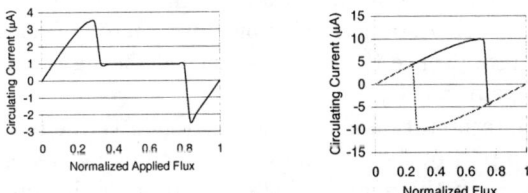

FIGURE 9. Predicted Transfer Functions for a) $\beta = 1.5, I_B = 1.5 I_J$, b) $\beta = 1.5, I_B = 0$

to some observations in SR.

We have developed an analytical model for the DC SQUID which can be used to calculate the TF when the SQUID is in the superconducting or resistive state [8]. The heart of the method is to solve equation 2 for the phase at the potential minimum. The value of the other phase variable is determined from the requirement for phase continuity. The circulating current is given by the expression:

$$I_{CIRC} = \frac{I_J}{2}(\sin\theta_2 - \sin\theta_1) - \frac{1}{2R}(V_1 - V_2) \qquad (3)$$

in which V_1 and V_2 are the voltages across junctions of equal resistance, R. In Figure 9a we show the predicted value of I_{CIRC} as a function of Φ_x for $I_B = 1.5 I_J$ and $\beta = 1.5$.

For small Φ_x, I_{CIRC} rises almost linearly until the critical current for the value of applied I_B is reached. The SQUID then enters the resistive state in which the phase oscillates with time about a non-zero average value. At a

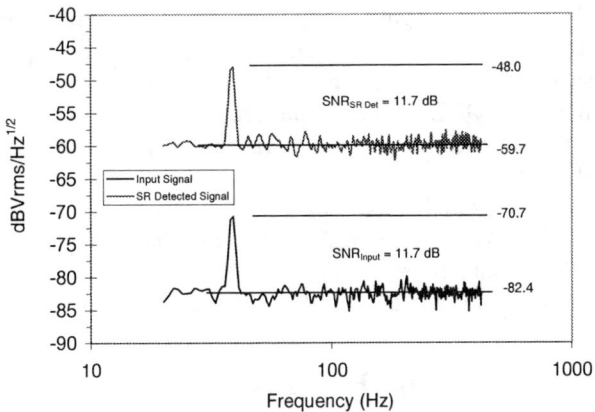

FIGURE 10. Effect of applied external noise. Upper trace is the output spectrum, lower trace the input spectrum.

value of Φ_x symmetric about 0.5, the SQUID leaves the resistive state and becomes fully superconducting again. The agreement between Figures 9a and 6b is good in both form and magnitude. Similarly good agreement is found in the fully superconducting state for both hysteretic and non-hysteretic TFs. A hysteretic TF for $\beta = 1.5$ and $I_B = 0$ is shown in Figure 9b. Comparison with Figure 6a shows the amplitude of the transitions to be almost the same.

We have also performed preliminary experiments on the effect of noise added from outside the system. Gaussian white noise with an upper frequency of 1 MHz was added to the signal. In the hysteretic regime ($I_B \leq I_{ZH}$) we observe the characteristic peak in the SNR ratio as a function of the amplitude applied noise, which is indicative of SR. The amplitude of this peak is always lower than the input SNR by approximately 10 dB. However, when the S-SQUID is biased close to the peak in the SNR we are able to configure the system so that the output SNR is exactly equal to the input SNR, a result not previously considered possible in conventional SR. In Figure 10 we show input and output spectra taken for S-4 at $I_B \approx 1.2 I_{ZH}$ for a signal frequency of 37 Hz.

We interpret the absence of a conventional stochastic resonance SNR peak to the limited bandwidth of our noise generator. For low $I_B (I_B \leq I_{ZH})$ the switching frequency spectrum caused by the internal thermal noise extends to approximately 100 kHz. Under these conditions, the applied noise is able to dominate the switching dynamics of the SQUID. However, for high I_B the highest switching rate may be of order 1 GHz and the applied noise has negligible overall effect. In this case the system acts as a simple amplifier.

The effects we observe shed new light on the effect of internal switching on the operation of DC SQUIDs and perhaps other related physical systems. The origin of the strong increase in signal amplitude at the output of the

SQUID is still under investigation. However, the ability to fully recover the signal while the system is switching internal states could greatly reduce the effects of noise interference and the need for the extensive shielding precautions usually required for SQUIDs. The enhancement in signal to noise ratio may offer the potential to increase basic SQUID sensitivity if multiple S- SQUIDs are connected in series.

This work was sponsored in part by the U. S. Navy under contract N66001-96-C-7009. We also thank Adi Bulsara, Mario Inchiosa, Hoke Trammell, Robert Penny, and Roger Koch.

electronic address: andy_hibbs@QM.com, brian_whitecotton@QM.com.

REFERENCES

1. Simon A., and Libchaber A., *Phys. Rev. Lett.*, **66**, 3375 (1992).
2. see for example, Barone A., and Paterno G., *Physics and Applications of the Josephson Effect*, (J Wiley, NY 1982).
3. Hibbs A. D., et al., *Proc. 12th Int. Conf. on Noise in Physical Systems and 1/f Fluctuations*, AIP Press, AIP Conference Proceedings **285**, 720 (1993).
4. Rouse R., Han S., and Lukens J. E., *Appl. Phys. Lett.* **66**, 108 (1995).
5. Hibbs A. D. et al., *J. Appl. Phys.*, **77**, 2582 (1995).
6. Tesche C. D., *J. Low. Temp. Phys*, **44**, 119 (1981).
7. β is defined to be $2\pi L I_J/\Phi_0$, where L = SQUID inductance.
8. Hibbs A. D., manuscript in preparation
9. Quantum Design, 11578 Sorrento Valley Road, San Diego CA 92121, U.S.A.

Quasicontinuous control using time-delay coordinates

A. Schenck zu Schweinsberg and U. Dressler

Daimler-Benz AG, Research Institute Frankfurt,
Goldsteinstraße 235, D-60528 Frankfurt

Abstract. We investigate quasicontinuous versions of the Ott-Grebogi-Yorke (OGY) control method in a bronze ribbon experiment using time-delay coordinates for the reconstruction of the attractor. We apply as quasicontinuous control methods the local control method and the minimal expected deviation method. As is known for the original OGY-method with time-delay coordinates, values of the control parameter at previous times appear in the linearized dynamics. We discuss two possible ways to derive from this linearization feedback control formulas. The robustness of the control methods with respect to noise is experimentally demonstrated.

INTRODUCTION

In this paper we combine the two following modifications of the Ott-Grebogi-Yorke (OGY) control method [1]. First, in the original OGY-control approach the control frequency was limited to the frequency of the piercings of the continuous trajectory through a Poincaré section. If now the instability of the unstable periodic orbit (UPO) is very high then the amplification of measurement noise can spoil the feedback control when controlling only once per period of the UPO. Therefore, for experiments with large instabilities quasicontinuous extensions of the OGY-control have been introduced by Reyl *et al.* [2] with the *minimal expected deviation* method (MED) and by Hübinger *et al.* [3,4] with the *local control* method (LC). For driven systems, which we consider in this paper, the control frequency is raised by introducing N equally spaced Poincaré sections Σ_n per period T as control stations.

The second modification which we address comes in when one uses time-delay coordinates [5] for the reconstruction of the attractor. In this case the OGY-feedback formula has to be modified [6], since the flow mapping also depends on all preceding control parameters which were changed during the time window $\tau_w = (d-1)\tau$ of the delay vector $(x(t), x(t-\tau), \ldots, x(t-(d-1)\tau))$.

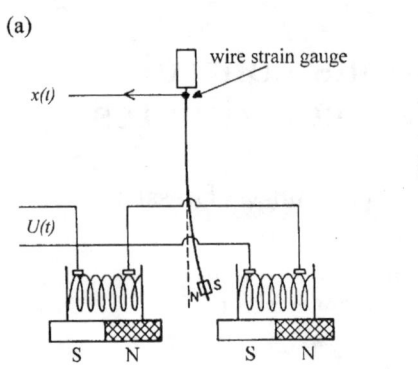 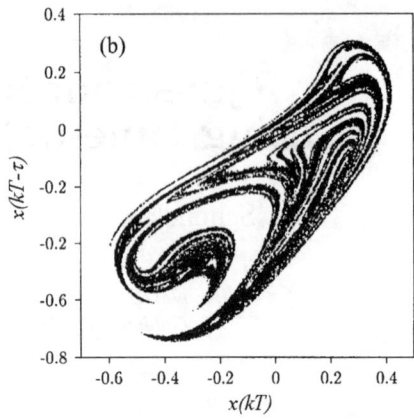

FIGURE 1. (a) Experimental setup of the chaotic bronze ribbon. (b) Chaotic attractor for the bronze ribbon experiment. 50 000 points $\mathbf{z}(kT) = (x(kT), x(kT - \tau))$ are shown in the Poincaré section Σ_1 with $\tau = \ell\Delta t$, $\Delta t = \frac{T}{64}$, and $\ell = 5$.

For a quasicontinuous control with a control frequency being, e.g., the sampling frequency $\frac{1}{\Delta t}$, the mapping $\mathbf{P}^{(n,n+1)} = \phi^{\Delta t}_{|\Sigma_n} : \Sigma_n \to \Sigma_{n+1}$ will depend on the last $w = \ell(d-1)$ parameter changes with ℓ being the time-delay in units of Δt and d the embedding dimension. We express this dependence as $\mathbf{z}^{n+1} = \mathbf{P}^{(n,n+1)}(\mathbf{z}^n, p^{n-w}, \ldots, p^{n-1}, p^n)$. Thus, to stabilize an UPO $\mathbf{z}_F^n$ we have to use the linearization

$$\delta \mathbf{z}^{n+1} = A^n \cdot \delta \mathbf{z}^n + \sum_{i=0}^{w} \mathbf{b}^{n,i} \delta p^{n-i} \qquad (1)$$

with $\delta \mathbf{z}^n = \mathbf{z}^n - \mathbf{z}_F^n$, $\mathbf{z}_F^n$ being the UPO in section Σ_n, and the linearizations $A^n = D_{\mathbf{z}^n}\mathbf{P}^{(n,n+1)}$ and $\mathbf{b}^{n,i} = \frac{\partial}{\partial p^{n-i}}\mathbf{P}^{(n,n+1)}$ around $(\mathbf{z}_F^n, p_0, \ldots, p_0)$.

EXPERIMENTAL SETUP

The experiment is a horizontally cantilevered elastic bronze ribbon equipped with two small permanent magnets (Fig. 1 (a)). The beam is located in an inhomogeneous magnetic field. To drive the system two coils are placed around the free end of the beam and supplied with an ac voltage $U(t) = U_A \sin(\frac{2\pi}{T}t) + p$ with $U_A = 0.6$ V and the driving period $T = 1$ s. The offset voltage p is used as control parameter.

As measurement signal $x(t)$ we use the voltage signal of a wire strain gauge. Setting the sampling time Δt to $T/64$ we introduce $N = 64$ Poincaré sections for the control. With the sampled measurement signal $x^n = x(t_0 + n\Delta t)$ the dynamics of the system is reconstructed using two time-delay coordinates $\mathbf{z}^n = (x^n, x^{n-\ell})$ and the phase of the periodic driving.

When the control parameter is set to $p_0 = -0.2\,\text{V}$ a chaotic attractor is found (Fig. 1 (b)). Embedded in the chaotic attractor four period-one UPOs, one period-two UPO, and one period-three UPO are detected using the method of best recurrent points described in [4]. The linearizations A^n and $\mathbf{b}^{n,i}$ of the mappings $\mathbf{P}^{(n,n+1)}$ around a UPO are extracted from the dynamics of the nearest neighbor points in the time-delay embedding space [7].

CONTROLLING THE BRONZE RIBBON

The control requirement of both quasicontinuous control methods considered in this paper (LC and MED) can be formulated in a unified way [7]. In the physical state space where the linearized dynamics depends only on the actual control parameter, i.e. $\delta\hat{\mathbf{z}}^{n+1} = \hat{A}^n \cdot \delta\hat{\mathbf{z}}^n + \hat{\mathbf{b}}^n \delta p^n$, both control methods require that the application of the control signal δp^n should diminish the projection of $\delta\hat{\mathbf{z}}^{n+1}$ on a vector $\hat{\mathbf{h}}^n$ by a factor of $(1-\rho)$ compared to the one which $\delta\hat{\mathbf{z}}^{n+1}_{\delta p^n = 0}$ would have if no control were applied, i.e.

$$\hat{\mathbf{h}}^{n\,\dagger} \cdot \delta\hat{\mathbf{z}}^{n+1} = (1-\rho)\,\hat{\mathbf{h}}^{n\,\dagger} \cdot \delta\hat{\mathbf{z}}^{n+1}_{\delta p^n = 0}. \quad (2)$$

For the LC method the vector $\hat{\mathbf{h}}^n$ is given by the singular vector corresponding to the largest singular value of $\hat{A}^{n+1}$ and for the MED method by $\hat{\mathbf{h}}^n = \hat{\mathbf{b}}^n$.

Now, to obtain a control formula in the case of the linearization (1) for a time-delay embedding, one could use the *extended states* $\mathbf{y}^n = (\mathbf{z}^n, p^{n-w}, \ldots, p^{n-1})$ and rewrite the dynamics as $\delta\mathbf{y}^{n+1} = C^n \cdot \delta\mathbf{y}^n + \mathbf{d}^n \delta p^n$ with

$$C^n = \begin{pmatrix} A^n & \mathbf{b}^{n,w} & \mathbf{b}^{n,w-1} & \cdots & \mathbf{b}^{n,1} \\ \hline & 0 & 1 & \cdot & 0 \\ 0 & \cdot & \cdot & \ddots & \cdot \\ & \cdot & \cdot & \cdot & 1 \\ & 0 & \cdot & \cdot & 0 \end{pmatrix} \quad \text{and} \quad \mathbf{d}^n = \begin{pmatrix} \mathbf{b}^{n,0} \\ 0 \\ \vdots \\ 0 \\ 1 \end{pmatrix}.$$

For the OGY-control with time-delay this was done in [8]. Since the linearized dynamics has the same form as in the physical state space, the control requirement can be chosen in complete analogy to (2):

$$\mathbf{h}^{n\,\dagger} \cdot \delta\mathbf{y}^{n+1} = (1-\rho)\,\mathbf{h}^{n\,\dagger} \cdot \delta\mathbf{y}^{n+1}_{\delta p^n = 0}. \quad (3)$$

with $\mathbf{h}^n$ being the singular vector corresponding to the largest singular value of C^{n+1} for the LC method or $\mathbf{h}^n = \mathbf{d}^n$ for the MED method.

The at first glance simple and convincing use of the extended states leads to an undesirable dependence on the units of the control parameter [7]. To verify this dependence in the experiment we replace all parameter p^n by p^n/σ and the corresponding vectors $\mathbf{b}^{n,i}$ by $\sigma\mathbf{b}^{n,i}$ and apply the LC method (Fig. 2). As can be seen successful control can only be achieved for $\sigma = 200$ and $\sigma = 2\,000$.

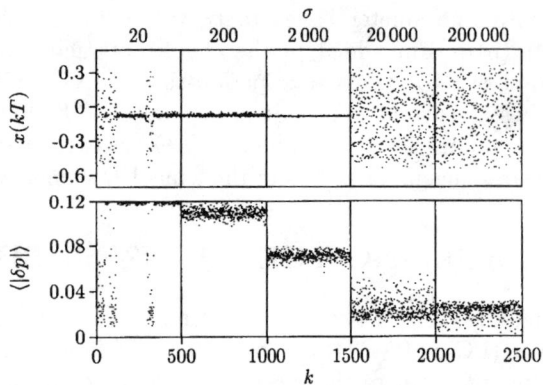

FIGURE 2. Local control of UPO 1 using extended states. Starting with the scaling factor $\sigma = 20$ every 500 driving periods the scaling factor σ is increased by a factor 10. $x(kT)$ is the stroboscopic measurement in Σ_1 and $\langle|\delta p|\rangle$ the averaged control amplitude.

Furthermore, the applied control signal clearly depends on the scaling factor σ. We did the same experiment for the MED method and observe that the appropriate value of σ depends on the method and even on the parameters d, τ of the time-delay embedding. The best choice with respect to control cannot be predicted in advance.

An alternative way to derive a feedback formula starting from the linearization (1) is to *modify the control requirement*. For the original OGY-method with time-delay coordinates this was done in [6]. We now generalize this modified control requirement for the quasicontinuous control. Because the parameter value p^n influences the trajectory $\mathbf{z}^{n+1}, \mathbf{z}^{n+2}, \ldots$ till $\mathbf{z}^{n+w}$, we require that the system stabilizes only after $(w+1)$ time steps for an appropriately chosen $\delta p^n \neq 0$ without further control interventions in between, i.e.,

$$\mathbf{h}_w^{n\,\dagger} \cdot \delta \mathbf{z}^{n+w+1} = (1-\rho)\, \mathbf{h}_w^{n\,\dagger} \cdot \delta \mathbf{z}^{n+w+1}_{\delta p^n = 0}, \qquad (4)$$

$$\delta p^{n+1} = \delta p^{n+2} = \cdots = \delta p^{n+w} = 0. \qquad (5)$$

The condition (4) is chosen in analogy to (2). For the LC method, $\mathbf{h}_w^n$ is given by the singular vector of A^{n+w+1} which refers to the direction of maximal stretching. To minimize the deviation in section Σ_{n+w+1} (MED), the vector $\mathbf{h}_w^n = \frac{\partial}{\partial p^n} \mathbf{P}^{(n,n+w+1)}(\mathbf{z}_F^n, p_0, \ldots, p_0)$ has to be chosen which can be calculated using A^{n+j} and $\mathbf{b}^{n+j,j}$, $j = 0, \ldots, w$. Both possibilities lead to a control formula which does not depend on a scaling factor of the control parameter.

Note that in the formulation of the modified control requirement the parameter perturbations are in principle calculated till δp^{n+w}. In the experimental realization, however, where noise is always present we calculate δp^n at every control step n once again.

To demonstrate that the modified control requirement (4) is indeed neces-

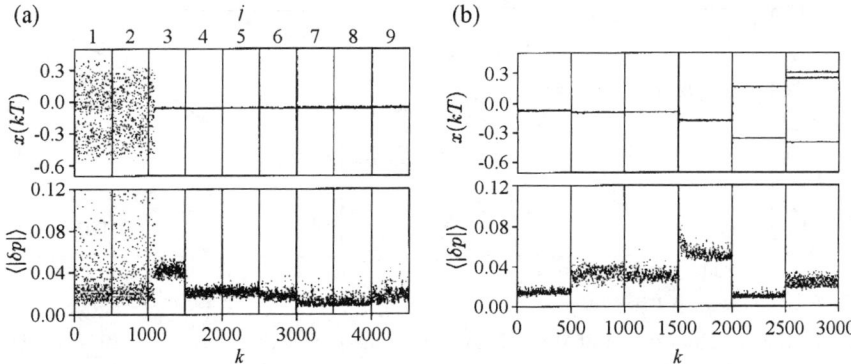

FIGURE 3. Local control using the modified control requirement. $x(kT)$ is the stroboscopic measurement in Σ_1 and $\langle|\delta p|\rangle$ the averaged control amplitude. (a) From 0 – 499 the stabilization is demanded after $j = 1$ time step, afterwards the value j is increased every 500 periods by 1. (b) The control ($j = w + 1$) is switched on at $k = 0$, every 500 driving periods we change the UPO to be controlled.

sary we apply the LC method using a feedback control formula which results if one requires stabilization after $j = 1, 2, \ldots$, till 9 time steps. Fig. 3 (a) shows the result of this experiment. The UPO can be only stabilized for $j \geq 3$ and the control signal is minimized for $j = 7$. A further investigation shows that the averaged distance between the UPO and the trajectory of the system has a minimum for $j = 5$. Both results confirm our theoretical choice $j = w + 1 = 6$. With this choice we could stabilize every UPO using the MED and LC method in (4) without qualitative difference between both control methods. In Fig. 3 (b) we show an example of a successful control experiment using the LC method. As can be seen the transient time to stabilize the next UPO is always very short.

Next, we investigate the robustness of the LC and the MED method with respect to measurement noise by adding small noise terms $\varepsilon \cdot \eta^n$ to the measurement x^n, η^n being identically distributed in $[-1, 1]$. Starting with $\varepsilon = 0$ every 100 periods the noise level is increased by 0.005. In Fig. 4 (a) the number N_S of stabilizied periods is shown for the LC and the MED method using two and three time-delay coordinates versus ε. For the MED method with a three-dimensional Poincaré section in the embedding space the control signal averaged over the stabilized periods N_S are shown in Fig. 4 (b).

Finally, we note that the position of the orbit used in the feedback control was always redetermined befor we started the experiment in question. For this purpose, the adaptive orbit correction [9,7] was applied which calculates a new estimate of the UPO during control by exploiting the periodicity of the control signal and the stabilized orbit.

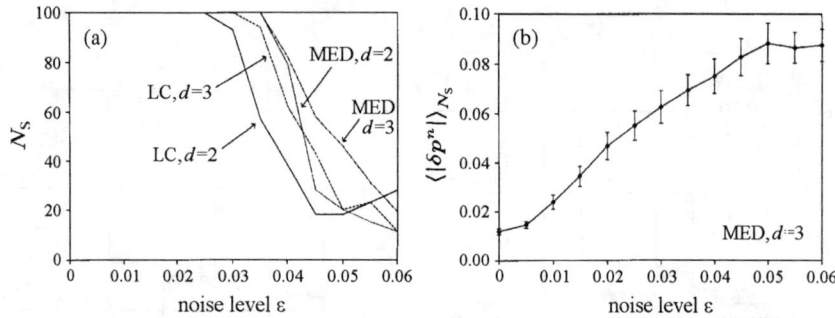

FIGURE 4. Robustness of the LC and MED method with respect to noise. (a) The number N_S of stabiliziedperiods and (b) the averaged control signal (exemplarily for the MED method) are plotted versus the noise level ε.

SUMMARY

In a bronze ribbon experiment we have implemented quasicontinuous versions of the OGY-control method using time-delay coordinates for the reconstruction of the attractor. As quasicontinuous methods we apply the local control method and the minimal expected deviation method. In order to handle the additional dependences on preceding control parameters two possible ways to derive feedback control formulas are discussed. These are the extended state space approach and a modified control requirement. We demonstrate in the experiment that the extended state space approach leads to an undesirable dependence on the units of the control parameter, while the modified control requirement shows satisfying results, even in the presence of noise.

REFERENCES

1. E. Ott, C. Grebogi, and J.A. Yorke, Phys. Rev. Lett. **64**, 1196 (1990).
2. C. Reyl, L. Flepp, R. Badii, and E. Brun, Phys. Rev. E **47**, 267 (1993).
3. B. Hübinger, R. Doerner, and W. Martienssen, Z. Phys. B **90**, 103 (1993).
4. B. Hübinger, R. Doerner, W. Martienssen, M. Herdering, R. Pitka, and U. Dressler, Phys. Rev. E **50**, 932 (1994).
5. F. Takens, in *Dynamical Systems and Turbulence*, edited by D. Rand and L.S. Young (Springer, Berlin, 1981).
6. U. Dressler and G. Nitsche, Phys. Rev. Lett. **68**, 1 (1992).
7. A. Schenck zu Schweinsberg, T. Ritz, U. Dressler, B. Hübinger, R. Doerner, and W. Martienssen, Phys. Rev. E **55**, 2145 (1997).
8. F.J. Romeiras, C. Grebogi, E. Ott, and W.P. Dayawansa, Physica D **58**, 165 (1992); P. So and E. Ott, Phys. Rev. E **51**, 2955 (1995); M. Ding, W. Yang, V. In, W.L. Ditto, M.L. Spano, and B. Gluckman, Phys. Rev. E **53**, 4334 (1996).
9. R. Doerner, B. Hübinger, and W. Martienssen, Int. J. Bif. Chaos **5**, 1175 (1995).

Coupled Brownian Rectifiers

Robert Häußler, Roland Bartussek, and Peter Hänggi

University of Augsburg, Department of Physics Memminger Str. 6, D-86135 Augsburg, Germany

Abstract. Periodic structures that lack reflection symmetry act – when periodically rocked by external forces – as rectifiers for Brownian motion. Here we investigate the role of a global coupling (of the 'Kuramoto-type') among such rectifiers. We demonstrate that the coupling strength K acts as a control on the sign (it yields a realization for current reversal) and the magnitude of the directed average velocity of Brownian particles. Moreover, raising the coupling strength K, results in an effective reduction of ambient noise. This intriguing effect is revealed in a striking manner for the mean velocity vs. load characteristics.

INTRODUCTION

In recent years, the study of directional Brownian motion in periodic potentials that lack reflection symmetry (*ratchets*) has attracted considerable attention (for the major ideas and a recent review with more references, see [1]). Such ratchets provide models for molecular motors, but in addition also carry a potential for novel technological applications in the worlds of micro- and nano-physics [1,2]. Due to the fact that motion is directed, particles can move uphill against a load with finite velocity; thus they serve as archetype systems that are able to act as rectifiers (i.e., identical direction of the velocity upon varying the bias around zero) on a level of Brownian noise. In the previous literature a variety of ratchet mechanisms have been identified [1]. Here, our focus is on the interplay of ratchet devices that are mutually coupled with each other. Thus far, such coupling effects have rarely been addressed [3,4]. The focus in this work will be on globally coupled rocking ratchets.

MODEL

We start from the *overdamped* dynamics of N interacting particles in a ratchet potential $V(x)$, i. e.

$$\dot{x}_i = -\frac{dV(x_i)}{dx_i} - \frac{K}{N}\sum_{j=1}^{N}\sin[2\pi(x_i - x_j)] + F + A\sin\Omega t + \sqrt{D}\xi_i(t), \quad (1)$$

where $i = 1, ..., N$. The dot denotes the derivative with respect to time t. The particles are forced by a static bias F and, additionally, by periodic driving with angular frequency Ω and amplitude A. Each particle experiences thermal fluctuations of strength D, modelled by independent Gaussian white noise sources $\xi_i(t)$ of zero average, and correlations $\langle \xi_i(t)\xi_j(s)\rangle = 2\delta_{i,j}\delta(t-s)$. We adopted the global interaction between particles of strength K/N in eq.(1) from the well known 'Kuramoto model' [5,6] describing excitable systems.

In this work we consider a ratchet potential being constructed from two Fourier modes [7], $V(x) \equiv -\frac{1}{2\pi}(\sin 2\pi x + \frac{1}{4}\sin 4\pi x)$. Equation (1) can be simulated numerically. This method has the disadvantage that for small noise strength D it is intrinsically rather time consuming. Thus, we look for an alternative route to the solution of our problem. An equivalent description of the noisy dynamics (1) is given by the N-dimensional Fokker-Planck equation, i.e.,

$$\frac{\partial}{\partial t}W(x_1, ..., x_N; t) = \Bigg\{\sum_{i=1}^{N}\frac{\partial}{\partial x_i}\Bigg[\frac{dV(x_i)}{dx_i} + \frac{K}{N}\sum_{j=1}^{N}\sin 2\pi(x_i - x_j)$$
$$-F - A\sin\Omega t\Bigg] + \sum_{i,j=1}^{N}\frac{\partial}{\partial x_i \partial x_j}D\Bigg\}W(x_1, ..., x_N; t). \quad (2)$$

For small N, this equation can effectively be solved by the method of *Matrix-Continued-Fractions* [2,7]. Further, to investigate the limit of many particles, $N \gg 1$, we turn to a *mean field description*: For the mean particle density $W(x;t) \equiv \lim_{N\to\infty}\frac{1}{N}\sum_{i=1}^{N}\delta(x - x_i(t))$ we find the one dimensional Fokker-Planck equation

$$\frac{\partial}{\partial t}W(x;t) = \Bigg\{\frac{\partial}{\partial x}\Bigg[\frac{dV(x)}{dx} + K\int_0^{2\pi}dx'W(x';t)\sin 2\pi(x - x') - F - A\sin\Omega t\Bigg]$$
$$+ \frac{\partial^2}{\partial x^2}D\Bigg\}W(x;t), \quad (3)$$

which is *nonlinear* in $W(x;t)$. We solve this equation numerically for periodic boundary conditions by expanding into Fourier modes, i.e.,

$$W(x,t) = \sum_{l=-\infty}^{\infty}c_l(t)e^{il2\pi x}. \quad (4)$$

This yields an infinite set of nonlinear ordinary differential equations for the expansion coefficients $c_l(t)$, which are solved by standard methods.

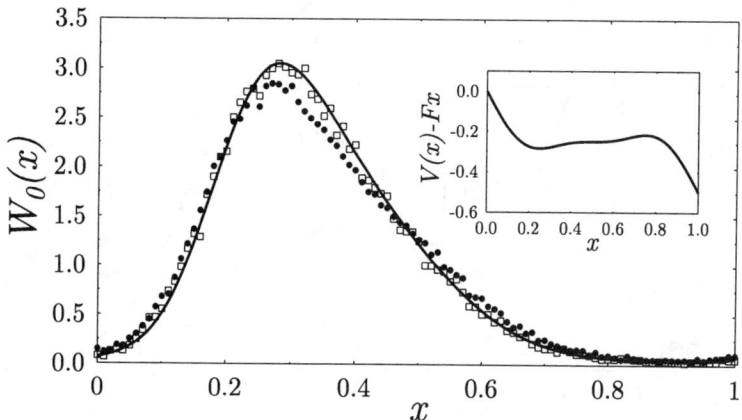

FIGURE 1. The asymptotic stationary mean field probability distribution $W_0(x)$ in a tilted potential (shown in the inset) is calculated within the mean field approximation (—) and compared to Langevin simulations for $N = 8(\bullet)$, $N = 256(\square)$ at model parameters $F = 0.5$, $D = 0.1$, and $K = 1$.

RESULTS

At this point we can ask, how large the particle number N must be, in order for the mean field to be a sensible valid approximation. For the undriven case $A = 0$ of a tilted potential $V(x) - Fx$ we compare in Fig. 1 the asymptotic stationary mean field solution $W_0(x) \equiv \lim_{t\to\infty} W(x;t)$ calculated by the solution of (3) with Langevin simulations, cf. Fig. 1.

Next, we turn to the investigation of the mean asymptotic particle velocity

$$\langle \dot{x} \rangle_{st} \equiv \lim_{t\to\infty, N\to\infty} \int_t^{t+2\pi/\Omega} dt' \sum_{i=1}^{N} \dot{x}_i(t')/N. \qquad (5)$$

Numerical results for the directed velocity versus coupling strength K are depicted with Fig.2. Clearly the interaction between particles can increase the current significantly, cf. Fig. 2(a). Moreover, the direction of transport can be *reversed* due to the coupling, cf. the large K-values in Fig. 2(a). We remark, that such a reversal emerges already for the periodically rocked thermal ratchet in absence of coupling [3,7]. In the limit of adiabatically slow driving, $\Omega \to 0$, we can apply an adiabatic approximation in order to calculate $\langle \dot{x} \rangle_{st}$ from (3), cf. [7], see Fig. 2(b).

In Fig. 3 we depict the dependence of $\langle \dot{x} \rangle_{st}$ in absence of a bias versus noise strength D for various coupling strengths K: The coupling can induce an increase for the absolute value of the current (e.g. $\langle \dot{x} \rangle_{st}$ at moderate-to-large

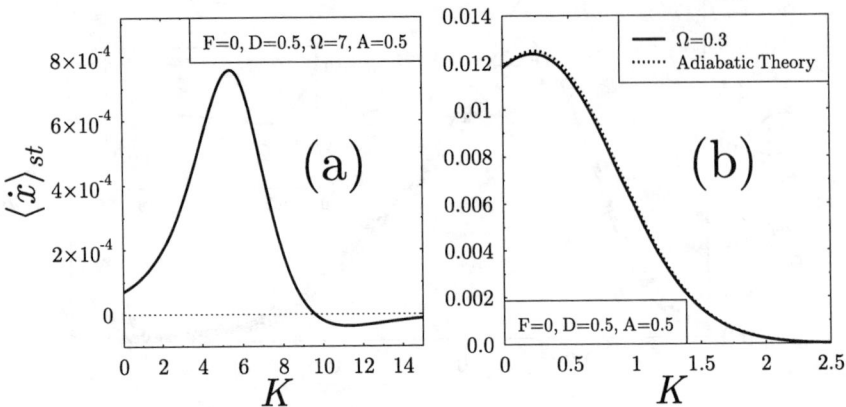

FIGURE 2. Numerically evaluated average particle velocity $\langle \dot{x} \rangle_{st}$ versus coupling strength K for parameter values given in the figures. (a) for non adiabatic driving, (b) for adiabatic rocking (—), which favourably compares with the adiabatic analytical theory ($\cdots$).

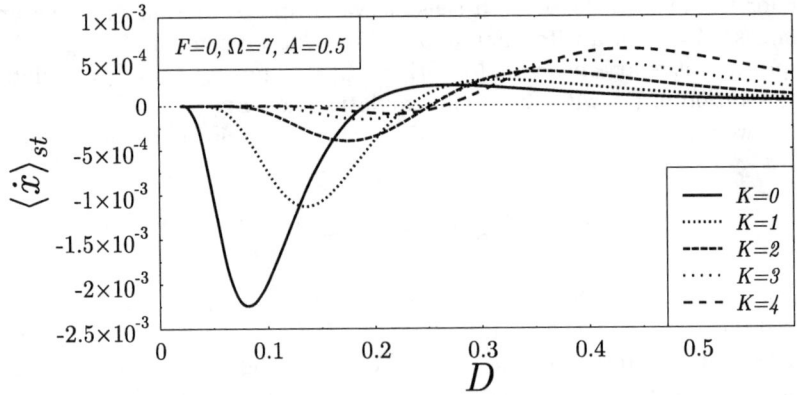

FIGURE 3. The numerically evaluated mean field particle velocity $\langle \dot{x} \rangle_{st}$ is shown versus the Gaussian white noise strength D, for various coupling strengths K.

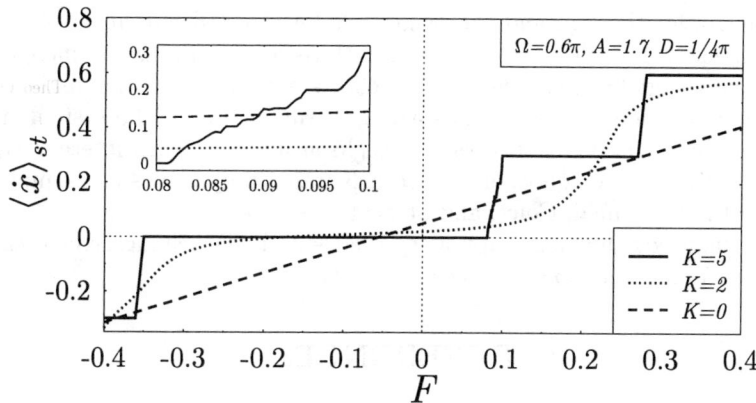

FIGURE 4. The numerically evaluated velocity $\langle \dot{x} \rangle_{st}$ in presence of an external bias force F versus different coupling strengths K. The inset presents an enlargement: The $\langle \dot{x} \rangle_{st}$-$F$ characteristics exhibits a devils staircase for large coupling K.

D in Fig. 3), as well as a decrease (at low noise level D). Increasing the coupling K shifts the point of current reversal to higher D-values.

In order to investigate the efficiency with which particles are performing work in a coupled rocking ratchet we study the mean velocity versus a nonvanishing external bias $F \neq 0$. In [2] we presented results for the velocity-bias characteristic for zero coupling K, which exhibits a devils staircase behavior, being increasingly smoothed out with increasing noise (not shown). In Fig. 4 we depict the load curves for the same parameters as in [2] for various values of coupling strength K. [1] We observe that an increase in K yields a more pronounced step-behavior. The coupling induced, extended plateaus can be utilized for the design of rectifiers operating with Brownian noise. Thus, the magnitude of the coupling strength can be used to control the smoothing level of load curves; a large value of K effectively reduces the bare noise level D.

CONCLUSIONS

In conclusion, we have investigated analytically (adiabatic limit) and numerically the physical role of mutual coupling among Brownian rectifiers that individually act as rocking ratchets. The effects of coupling are multifacetted: The presence of finite coupling K can induce a reversal of velocity; K acts

---

[1] The potential $\tilde{U}(\tilde{x})$ chosen in [2] (denoting all variables in [2] with a tilde) has a different periodicity and height than in this work: Rescaling the variables in [2] by $\tilde{x} = 2\pi x$, $\tilde{U}(\tilde{x}) = 2\pi V(x/2\pi)$, $\tilde{A} = A$, $\tilde{F} = F$, $\tilde{t} = 2\pi t$, $\tilde{\Omega} = \Omega/2\pi$, $\tilde{D} = 2\pi D$ yields $\left\langle \frac{d\tilde{x}}{d\tilde{t}} \right\rangle_{st} = \left\langle \frac{dx}{dt} \right\rangle_{st} = \langle \dot{x} \rangle_{st}$, when $K = 0$.

as a control for the sign and the magnitude for directed movement of Brownian particles. Its role for the load-characteristics is particularly intriguing: Increasing the global coupling strength effectively diminishes the influence of strong ambient noise forces, such as strong thermal noise D. Our system in (1) can likely be realized with arrays of Josephson junctions with internal asymmetry [2] due to the equivalence with a Kuramoto-dynamics as exemplified recently by Wiesenfeld, Colet, and Strogatz [6].

We acknowledge financial support by the Deutsche Forschungsgemeinschaft, Az. Ha1517/13-1, and BMBF, Az.13N7121/1.

REFERENCES

1. A. Ajdari and J. Prost, C. R. Acad. Sci. Paris **315**, 1635 (1992); M. O. Magnasco, Phys. Rev. Lett. **71**, 1477 (1993); C. R. Doering, W. Horsthemke, and J. Riordan, Phys. Rev. Lett. **72**, 2984 (1994); P. Hänggi and R. Bartussek, in *Nonlinear Physics of Complex Systems, Lecture Notes in Physics*, Vol. **476**, edited by J. Parisi, S. C. Müller, and W. Zimmermann (Springer, Berlin Heidelberg, 1996), pp. 294–308.
2. I. Zapata, R. Bartussek, F. Sols, and P. Hänggi, Phys. Rev. Lett. **77**, 2292 (1996).
3. I. Derényi and T. Vicsek, Phys. Rev. Lett. **75**, 374 (1995).
4. F. Marchesoni, Phys. Rev. Lett. **77**, 2364 (1996).
5. S. Shinomoto and Y. Kuramoto, Prog. Theor. Phys. **75**, 1105 (1986).
6. K. Wiesenfeld, P. Colet, and S. H. Strogatz, Phys. Rev. Lett. **76**, 404 (1996).
7. R. Bartussek, P. Hänggi, and J. G. Kissner, Europhys. Lett. **28**, 459 (1994).

Generation and Processing of Chaotic Signals in Range Measurement Systems

Andreas Bauer

Technical University Dresden
Institute of Fundamentals of Electrical Engineering and Electronics
Mommsenstr. 13, 01069 Dresden, Germany
email: bauer@iee.et.tu-dresden.de

Abstract. In this paper the application of chaotic signals in *continuous wave* (CW) ranging systems is addressed. Different chaos generators that can be applied in this application are presented and analysed with respect to their signal properties.

In the case of a CW ranging system the important requirement is an unambiguous range estimation. If classical correlation processing is applied in the CW ranging system, the capability of the ranging signal for unambiguous range estimation is analysed by means of the *Woodward Ambiguity Function* (WAF). A discrete time chaotic ranging system is presented and it performance with respect to multi user operation (CDMA) is analysed.

CHAOS GENERATORS AS CW RANGING SIGNAL SOURCES

A CW ranging system emits a ranging signal x_t into a medium where targets at radial range from the ranging system r_i, which are possibly moving with radial velocity v_i relative to the ranging system, are present. These signal waves propagate

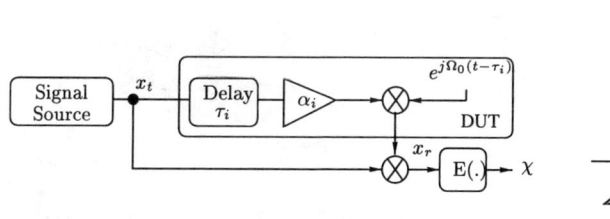

FIGURE 1. System structure of a CW ranging setup with correlation receiver

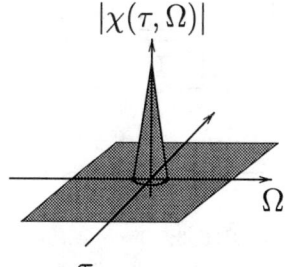

FIGURE 2. Shape of an ideal ambiguity function

with propagation velocity c. The targets reflect shares of this wave back to the ranging system. Due to the target range r_i and the finite propagation velocity c the signal is subject to the delay $\tau_i = \frac{2r_i}{c}$. Moreover if $v_i \neq 0$ the Doppler effect shifts the frequency of x_t by $\Omega_i = \frac{2v_i}{c}\omega$. The received signal x_r is cross-correlated with a copy of $T_{\tau_0}x_t$ where T_{τ_0} denotes the delay operator that delays the signal which it is applied to τ_0 time units. A system representation of these principles is depicted in Fig. 1 with $\tau_0 = 0$ and a linear channel attenuation α_i. The output signal of a correlation based ranging system therefore is:

$$y_i(\tau_i, \Omega_i) = \lim_{t_c \to \infty} \frac{1}{t_c} \int_{t_0}^{t_c} x_t(t) x_{r_i}^*(t) dt \text{ with } x_{r_i}(t) = x_t(t - \tau_i) e^{j\Omega_i(t-\tau_i)} \qquad (1)$$

where the * denotes the complex conjugate.

The quantity $\chi(\tau, \Omega) = \left|\frac{y(\tau,\Omega)}{y(0,0)}\right|$ is the so called Woodward Ambiguity Function [?]. It is a important performance criteria for the suitability of a signal x_t for the range and range rate estimation.

The shape of an ideal WAF is depicted in Fig. 2. It should posses only one significant main lobe, because the appearance of side lobes exceeding a certain threshold level causes ambiguities in the estimation of the target parameters. The remaining uncertainty in the target parameter estimation is as lower as steeper the slopes of the main lobe of the WAF are and as narrower the main lobe is. Since $\chi(\tau = 0, \Omega = 0) = 1$ and the volume under $V(\chi(\tau, \Omega)) = \text{const.} \neq 0$ a WAF that is $\chi(\tau, \Omega) = \delta(\tau, \Omega)$ is impossible. There is in any case a remaining uncertainty in the target parameter estimation.

Chaotic signals are *aperiodic broadband* and *noise-like* signals possessing a sophisticated phase structure. The aperiodicity of the signal causes the autocorrelation (ACF), which is $\chi(\tau, 0)$ to be aperiodic too. The broad bandwidth of the signal causes the ACF to decrease rapidly as τ increases. Moreover, the correlation between the signal spectrum and its Doppler-shifted copies, which is $\chi(0, \Omega)$, has also only one narrow main lobe. This is due to the complicated phase structure of the chaotic ranging signal.

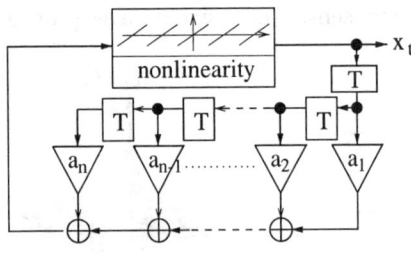
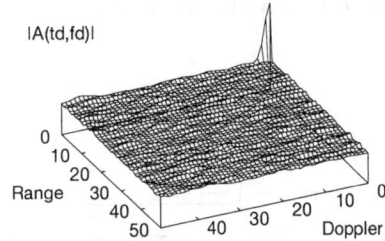

FIGURE 3. Discrete time chaos generator based on digital filter structure

FIGURE 4. WAF of a discrete time chaotic signal

In Fig. 3 a simple discrete time chaos generator based on a digital filter structure is depicted. The nonlinearity is a two's compliment overflow characteristic. Such structures used as chaos generators are thoroughly studied and investigated in [?].

Fig. 4 shows a simulated ambiguity function of a randomly chosen signal section x_t of the chaos generator in Fig. 3. Since the power density spectrum (PDS) of

x_t tends to be white with increasing system dimension n $\chi(\tau,0)$ is an almost $\delta(\tau)$ shaped ACF. But also in the Ω direction $\chi(0,\Omega)$ exhibits almost ideal behaviour. No appearance of significant side lobes can be observed. Therefore, a simultaneous estimation of the target parameters range and velocity without ambiguities is possible when suitable chaotic signals are used.

Comparable behaviour can be expected from a transmission line based continuous time chaos generator as depicted in Fig. 5. Such a chaos generator has been

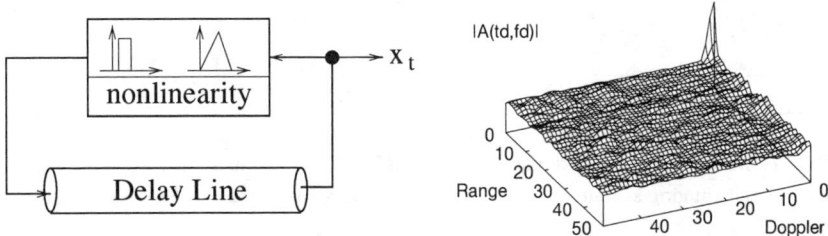

FIGURE 5. Continuous time chaos generator based on a analogue transmission line delay

FIGURE 6. WAF of a continuous time chaotic signal from a infinite dimensional generator

implemented and x_t has been sampled. The bandwidth of x_t is approximately 500MHz. From the sampled signal the WAF has been computed and the result is depicted in Fig. 6. The WAF of this infinite dimensional chaos generator exhibits a shape comparable to the WAF of the discrete time generator. Therefore, these signals can also be used for simultaneous target range and range rate estimation. However, some small side lobes of $\chi(\tau,0)$ which result from periodicities in x_t appear.

In the following only discrete time systems are discussed because of their better WAF and the possibility of the application of simple digital signal processing schemes. Moreover, the discussion is restricted to stationary targets and therefore Doppler processing of chaotic signals is not discussed in this paper.

EXAMPLE BASEBAND MODEL OF A CHAOTIC RANGING SYSTEM

In Fig. 7 a baseband model of a chaos based ranging system is depicted. It contains a chaos generator as introduced in the previous section.

The channel in the baseband model is represented simply by a FIR filter and additive white gaussian noise (AWGN). Therefore the output to the k-th correlator is given by:

$$R_{x_t x_r}(k\tau) = \mathrm{E}(x_r(t)x_t^*(t-k\tau)) = \mathrm{E}\left(\left(\sum_{i=1}^{i_{max}} c_i x_t(t-i\tau) + n_c\right) x_t^*(t-k\tau)\right) \quad (2)$$

The system in Fig. 7 together with the channel model contained in (2) has been simulated. All parameters c_i are zero except for $c_{20} = 0.5$ and $c_{28} = 0.4$ where the index of the coefficient represents the target range and the magnitude of the c_i models the target reflection strength. As a noise source a tent map with half

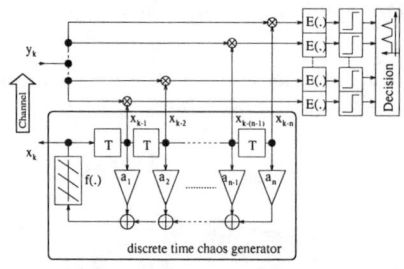

FIGURE 7. Baseband model of a discrete time chaos based ranging system

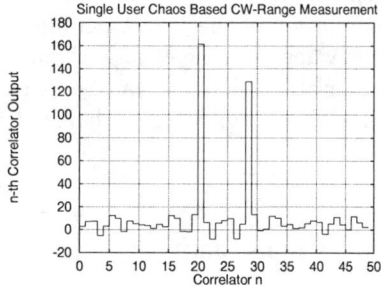

FIGURE 8. Simulated range measurement using the system in Fig. 7

the variance of the ranging signal has been used. As necessary in every practical realisation the expectation E is replaced by a finite averaging. In the simulation example in Fig. 8 the averaging time is 1000. The sound peaks at $k = 20$ and $k = 28$ clearly indicate the presence of targets at ranges according to these k.

REALISATION ISSUES.

For practical realisations a chaos generator as in Fig. 3 has been implemented as an IC [?]. This generator is capable to produce discrete time chaotic signals with prescribed statistical properties at a chip clock of 500kHz.

In order to adapt the chaotic baseband code to the channel $x_t \in (-1, 1)$ is mapped onto a uniform complex code $\mathbf{x_t}$ on the unit circle:

$$\mathbf{x}_t = e^{j\pi x_t} = \cos(\pi x_t) + j\sin(\pi x_t) = \Re(\mathbf{x}_t) + j\Im(\mathbf{x}_t) \quad (3)$$

An IQ-modulator modulates the complex code $\mathbf{x}_t$ onto a carrier of frequency ω_c. Therefore the transmitted signal $x_{ct}(t)$ is obtained:

$$x_{ct}(t) = |\mathbf{x}_t(t)|\cos(\omega_c t + \arg(\mathbf{x}_t(t))) = |\mathbf{x}_t(t)|\cos(\omega_c t + \pi x_t(t)) \quad (4)$$

On the channel this $x_{ct}(t)$ has a constant envelope and appears to be a APSK signal. This IQ modulation scheme retains the linear properties of the channel and therefore allows correlation processing in the baseband. The correlation processing is done with the complex signal obtained on the IQ-demodulator output:

$$\mathbf{x}_r(t) = \mathbf{x}_t(t + \phi_c) + \mathbf{n}_c(t) = e^{j(\pi x_t(t) + \phi_c)} + \mathbf{n}_c(t) \quad (5)$$

where $\mathbf{n}_c$ denotes the complex representation of the channel noise and ϕ_c is a phase error due to the phase difference between the carriers in the modulator and demodulator. Under the assumption that $\mathbf{n}_c$ and $\mathbf{x}_r$ are independent the cross-correlation between $\mathbf{x}_t$ and $\mathbf{x}_r$ gives the expected result:

$$r_{\mathbf{x}_t(\tau)\mathbf{x}_r} = \mathrm{E}(\mathbf{x}_t(t-\tau)\mathbf{x}_r^*(t)) = E_x e^{j\phi_c}\delta(\tau - \tau_c) \quad (6)$$

where τ_c is the signal delay due to the target range and E_x is the the power of $\mathbf{x}_t$.

If the signal is to be delayed longer than than $n\tau$ where n is the dimension of the generator, additional delay can be introduced either by sample and hold steps, or by means of binary delay exploiting nonlinear digital waveform coding [?]. The latter method uses the generator structure in order to reconstruct an approximation of the chaotic signal from a binary signal representing the symbolic dynamics of the chaos generator. The principle of a nonlinear digital waveform coding based ranging baseband system is depicted in Fig. 9.

MULTI USER CAPABLE CHAOTIC RANGING SYSTEM

A chaotic signal is not only an excellent ranging signal. It can also be considered a code sequence. Since chaos generators have the favourable property to produce cross-uncorrelated signals, even if identical chaos generators are used, the chaotic signals can be used for a *code division* (CDMA) on the measurement channel. A

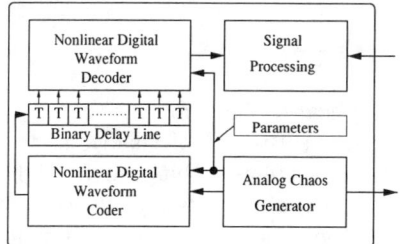

FIGURE 9. Baseband model of a chaotic ranging system using nonlinear digital waveform coding

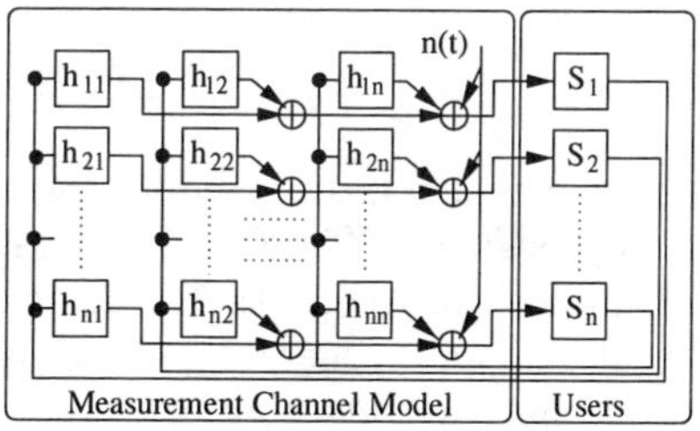

FIGURE 10. Baseband model of a multi user range measurement setup based on chaotic codes

baseband model for this situation when n users measure in a common environment (channel) is depicted in Fig. 10. In this figure every box S_i represents a system as depicted in Fig. 7. The h_{ij} resembles the FIR channel model between the transmitting system S_j and the receiving system S_i. $n(t)$ denotes the channel noise. Therfore the received signal of the k-th measurement system is:

$$x_{rk}(t) = \sum_{j=1}^{n} h_{kj} * x_{tj}(t) + n(t) \qquad (7)$$

where $*$ denotes the convolution. The CDMA capability of the chaotic codes allows the k-th system to estimate the ranges of reflecting targets in the channel h_{kk} regardless the cross talking of the other S_i via h_{ki} and the noise $n(t)$.

The error rate of a system as in Fig. 10 has been estimated by simulation means. The setup has been simplified such that all $h_{ij} = m, i \neq j$ and $h_{kk} = a_l$ with randomly chosen $l > 0$ representing the range and $a_l = 1$. The simulation parameters are the number of users n and the cross talking coefficient m. Both, a target miss

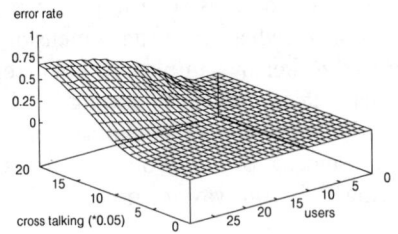 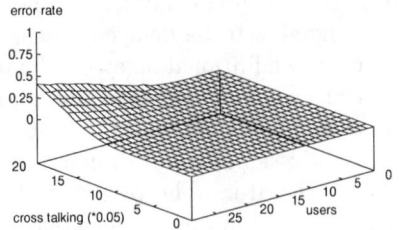

FIGURE 11. CDMA Simulation Results, Correlation Time 100 Samples

FIGURE 12. CDMA Simulation Results, Correlation Time 200 Samples

and a false alarm are counted as an error. For finite correlation time the error rate depends on the number of correlated samples T_c. Therefore two results for $T_c = 100$ and $T_c = 200$ are plotted in Fig. 11 and in Fig. 11 respectively.

As expected, the chaotic codes allow multi user ranging. The error rate increases with the number of users and the cross talking between the users but can be reduced by increasing T_c. In the simulated setup the error rate tends to zero for $T_c > 500$.

CONCLUSIONS

Ranging is a relatively new prospective application field for chaos generators and their signals. The CDMA capability suggests the use of chaotic signals in mass applications such as car collision warning systems. But also in high-end applications chaos generators can be applied. Multi rate processing allows Doppler shift estimation and chaotic beam forming is possible too. Another impact of chaotic signals can be expected if higher order signal processing is used. This is due to the fact, that discrete time chaos generators can be designed to produce signal with prescribed higher order statistics.

ACKNOWLEDGEMENTS

The author would like to thank the Deutsche Forschungsgemeinschaft (DFG) and the Willkomm Stiftung for their support. Furthermore, thanks to Prof. W. Schwarz and the Dresden Nonlinear Systems Research Group.

Spatio-Temporal Simulation in Subthreshold CMOS

John Neeley and John G. Harris
Computational Neuro-Engineering Laboratory
University of Florida

Abstract. This paper reports on the design and chip measurements from a CMOS chaotic oscillator operating by itself and connected in a ring of four similar oscillators. The oscillator is autonomous and generates signals with three state variables analogous to Chua's circuit. For commensurate bandwidth, this design utilizes currents and capacitors over 200 times smaller than above threshold CMOS realizations. Also, all circuit elements are on chip. The resulting voltage-controlled bifurcation parameters simplify exploration of the circuit's dynamics, alleviating the need to interchange physical components. This combination of reduced size and variable parameters make the design suitable for single-chip VLSI synthesis of higher dimensional chaotic circuits, including coupled maps generating spatio-temporal chaos and systems exploiting chaos synchronization.

I INTRODUCTION

Electronic circuits exhibiting well-defined bifurcations and chaotic behavior may be utilized in networks that produce spatial and temporal chaos and other systems exploiting their unique spectra. Chua's circuit [2] [6] [7] is the simplest autonomous circuit that exhibits chaos and has become a research vehicle and circuit building block for studying chaotic dynamics. Chua's circuit has been implemented with discrete components [5], as an integrated circuit with an external resistor [3], and with all components completely integrated [12]. Since not all phenomena of the real circuit may be explained by a piecewise-linear model [9], Chua's circuit has also been implemented with a smooth (cubic) nonlinearity in discrete components [13].

Unlike previous voltage-mode IC simulations, this design utilizes CMOS circuits operating in weak-inversion (subthreshold). Currents in the subthreshold region of CMOS operation can be over two orders of magnitude smaller than currents in above threshold CMOS. The immediate result of reduced current is a proportional decrease in the capacitor size necessary for IC realizations of dynamic simulation. Since capacitor area dominated previous voltage-mode

IC realizations, this decrease in capacitor size enables a simple extension to higher dimensional IC-based simulation.

In this paper, we summarize the implementation details, simulations, and experimental data of a subthreshold CMOS design exhibiting a predictable bifurcation sequence analogous to that of Chua's circuit. This design has been fabricated and tested both by itself and connected in a ring of four such oscillators. In section II, we describe the design from a network element level, then briefly describe the interactions of these elements from a state-variable perspective. In section III, we show bifurcation sequences for a simulated and experimental attractor. Finally, in section IV, we present experimental results for a ring of four oscillators.

II NETWORK STRUCTURE

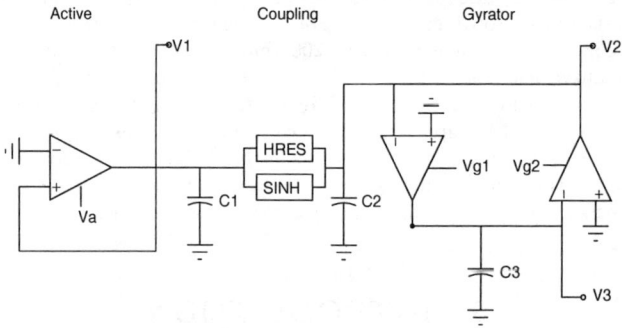

FIGURE 1. *Network schematic of subthreshold IC implementation.*

This design, like existing IC versions of Chua's circuit, is a third-order circuit. A network schematic of the subthreshold design is shown in Figure 1. The network contains a locally active element coupled resistively to a linear oscillator. In the Chua circuit, the nonlinear driving point characteristics of the Chua diode provide three regions of linear dynamic evolution. The evolution is specified by the Jacobian or conductance matrix:

$$J = \begin{bmatrix} G'/C_1 & G_c/C_1 & 0 \\ G_c/C_2 & -G_c/C_2 & G_{L2}/C_2 \\ 0 & -G_{L1}/C_3 & 0 \end{bmatrix} \quad (1)$$

where

$$G' = G_a - G_c \quad (2)$$

G_a and G_c are the transconductances of the active and coupling elements respectively. The Chua diode [2] defines three linear regions for G_a in Chua's circuit. The associated eigenvalues and eigenplanes of these three matrices completely characterize the circuit's dynamics. When the magnitude of V_1 is greater than some voltage threshold (V_{th}) the conductance matrix produces two unstable complex eigenvalues and one stable real eigenvalue, if the magnitude of V_1 is less than V_{th} the three eigenvalues are comprised of two stable complex and one unstable real eigenvalue.

In normal operation, the conductance matrix remains invariant. However, to produce the aforementioned eigenvalues/eigenplanes we draw further attention to the relation between the conductance of the locally active element (G_a), the Chua diode, and the conductance of the passive coupling (G_c), variable resistor between the active element and the gyrator. When the magnitude of V_1 is greater than V_{th}, the conductance of the active element (G_a) is less than the coupling conductance (G_c) ($G_a < G_c, G' < 0$); conversely, if the magnitude of V_1 is less than V_{th}, the conductance of the active element is greater than the passive connection ($G_a > G_c, G' > 0$). In Chua's circuit G_c is constant while G_a varies with respect to V_1. With appropriate scaling of the energy elements, these conditions produce the necessary eigenplanes for chaos.

The subthreshold design utilizes resistive nonlinear connections between a locally active element and a gyrator to provide the nonlinearity necessary for the required dynamics. Thus, G_c varies while G_a is approximately constant. This nonlinearity is a function of the difference $|V_1 - V_2|$ rather than only a function of V_1, but distributes charge between V_1 and V_2 analogous to the Chua's circuit with a cubic nonlinearity. The resistive nonlinearity is odd-symmetric and preserves the above qualitative ratios with conductance $G_a < G_c$ for $|V_1 - V_2| > V_{th}$ and $G_a > G_c$ for $|V_1 - V_2| < V_{th}$. In this way, we produce a nonlinear transconductance matrix having eigenfunctions necessary for chaotic dynamics.

The locally active element is a standard five-transistor transconductance amplifier [10]. The nonlinear coupling between the active element and gyrator are sinh [8] and hres [10] circuits. These circuits produce a voltage input to current output relationship equivalent to the hyperbolic sine and hyperbolic tangent respectively. The realization of these functions requires fourteen transistors each. All elements of the circuit are implemented on chip and require bias voltages, permitting further modification of the standard Chua parameters: a, b, and β [6] [7]. Specifically, the coupling transconductance (G_c) may be modified, affecting a and b, even in such a manner that G_a becomes asymmetric. Furthermore, since the oscillator is implemented as a gyrator or second order transconductance matrix with an inductor replaced with a capacitor, this permits individual variation of another bifurcation parameter β.

III SINGLE CELL: SIMULATED AND MEASURED RESULTS

We have simulated the complete design of this oscillator in the circuit simulation package ANALOG. The α bifurcation sequence is specified by the on-chip capacitors. For the simulation shown below $\alpha = 10$ with the largest capacitor equal to 7.5pF (Previous designs required capacitances in the nF range). A 450μm by 350μm (0.1575 mm^2) prototype was fabricated in 2μm CMOS process. Both the simulated and measured spiral attractor from this chip are shown in Figure 2. Some of the apparent variation of the eigenfunctions can be attributed to quantization error and the time delay between the data acquisition channels.

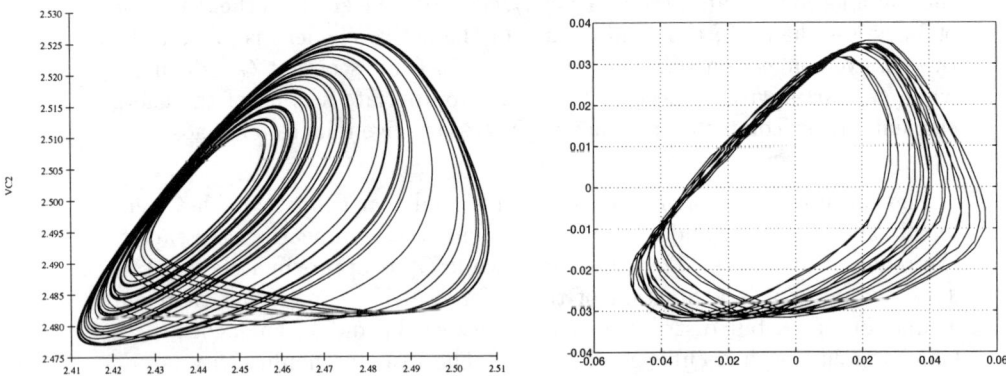

FIGURE 2. *Simulated β spiral. Measured β spiral.*

IV MULTIPLE CELLS: MEASURED RESULTS

One phenomena of coupled chaotic circuits generating considerable interest is synchronization. Synchronization has been demonstrated for piecewise linear Chua's circuits [11], Lorenz oscillators [4], and in other nonautonomous nonlinear systems [1].

The fabricated circle map consists of four oscillators with a parallel combination of hres and sinh elements between adjacent oscillators. Four cells with such coupling fit comfortably on a MOSIS 2mm X 2mm die (TINY chip.) One concern was that the synchronization be dependent on circuit (designed) coupling and not unintentionally couple through the substrate. For similar parameters, the asynchronous Lissjous pattern generated with no (extremely low) intercell coupling, and the synchronous Lissjous pattern with high coupling (greater than approximately one-third of the passive coupling found

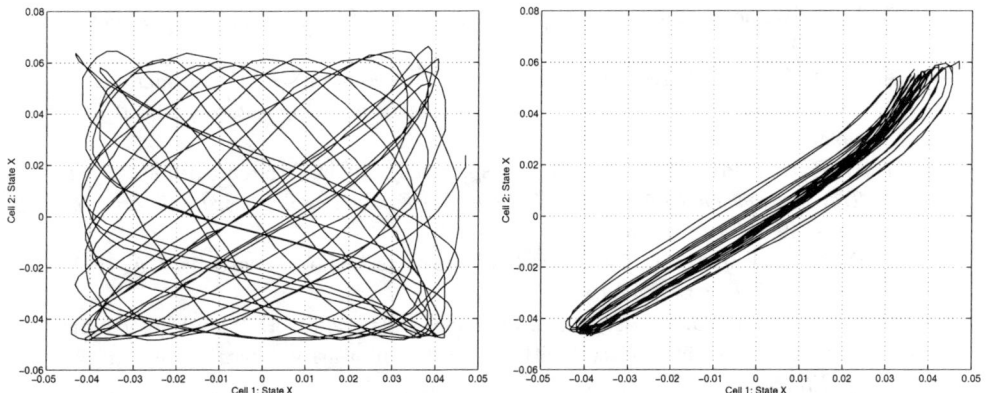

FIGURE 3. *Asynchronous and (Nearly) Synchronous Ring Operation*

within each oscillator) are shown in Figure 3. The plots show that the ring will synchronize with the designed rather than accidental coupling.

V CONCLUSION

This work demonstrates the viability of compact, low-power subthreshold CMOS simulations that serve as robust sources of chaotic signals while incorporating all necessary circuit elements on chip. In this way, we provide a basic circuit elements that can be used as a self-contained building block for massively parallel arrays that exploit desirable aspects of chaos.

Acknowledgments: This work was supported by an NSF CAREER award #MIP-9502307.

REFERENCES

1. Thomas L. Carroll and Louis M. Pecora. Synchronizing nonautonomous chaotic circuits. *IEEE Transactions of Circuits and Systems - II: Analog and Digital Signal Processing*, 40:646–650, 1993.
2. Leon O. Chua. A universal circuit for studying and generating chaos - part I: Routes to chaos. *IEEE Transactions on Circuits and Systems - I: Fundamental Theory and Applications*, 40:732–744, 1993.
3. Jose M. Cruz An IC chip of Chua's circuit. *IEEE Transactions on Circuits and Systems - II: Analog and Digital Signal Processing*, 40:614–625, 1993.
4. Kevin M. Cuomo, Alan V. Oppenheim, and Steven R. Strogatz. Synchronization of lorenz-based chaotic circuits with applications to communications. *IEEE Transactions on Circuits and Systems - II: Analog and Digital Signal Processing*, 40:626–633, 1993.

5. M. P. Kennedy. Robust op amp realization of Chua's circuit. *Frequenz*, 46:66–80, 1992.
6. M. P. Kennedy. Three steps to chaos - part I: Evolution. *IEEE Transactions on Circuits and Systems - I: Fundamental Theory and Applications*, 40:640–656, 1993.
7. M. P. Kennedy. Three steps to chaos - part II: a Chua's circuit primer. *IEEE Transactions on Circuits and Systems - I: Fundamental Theory and Applications*, 40:657–674, 1993.
8. D. Kerns. PhD thesis, California Institute of Technology, Pasadena, CA, 1991. Dept. of Electrical Engineering.
9. A. I. Khibnik, D. Roose, and L. O. Chua. On periodic orbits and homoclinic bifurcations in Chua's circuit with a smooth nonlinearity. *International Journal of Bifurcation and Chaos*, 3:363–384, 1993.
10. C. Mead. *Analog VLSI and Neural Systems*. Addison-Wesley, 1989.
11. M. J. Ogorzalek. Part i, synchronization. *IEEE Transactions of Circuits and Systems - I: Fundamental Theory and Applications*, 40:693–699, 1993.
12. Angel Rodriguez Vazquez and Manuel Delgado Restituto. Cmos design of chaotic oscillators using state variables: a monolithic chua's circuit. *IEEE Transactions on Circuits and Systems - II: Analog and Digital Signal Processing*, 40:596–613, 1993.
13. Guo-Qun Zhong. Implementation of Chua's circuit with a cubic nonlinearity. *IEEE Transactions on Circuits and Systems - I: Fundamental Theory and Applications*, 41:934–941, 1994.

Non-linear Quantum Mechanical Ground State SQUID Magnetometer Dynamics

R.R.Whiteman, J.Diggins, V.Schöllmann, T.D.Clark,
R.J.Prance, H.Prance, J.F.Ralph* and M.Everitt

Physical Electronics Group, School of Engineering
University of Sussex
Brighton, Sussex BN1 9QT, U.K.
*DRA, Farnborough, U.K.

Abstract. We consider the dynamical behaviour of a coupled SQUID ring-tank circuit system when the ring is in its quantum mechanical ground state. As we show, the underlying quantum evolution of the ring generates a non-linear dynamics in the classical tank circuit. In this sense the SQUID ring-tank circuit is a model system for studying the quantum-classical interface.

Over the years the quantum mechanics of the simple SQUID ring (a thick superconducting ring, inductance Λ, enclosing a weak link of effective capacitance C) has attracted much attention. As a macroscopic object, coupled to a macroscopic circuit apparatus, it offers perhaps a unique experimental system through which to study the quantum-classical interface. The standard quantum description of the ring is made in terms of the integrated electromagnetic field variables Φ, the magnetic flux through the ring, and Q, the (total) displacement flux threading between the electrodes of the weak link, such that $[\Phi, Q] = i\hbar$ with $\Delta\Phi\Delta Q \geq \hbar/2$. The time independent SQUID ring Hamiltonian then takes the form[1] $H = \frac{Q^2}{2C} + \frac{(\Phi-\Phi_x)^2}{2\Lambda} - \hbar\nu\cos\left(2\pi\frac{\Phi}{\Phi_o}\right)$, where $\hbar\nu/2$ is the pair tunnelling matrix element (for a weak link critical current $I_c = 2e\nu$), Φ_x is the applied magnetic flux, $\Phi_o = h/2e$ and C acts as the effective mass of the SQUID ring. Given typical experimental values of $\hbar\nu$ and Λ, a value of C in the range of 5 to $1 \times 10^{-16} F$ is needed for quantum properties to become manifest at a few K. Using the above Hamiltonian we can solve the time independent Schrödinger equation (TISE) to yield the energy eigenvalues $E_\kappa(\Phi_x)$ for SQUID states $\kappa = 0$ (ground), $\kappa = 1$ (first excited state), etc., each Φ_o-periodic in Φ_x. These small values of capacitance can be attained

CP411, *Applied Nonlinear Dynamics and Stochastic Systems Near the Millenium*
edited by J. B. Kadtke and A. Bulsara
© 1997 The American Institute of Physics 1-56396-736-7/97/$10.00

(and measured) straightforwardly using niobium point contact weak links enclosed in monolithic niobium (Zimmerman-type) SQUID rings. We also note that by adopting a more complete quantum electrodynamic description of a SQUID ring[2], it can be shown that any geometric capacitances of the ring are in series with the capacitance of the weak link, meaning that for the rings we use, the latter capacitance always dominates. In figure 1 we show, as an

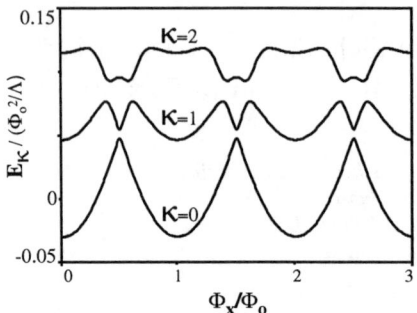

Figure 1. *The first three SQUID energy levels found by solving the time independent Schrödinger equation with $\hbar\omega_o=0.043\Phi_o^2/\Lambda$ and $\hbar\nu=0.070\Phi_o^2/\Lambda$.*

Figure 2. *The SQUID potential $U(\Phi,\Phi_x)$ at $\Phi_x=n\Phi_o$, with the wavefunctions of the first three states. ($\kappa=1,2$ shown vertically offset).*

example, the first three SQUID ring energies, found by solving the TISE[1] for easily accessible ring parameters [here, where $\hbar\omega_o = \hbar/\sqrt{\Lambda C} = 0.043\Phi_o^2/\Lambda$ for $\Lambda = 3 \times 10^{-10}H$, $C = 1 \times 10^{-16}F$, and $\hbar\nu = 0.07\Phi_o^2/\Lambda$, i.e. $I_c = 3\mu A$]. In figure 2 we show the wave function $\Psi_\kappa(\Phi, \Phi_x)$ of the ring for these first three states at a particular value of Φ_x ($n\Phi_o$, n integer) together with the SQUID potential $U(\Phi, \Phi_x)$ at this value of flux. Now, since the SQUID ring wavefunction contains all physically accessible information concerning its behaviour, it would appear desirable to preserve its structure in any experimental technique used to probe the ring. Since the ring is a quantum *circuit* object this is straightforward as is clear when we take derivatives of its energy with respect to Φ_x. Thus, $-\partial E_\kappa(\Phi_x)/\partial\Phi_x$ gives the ring screening supercurrent $\langle I_s(\Phi_x)\rangle_\kappa$ while $-\Lambda\partial^2 E_\kappa(\Phi_x)/\partial\Phi^2$ yields the ring magnetic susceptibility $\chi_\kappa(\Phi_x)$. In general these SQUID rings are probed via an ac resonant circuit. Almost invariably (as here) this takes the form of a radio frequency (rf), parallel LC (tank) circuit, inductively coupled to the ring. In real experimental situations this coupling is never negligible. This means that the $\langle I_s\rangle_\kappa$ in the ring due to a time varying magnetic flux φ in the tank circuit coil feeds back into the tank circuit, so changing its dynamical behaviour. As we shall see, this is the route whereby the quantum evolution of the SQUID ring is monitored. The equation of motion for this coupled system is[3]

$$C_T\ddot{\varphi} + \frac{\dot{\varphi}}{R_T} + \frac{\varphi}{L_T} = I_{in}(t) + \mu\langle I_s(\Phi_{xstat} + \mu\varphi)\rangle_\kappa \tag{1}$$

where L_T and C_T are the tank circuit inductance and capacitance, respectively, R_T is the resistance of the tank circuit at (parallel) resonance, $\mu\,(=\Lambda/L_T)$ is the fraction of rf flux coupling between the SQUID ring and the tank circuit coil, $I_{in}(t)$ is the drive current which is used to excite the tank circuit and Φ_{xstat} is the static flux applied to the ring. As we have demonstrated previously[4], the correspondence between the solutions of (1) and experiment can be extremely good.

In our work we make a distinction between weakly and strongly driven situations, i.e. where $\mu\varphi$ is either very small or comparable with Φ_0. In both regimes the screening current response in the ring manifests itself in the growth of a classical non-linear dynamics in the tank circuit which constitutes a record of the underlying quantum evolution of the SQUID ring. We note that, although this non-linear dynamics is derived from the functional form of $\langle I_s(\Phi_x)\rangle_\kappa$, neither this, nor the underlying SQUID energy, is measured in any conventional sense. However, the development of this (classical) non-linear dynamics is a record from which we can infer the underlying quantum behaviour in the SQUID ring. For this particular system, therefore, we might adopt the term *recordable* when considering the problem of the measurement of a quantum object by a classical apparatus.

This view that the quantum object (the SQUID ring) renormalizes its classical environment (the tank circuit), and in so doing creates a record of its own quantum evolution, turns convention on its head, i.e. there is no projection; the SQUID ring continuously maintains its superposition state. This is best illustrated by example. As we have shown[1], the non-linear form of $\langle I_s(\Phi_x)\rangle_\kappa$ [or $\chi_\kappa(\Phi_x)$] acts to produce shifts in the resonant frequency of the coupled system. For arbitrary rf flux coupling $K = \sqrt{M^2/\Lambda L_T}$, where M is the ring-tank circuit mutual inductance, the dynamical behaviour of the system has to be found by solving (1). The general result is that both the peak frequency $f_{r\kappa}$ and the peak amplitude $A_{r\kappa}$ of the system resonance change as a function of the bias flux Φ_{xstat}. In an experiment these changes with bias flux constitute the (if required, permanent) record of the processes occurring at the quantum level in the SQUID ring.

Experimentally, frequency (and amplitude) changes are followed on spectrum analyzers. With commercial analyzers the technique is to use the tracking generator output of the analyzer (producing a fixed current drive over a preset range of frequencies) to excite the tank circuit and hence the SQUID ring. However, commercial machines only sweep from low to high frequency. For much of the work reported here we needed to sweep both up and down in frequency. We achieved this by using a spectrum analyzer of our own design[4]. The voltage response across the tank circuit was first amplified at 4.2K by a very low noise GaAsFET amplifier, with subsequent amplification and detection being provided by a low noise, room temperature receiver. The signal was then fed either into the commercial analyzer or into a low noise linear mixer in our system, followed by second order low pass filter to set the measurement bandwidth. In our analyzer the variable frequency rf current source and the data acquisition from the low pass filter were controlled by a purpose designed LabView virtual instrument, allowing us to choose the direction(s) of the frequency sweep. As stated, the SQUID rings used were of the niobium point contact type, with the weak link made in situ at 4.2K. Using this technique SQUID rings with almost sawtooth screening current responses to applied flux (corresponding to the ground state in figure 1) could be made with ease[4]. As

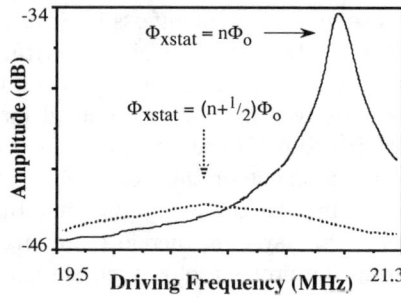

Figure 3. Experimental weakly driven resonance lineshapes for $\Phi_{xstat}=n\Phi_o$ and $(n+1/2)\Phi_o$.

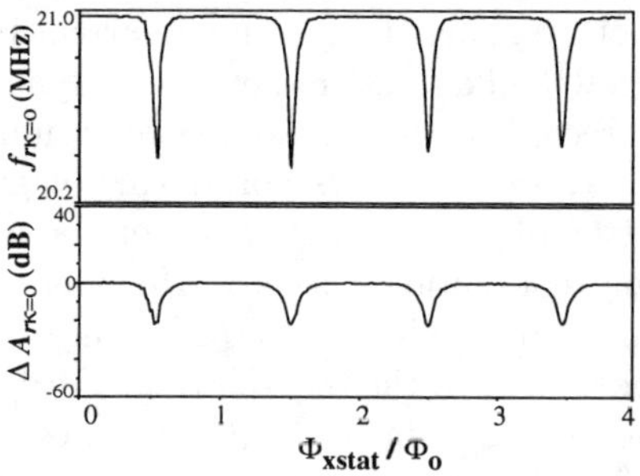

Figure 4. Experimental variation of the peak frequency and amplitude of the coupled system resonance as a function Φ_{xstat} for the weakly driven case.

we have discussed above, the changing quantum state of a SQUID ring can be probed by following the renormalized dynamics of a coupled tank circuit. By way of illustration, we show in figure 3, for $\mu\varphi = 0.014\Phi_o$, the effect on the resonance lineshape of an experimental ground state SQUID ring-rf tank circuit system when the static bias flux Φ_{xstat} changes from $n\Phi_o$ to $(n+1/2)\Phi_o$. Here, $K^2 = 0.038$, $\omega_{rf}/2\pi = 20.9 MHz$ and the system quality factor (Q) at integer bias flux is approximately 1000. As can be seen, this lineshape is altered quite drastically (the experimental record) at half integer flux bias, where $\chi_{\kappa=0}$ is a maximum. The examples given in figure 3 are, of course, for just two values of Φ_{xstat}. Much more information can be extracted by plotting both $f_{r\kappa=0}$ and $A_{r\kappa=0}$ over several Φ_o periods of Φ_{xstat}. In figure 4 we present these plots for the system of figure 3. It is apparent that $f_{r\kappa=0}$ peaks negatively sharply at $\Phi_{xstat} = (n + 1/2)\Phi_o$, indicating both a very strong non-linearity and an almost sawtooth screening current response to external flux. This current $\langle I_s(\Phi_x)\rangle_{\kappa=0}$ can be inferred from the $f_{r\kappa=0}(\Phi_{xstat})$ pattern of figure 4 by solving (1), which is the course we have adopted here. With this $\langle I_s(\Phi_x)\rangle_{\kappa=0}$ inferred from the weakly driven limit, we can then solve (1) for the strongly driven regime, using the same screening current pattern. Again, for the ring-tank circuit system of figures 3 and 4, we show in figure 5 the experimental resonance lineshapes (using our own analyzer), for three different values of $\Phi_{xstat}[= n\Phi_o, (n+1/2)\Phi_o, (n-1/4)\Phi_o]$, for $\mu\varphi = 0.7\Phi_o$. To the best of our knowledge, the SQUID fold bifurcations shown in figure 5 have never been observed before. In particular, the opposed bifurcations shown in figure 5c appear to be unique. The predictive power of our model for this quantum-classical system can be seen from the theoretical resonance curves of figure 6. These resonance curves were found by solving (1) using the measured system circuit parameters together with the inferred $\langle I_s(\Phi_x)\rangle_{\kappa=0}$, for the three flux states of figure 5. The correspondence between experiment and theory is clear.

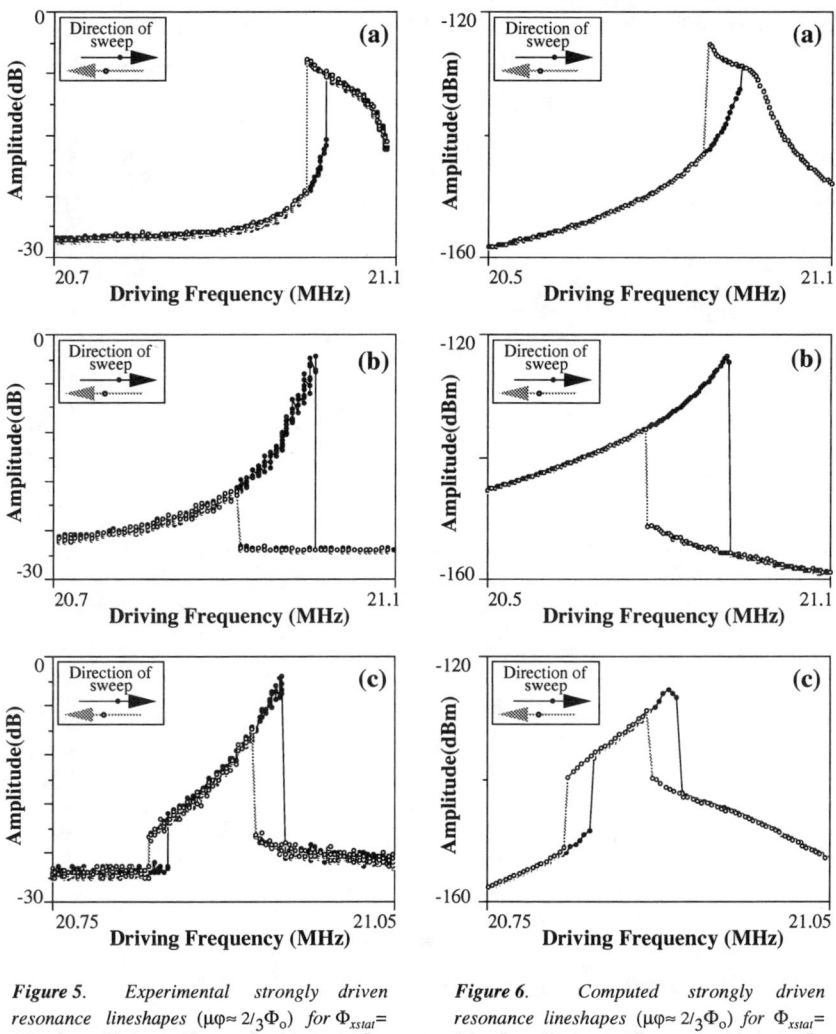

Figure 5. Experimental strongly driven resonance lineshapes ($\mu\varphi \approx 2/3\Phi_o$) for $\Phi_{xstat}=$ *(a)* $n\Phi_o$, *(b)* $(n+1/2)\Phi_o$ and *(c)* $(n-1/4)\Phi_o$.

Figure 6. Computed strongly driven resonance lineshapes ($\mu\varphi \approx 2/3\Phi_o$) for $\Phi_{xstat}=$ *(a)* $n\Phi_o$, *(b)* $(n+1/2)\Phi_o$ and *(c)* $(n-1/4)\Phi_o$.

Reprinted from *Physics Letters A*, R. Whiteman, "Opposed (Hammerhead) Bifurcations in the Resonant Lineshape of a Strongly Driven SQUID Ring-Tank Circuit System," in press 1997 with kind permission of Elsevier Science - NL, Sara Burgerhartstreet 25, 1055 KV Amsterdam, The Netherlands.

REFERENCES

1. R.J.Prance, T.D.Clark, R.Whiteman, J.Diggins, J.F.Ralph, H.Prance, T.P.Spiller, A.Widom and Y.N.Srivastava, Physica **B 203** 381 (1994)
2. J.F.Ralph, T.D.Clark, R.J.Prance, H.Prance and J.Diggins, J.Phys. Cond. Matt., **8** 10573 (1996)
3. J.Diggins, J.F.Ralph, T.P.Spiller, T.D.Clark, H.Prance, R.J.Prance, Phys. Rev. **E 49** 1854 (1994)
4. R.Whiteman, J.Diggins, V.Schöllmann, G.Buckling, T.D.Clark, R.J.Prance, H.Prance, J.F.Ralph and A.Widom, Phys. Lett. **A 226** 275 (1997)

Nonlinear Signal Transformation in the Regime of SR

Igor A. Khovanov and Vadim S. Anishchenko

Department of Physics, Saratov State University, Saratov, 410026, Russia

Abstract. We discuss the peculiarities of stochastic resonance in a bistable system for different levels of the amplitude of a regular force. It is shown that for a certain class of signals the output signal-to-noise ratio, as conventionally defined, can substantially exceed the signal-to-noise ratio at the input. The connection between the phenomenon of switching synchronization by regular force and the improvement of the signal-to-noise ratio is considered.

INTRODUCTION

Recently, the results in the research field of "noise in nonlinear systems" give evidence that noise may enhance a system response and may cause new regimes which do not take place in the absence of noise. Several surprising non-trivial phenomena have been observed and studied. Among them a phenomenon of stochastic resonance (SR) [1,2] is indeed of great interest and importance. The SR is a nonlinear effect which is observed in a wide class of nonlinear systems driven by a combination of periodic signal and noise. The response of a nonlinear system to a periodic signal can be optimized by varying noise intensity. It has been shown, that the output signal-to-noise ratio (SNR) and the coefficient of amplification have maximum values at non-zero noise level.

In the present paper we discuss the phenomenon of SR for different levels of signal amplitude and study the effect of the improvement of SNR in a bistable system.

SR FOR DIFFERENT AMPLITUDE OF SIGNAL

As is well known, in addition to the amplification and the SNR for the manifestation of SR a group of characteristics, which define the process of switchings between states, can be used [3,4]. Such characteristics are the mean switching frequency (MSF) $\langle f \rangle$, the residence time distribution $p(\tau)$ showing the different residense time probability between two subsequent switching events, and the cycle distribution $p(\varphi)$ that estimates the probability of switchings at different phases of a periodic signal [5]. For a very weak signal these characteristics are similar to those in the absence of noise (curve *1* in Fig. 1). Note that this conclusion is justified for the Gaussian noise and for another type of random force may not be valid. For example, in a chaotic bistable system the mechanism of the SR is defined by the coherence between a periodic signal and switchings regardless of the signal amplitude [6].

Therefore, for a very weak periodic signal in a noisy bistable system the SR manifests itself in the amplification and SNR, i.e. the spectrum characteristics. Besides, if the signal is weak, the physical nature of the SR is the same for both harmonic signal and complex nonharmonic signal [7-12]. Thus, for a weak complex signal the addition of noise in a bistable system leads to two effects. The first one is that the response of the bistable system tends to the linear one, i.e. the output SNR tends to the input value. The second one is that the bistable system amplifies the signal regardless of its complexity.

Let us now consider the SR in a bistable system for a strong harmonic signal whose amplitude does not exceed the height of the barrier between the states. It is known, that the increase of the signal amplitude leads to the decrease of the amplification and ΔSNR (the difference between the output and input SNR) and to the growth of the coefficient of nonlinear distortion [7]. The characteristics of switchings undergo the qualitative changes for the strong force (curve *2* in Fig. 1). The structure of the residence time and cycle distributions have a form of peaks. The dependence of the MSF versus noise amplitude becomes anomalous compared to Kramer's law for large signal amplitudes. Under a certain condition [13] a phenomenon of locking of the MSF by external periodic force is observed (curve *3* in Fig. 1). This phenomenon

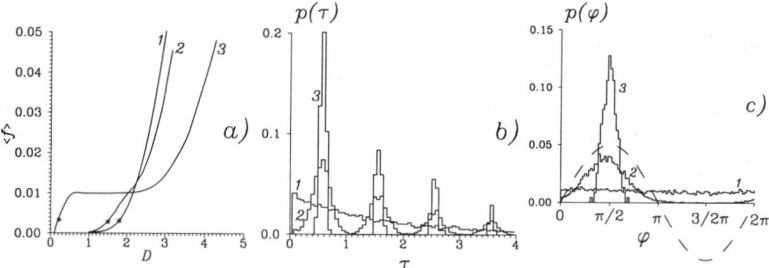

FIGURE 1. The results of numerical simulation of the bistable overdamped oscillator driven by harmonic signal and white noise: $\dot{x} = 5x - x^3 + a\sin\Omega t + \sqrt{D}\xi(t)$. The plots of the MSF as a function of noise intensity (a), $p(\tau)$ (b) and $p(\varphi)$ (c) for three values of the signal amplitude (curve *1* - a=0.05, *2* - a=1, *3* - a=3). The signal frequency is fixed as $\Omega = 0.01$. The distributions in figures (b) and (c) were obtained for noise intensities which are marked by asterisks in figure (a). The time at the residence time distribution is normalized by the period of the signal.

is called also the "switching synchronization" and is as follows: there is an interval of noise intensities in which the MSF remains constant and coincides with the frequency of periodic signal.

Obviously, for an arbitrary (complex) signal with a large amplitude the picture of the SR will be more complex. Nevertheless, for a complex signal we can indicate the changes of the signal form on which a bistable system will react the best way. And we may say that the response of the system to a certain signal is more efficient. Such signals are pulsed ones which have a square shape. Besides, the nonlinear distortion of such signals is minimal since the form of the signal and the response of the bistable system are similar.

IMPROVEMENT OF SNR IN A BISTABLE SYSTEM

The response of a system driven by pulsed signal and showing the SR was considered in a few works [14–16]. The most interesting results were obtained by L.B. Kiss and co-authors [16]. It was found that a level-crossing detector does greatly improve SNR of a periodic spike train (pulsed signal) with additive white noise. The main condition which provides an improvement of SNR is as follows: the duration of the periodically repeated input signal spikes is equal to the spike firing duration of the level-crossing detector. In other words, for the improvement of SNR it is essential to know the parameters of the input signal. Obviously, it is practically impossible in reality.

Notice that the article [17] contains the most detailed consideration of SNR improvement. We present some new results in this field.

Consider a simple bistable system driven by periodic rectangular signal and noise. Let us select the Schmitt trigger, which is an two-state electronic device demonstrating pure hopping dynamics, as a bistable system. The input of this device is an arbitrary function of time $V_{inp}(t)$ but the output has only two possible values $\pm V_{out}$. The ideal Schmitt trigger circuit driven by periodic rectangular signal and noise $\xi(t)$ obeys the equation:

$$y = U_{out}\text{sgn}\left[\Delta U y - a \cdot \text{sgn}(\sin(\Omega t + \varphi_0) - \xi(t)\right], \tag{1}$$

where ΔU refers to threshold level of the trigger, $\xi(t)$ is noise with a cutoff frequency f_c and amplitude σ. Let us select the signal amplitude a is large but it is small to prevent the switching of the trigger in the absence of noise.

The plots of MSF and SNR as a function of noise amplitude σ are shown in Fig. 2[1]. As it is seen, the phenomenon of switching synchronization is observed and in some interval of noise amplitude the output SNR exceeds the input value, i.e. there is the improvement of the signal-to-noise ratio! The physical

---

[1] The SNR is here defined, by analogy with conventional definition [2] used for a harmonic signal, as the ratio between the intensity of the delta function in the power spectrum at Ω to the intensity of the noisy background at the same frequency.

mechanism of the improvement of SNR is connected with the phenomenon of switching synchronization and with the peculiarity of rectangular shape of signal which imposes the moment of switching events in a bistable system. Notice that the improvement of SNR is found only for signal amplitudes when the effect of switching synchronization is possible.

The residence time distributions and the cycle distributions in the synchronization area are shown in Fig. 3, a, b. In the moment of the locking of the MSF the distributions have a single sharp peak (Fig. 3, a). The residence

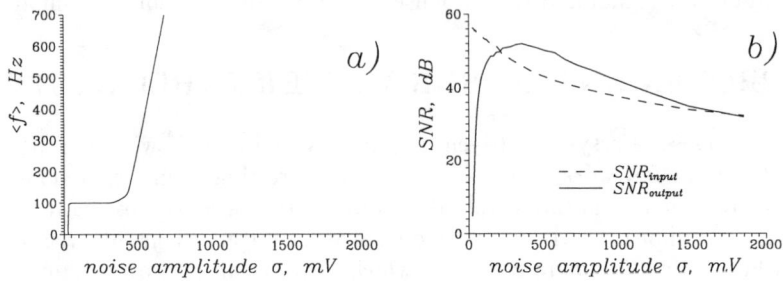

FIGURE 2. The experimental dependencies of MSF (a), input and output SNR (b) versus noise amplitude σ ($\Omega = 2\pi 400 Hz$, $a = 410 mV$, $\Delta U = 470V$, $f_c = 100 kHz$).

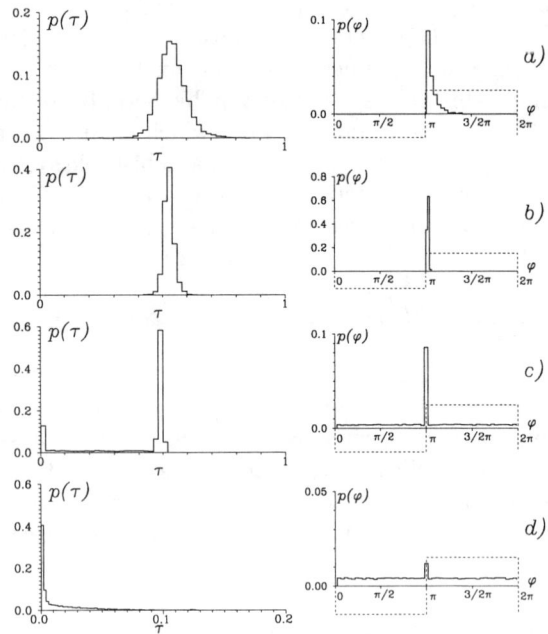

FIGURE 3. The experimental residence time and cycle distributions.

times and the switching phases are confined in a narrow range. The width of the peak is decreased by increasing noise amplitude (Fig. 3, b). Consequently, the regularity of switching events is significantly increased with the growth of noise amplitude. The switching events and the moments of changes of signal shape are practically coincide that leads to the improvement of the SNR.

When the switching synchronization is destroyed (Fig. 3, c), the improvement of the SNR is also observed since the large path of the switching events is occurred synchronously with regular force. If the switching probability is the same for any time (Fig. 3, d), the output SNR becomes smaller than the input one.

Thus, the simple bistable system does improve the signal-to-noise ratio. The improvement of the SNR is observed in a bistable system driven by noise and periodic rectangular signal.

The next step of our consideration is the investigation of the influence of a signal shape on the improvement of the SNR. Let us consider the Schmitt trigger driven by a noise and a signal $x(t)$ which is generated by Van-der-Pol oscillator

$$\ddot{x} - \epsilon(1 - x^2)\dot{x} + \omega^2 x = 0 \qquad (2)$$

As it is well known, the shape of the generated signal depends on the oscillator parameter ϵ and the shape of the signal becomes close to the rectangular shape when increasing the parameter ϵ.

The dependences of ΔSNR on noise amplitude σ are displayed in Fig. 4 for different values of the parameter ϵ. It is seen that for the signal shapes that are similar to harmonic ones the improvement of the SNR does not occur and ΔSNR is increased for the large values of the parameter ϵ. Consequently, the improvement of the SNR takes place only for the certain shapes of signals which are similar to the rectangular shape.

The same results of the improvement of the SNR have been found for the bistable overdamped oscillator that prove the community of the phenomenon of the SNR improvement.

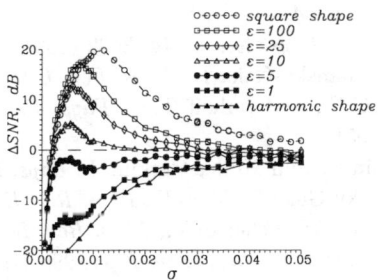

FIGURE 4. The plots of ΔSNR versus σ for different signal shapes obtained by numerical simulation of the equation (1) ($\Omega = 1$, $a = 0.22$, $\Delta U = 0.25$, $f_c = 100$).

CONCLUSION

Our investigations have shown that the physical picture of SR is quite different for different levels of regular signal amplitude. From the point of view of the practical applications of the SR we make the summary:

1) for the case of weak signal the application of the effect of SR for the amplification of a complex signal is efficient;

2) in the regime of large amplitudes the signal transformation is more effective in the regime of switching synchronization;

3) for the signal of the rectangular shape the improvement of the signal-to-noise ratio is observed.

The authors thank Dr. A. Neiman and Prof. F. Moss for fruitful discussions. This work was supported by the grant of DFG and RFFI 436 RUS 113/334/0(R).

REFERENCES

1. Benzi R., Sutera A. and Vulpiani A., *J. Phys. A* **14**, L453 (1981); Benzi R., Parisi G., Sutera A., and Vulpiani A. *Tellus* **34**, 10 (1982); Nicolis C, *Tellus* **34**, 1 (1982).
2. Proc. NATO Adv. Research Workshop on Stochastic Resonance in Phys. and Biology, eds. F. Moss, A. Bulsara, and M.F. Shlesinger *J. Stat. Phys.* **70** (1993); Proc. Int. Workshop on Fluctuations in Physics and Biology: Stochastic Resonance, Signal Processing and Related Phenomena, eds. R. Mannella, and P.V.E. McClintock, *Il Nuovo Cim. D* **17** (1995);
3. Zhou T., Moss F., and Jung P., *Phys. Rev. A* **42**, 3161 (1990).
4. Gammaitoni L., Marchesoni F., Santucci S., *Phys. Rev. Lett.* **74**, 1052 (1995).
5. French A.S., Holden A.V., and Stein R.B., *Kybernetik* **11**, 15 (1972).
6. Khovanov I., Anishchenko V., *Tech. Phys. Lett.* **2**, 854 (1996).
7. Anishchenko V., Neiman A., Safonova M., Khovanov I., in: *Chaos and Nonlinear Mechanics: Proc. Euromech Colloquium* Eds. T. Kapitaniak, J. Brindley, p.45-53 (World Scientific, 1995); Anishchenko V., Safonova M., and Chua L.O., *Int. J. Bifurc. and Chaos* **4**, 441 (1994).
8. Jung P., and Hänggi P., *Phys. Rev. A* **44**, 8032 (1991).
9. Neiman A., and Schimansky-Geier L., *Phys. Rev. Lett.* **72**, 2988 (1994).
10. Collins J.J., Chow C.C., and Imhoff T.T., *Phys. Rev. E* **52**, R3321 (1995); *Nature* **376**, 236 (1995).
11. Chivalo D.R., Longtin A., and Müller-Gerking J., *Phys. Rev. E* **55**, 1798 (1997).
12. Neiman A., Schimansky-Geier L., Moss F., *Phys. Rev. E* **56**, (1997) (to appear).
13. Shulgin B., Neiman A., Anishchenko V., *Phys. Rev. Let.* **75**, 4175 (1995).
14. Morillo M., Gomezordonez J., *Phys. Rev. E* **51**, 999 (1995).
15. Chapeaublondeau F., Godivier X., Chambet N., *Phys. Rev. E* **53**, 1273 (1996).
16. Loerincz K., Gingl Z., Kiss L.B., *Phys. Let. A* **224**, 63 (1996).
17. Inchiosa M.E., Bulsara A.R. *Phys. Rev. E* **52**, 327 (1995).

NEUROSCIENCE AND BIOMEDICAL ENGINEERING

Noise in Any Frequency Range Can Enhance Information Transmission in a Sensory Neuron

Jacob E. Levin

Department of Brain and Cognitive Sciences
Massachusetts Institute of Technology, Cambridge, MA 02139

Abstract. The effect of noise on the neural encoding of broadband signals was investigated in the cricket cercal system, a mechanosensory system sensitive to small near-field air particle disturbances. Known air current stimuli were presented to the cricket through audio speakers in a controlled environment in a variety of background noise conditions. Spike trains from the second layer of neuronal processing, the primary sensory interneurons, were recorded with intracellular Electrodes and the performance of these neurons characterized with the tools of information theory. SNR, mutual information rates, and other measures of encoding accuracy were calculated for single frequency, narrowband, and broadband signals over the entire amplitude sensitivity range of the cells, in the presence of uncorrelated noise background also spanning the cells' frequency and amplitude sensitivity range. Significant enhancements of transmitted information through the addition of external noise were observed regardless of the frequency range of either the signal or noise waveforms, provided both were within the operating range of the cell. Considerable improvements in signal encoding were observed for almost an entire order of magnitude of near-threshold signal amplitudes. This included sinusoidal signals embedded in broadband white noise, broadband signals in broadband noise, and even broadband signals presented with narrowband noise in a completely non-overlapping frequency range. The noise related increases in mutual information rate for broadband signals were as high as 150%, and up to 600% increases in SNR were observed for sinusoidal signals. Additionally, it was shown that the amount of information about the signal carried, on average, by each spike was INCREASED for small signals when presented with noise - implying that added input noise can, in certain situations, actually improve the accuracy of the encoding process itself.

INTRODUCTION

Traditionally in signal analysis noise is considered detrimental to the process of signal encoding, and merely a necessary evil to be avoided. Biological sensory systems, however, are complex, non-linear information transmission channels, and in certain signal and noise regimes can exhibit some initially non-intuitive behavior. It has been reported in a number of physical and biological systems that for sinusoidal input signals the output signal-to-noise ratio (SNR) of the system can actually be increased by the addition of a certain amount of broadband white noise input. This effect, termed stochastic resonance (SR), has previously to 1996 been investigated experimentally only for sinusoidal signals embedded in a broadband noise background. Recently, two experimental studies in sensory systems have demonstrated a similar noise-induced enhancement of output SNR for broadband signals in the presence of additive external noise [1,2]. We describe here some of the key results of one of those studies. Specifically, that additive noise anywhere within a neuron's frequency sensitivity range can serve to enhance the cell's encoding of a small amplitude input signal, provided that the signal and noise amplitudes are in the appropriate range.

Methods

Intracellular recordings were made from the primary ascending sensory interneurons (10-2 and 10-3) of adult female crickets, *Acheta domestica* [3,5]. Known sinusoidal and white noise signal waveforms of a variety of bandwidths were presented at a range of amplitudes spanning the amplitude operating range of the cells, in the presence of a similar range of uncorrelated white noise background waveforms, also in a variety of bandwidths. Sine signals embedded in broadband noise, broadband signals with broadband noise, and several combinations of narrowband signal with narrowband noise, including partially and completely non-overlapping frequency ranges were analyzed. Signal and noise waveform presentations were randomized, and multiple presentations of each signal/noise pair were made over each experiment in order to account for neuronal variability. Sinusoidal signal experiments were analyzed in the traditional manner, by calculating SNR from the output power spectra. White noise experiments were analyzed by calculating the amount of information contained in the evoked spike train about the particular signal waveform, known as the transinformation or mutual information, which is proportional to both the SNR and the normalized coherence function of the evoked spiking response with the signal waveform [3][1]. The use of information theory to characterize multifrequency signal encoding is a natural choice, and becoming popular in the analysis of neural systems, as measures such as the SNR derived from simple power spectra cannot be used [1,6,7].

---

1)

Transinformation, $T = \int_0^\infty df \log_2\left(\frac{1}{1-\gamma(f)^2}\right)$, where $\gamma^2 = \frac{\langle S^*(f)R(f)\rangle \langle R^*(f)S(f)\rangle}{\langle S^2(f)\rangle \langle R^2(f)\rangle}$

is the coherence function between the stimulus S and the spiking response R at frequency f.

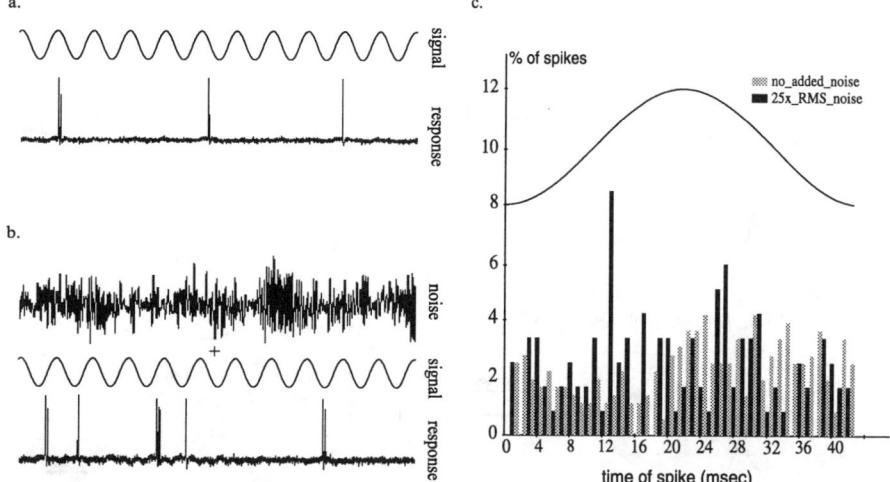

FIGURE 1. Effect of added noise on phase locking response to a perithreshold signal a) Segment of spike train evoked by a very low amplitude 23Hz sine wave stimulus presented alone. b) Same segment of low amplitude 23Hz sine wave presented with a broadband noise background at 25 times the RMS amplitude. c) Histograms of spike occurrence times relative to the 23Hz driving signal over the entire experiment with & without added noise, 1 msec bins. Note the peaks in the added noise case near the maximal slope of the signal waveform.

RESULTS

The standard SR effect [9] was replicated for the cricket cercal sensory interneurons: for small amplitude sinusoidal driving signals output SNR increased as a function of added white noise amplitude to a maximum, then fell beyond the SR peak (see [1]). In addition to the expected peak in the SNR, evidence that the timing accuracy of individual spikes was enhanced was also seen. Like many sensory neurons, 10-2 and 10-3 encode input signals by phase locking to their waveforms. For very weak sinusoidal signals, phase locking is enhanced significantly by the addition of broadband white noise, as can be seen in Fig.1. This enhancement is primarily due to a fraction of the spikes that phase lock extremely well in the presence of noise, and are not present in the no noise case.

For the broadband signal experiments an SR type effect was also observed for a range of low signal amplitudes, spanning approximately an order of magnitude in signal strength. Fig.2a shows the mutual information rate (directly proportional to the SNR) as a function of noise amplitude for three very low amplitudes of broadband (5-400Hz) signal presented with broadband (5-400Hz) noise. Note the peak in the information rate at a non-zero input noise amplitude, as well as the fact that this peak shifts to higher noise levels

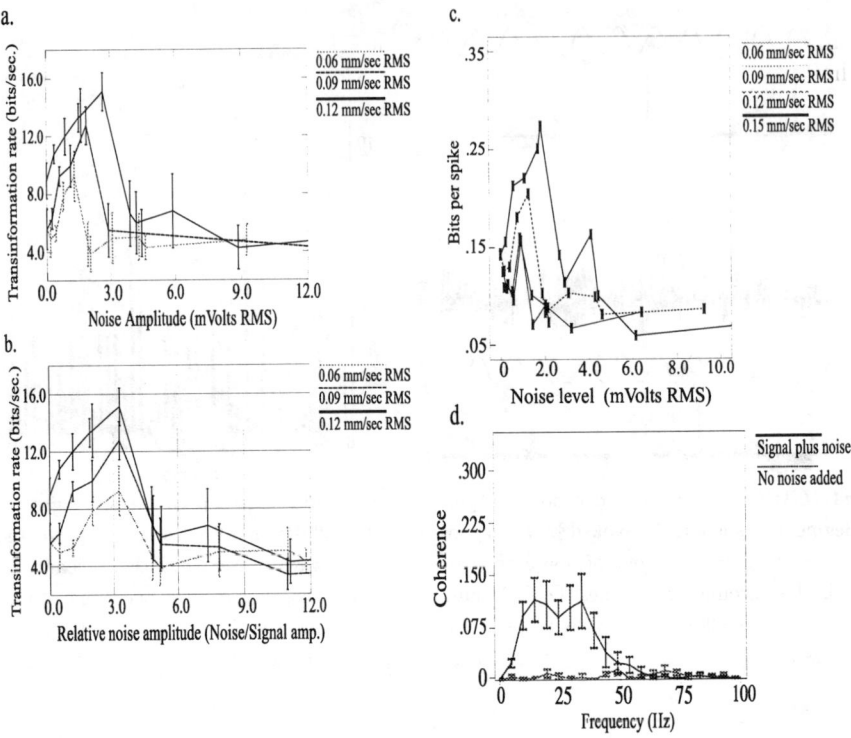

FIGURE 2. Effect of broadband noise on the encoding of broadband signals. a) Information rate as a function of input noise amplitude for three very low signal amplitudes. b) Same data as a), plotted as a function of *relative* noise amplitude, that is, noise amplitude normalized by signal amplitude for each signal strength. c) Bits of information carried per spike, on average, for four low signal amplitudes. d) Example of the coherence of the output spike train with the signal with no added noise and at the SR peak.

for higher signals. This shift is directly opposite to predictions of SR theory (i.e. [9,10]), and can be understood as a consequence of the biological phenomenon of adaptation. This can be seen more clearly in Fig.2b, where info rate is plotted vs. noise in units of the stimulus. The SR peak occurs at a constant noise/signal ratio, implying that the cell has adapted its threshold to the overall input power. For higher signal amplitudes, where encoding is good, added noise serves only to degrade signal encoding (see [1]). To quantify the effect of additive noise on spike placement accuracy, the average number of bits of information carried per spike, calculated by dividing the mutual information by the spiking rate in each situation, is plotted in Fig2c. A similar peak is seen here as well - a certain amount of added noise serves to maximize information carried per spike. As with the sine signal case, spiking accuracy has been improved by noise.

Contrary to predictions of SR theories [8,9] and so-called 'ASR' [4], enhancements to signal encoding were observed across most or all of the frequency operating range of all cells studied, in addition to the clear peak in encoding vs. noise, as evidenced by enhancements of the coherence function, plotted for one cell in Fig.2d. Thus, it is more than just "a non-linear cooperative effect" at a single frequency, i.e. the Kramer's rate. A 'resonance' exists, but it is as broad as the frequency response of the cell.

Several similar experiments were performed (9 cells in 8 animals) with white noise stimuli bandpassed so that signal and noise frequency ranges were partially or completely non-overlapping. In all situations, an SR like effect was observed: Addition of a certain amount of noise served to maximize the information carried by the spike train about the stimulus. Although the level at which this effect was maximized varied depending on the specific frequency ranges employed on account of the differential frequency sensitivity of the cells, a maximum was seen in every case. As with the fully overlapping broadband case, the encoding enhancements, as measured by the coherence, were observed over most or all of the frequency range of the signal, *even when there was no noise power at that frequency*, as seen in Fig.3.

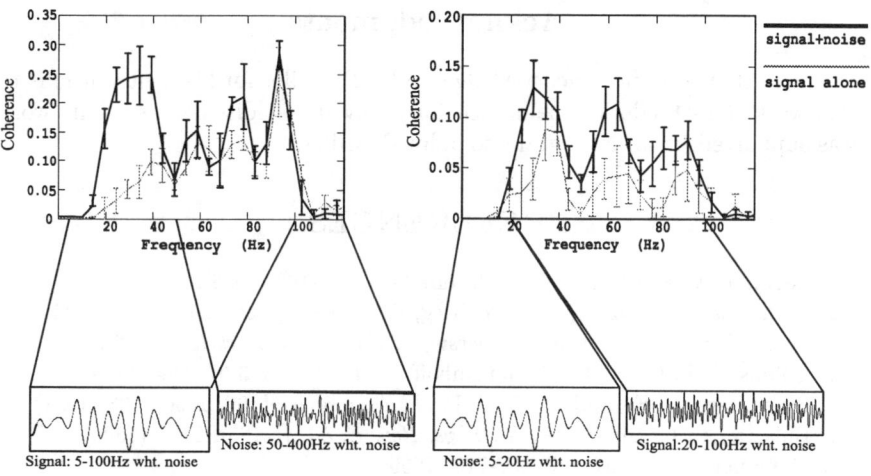

FIGURE 3. Coherences improved across the frequency operating range of the cells, even when noise and signal frequency ranges are partially or completely non-overlapping. a) Low frequency signal presented with high frequency noise. b) High frequency signal presented with low frequency noise.

Discussion

Enhancements to signal encoding through the addition of ambient noise were observed in sensory interneurons of the cricket for a wide range of stimulus and noise amplitudes, and for all combinations of signal and noise frequency ranges investigated. Although this result apparently contradicts theories of SR [8–10], it is consistent with current views of neuronal function. A neuron is not a simple threshold crossing detector, or even a thresholded charge collector, but rather, a stochastic frequency and time weighted integrating device. It has, at any moment, a probability of firing which is modulated by its inputs, past history, and random synaptic events. Any additive noise presented to the cell in the frequency range to which it is sensitive during the integration time preceding a spike will have the effect of altering its firing probability. In the proper regime, this added noise can serve to bias the cell to encode more accurately weak signals of interest.

Ambient noise is an unavoidable consequence of a system operating in the physical world. It seems that, rather than have its performance degraded by this noise, the cricket cercal system, and perhaps all sensory systems, utilize this broadband noise background to enhance their sensitivity to extremely weak signals, by adjusting the response properties of their neurons on both the behavioral and evolutionary time scales.

Acknowledgments

The author gratefully acknowledges John P. Miller for his contributions to this work, and Frederic Theunissen for many useful discussions. This work was supported by an NIH grant to John P. Miller.

REFERENCES

1. Levin, J. E., and Miller, J. P., *Nature* **380**, 165-168 (1996).
2. Collins, J. J., Imhoff, T. T. and Grieg, P., *J. Neurophysiol.* **76**, 642-645 (1996).
3. Levin, J. E., *Ph.D Thesis, University of California, Berkeley* (1996).
4. Collins, J. J., Chow, C. C. and Imhoff, T. T., *Nature* **376**, 236-238 (1995).
5. Theunissen, F. E., and Miller, J. P., *J. Neurophysiol.* **66**, 1690-1703 (1991).
6. Bulsara, A., R. and Zador, A., *Phys. Rev. E.* **54**, R2185-R2188 (1996).
7. Stemmler, M., *Network* **7**, 687-716 (1996).
8. Chialvo, D. R., Longtin, A. and Muller-Gerking, J., *Phys. Rev. E* **55**, 1798-1808 (1997).
9. Wiesenfeld, K. and Moss, F., *Nature* **373**, 33-36 (1995).
10. McNamara, B. and Wiesenfeld, K., *Phys. Rev. A* **39**, 4854-4869 (1989).
11. Heneghan, C., Chow, C. C., Collins, J. J., Imhoff, T. T., Lowen, S. B. and Teich, M. C., *Phys. Rev. E* **54** R2228-R2231 (1996).

The Neuromodulatory Properties of "Noisy Neuronal Oscillators"

Hans A. Braun *, Martin T. Huber [†],
Mathias Dewald * and Karlheinz Voigt *

*Institute of Physiology[1]
University of Marburg, Deutschhausstr.2, D-35037 Marburg
[†]Department of Psychiatry

Abstract. Many neurons, in various areas of the nervous system, exhibit spontaneous and intrinsic oscillations of their membrane potential or can develop oscillations on depolarisation. Here we consider so-called subthreshold oscillations which operate close to the threshold of spike-generation. To analyze the principle neuromodulatory properties of such neuronal mechanisms we developed a computer model consisting of only a minimal set of ionic conductances but which nevertheless attains subthreshold membrane potential oscillations of variable frequency and amplitude. With the addition of stochastic components, the oscillations generate typical impulse patterns which originate from a mixture of spike-triggering and subthreshold oscillations. The spiking probability thereby depends on the oscillation and noise parameters. According to the experimental data, the model can be tuned from complete silence to tonic firing with maximum sensitivity between these two extremes. Range and steepness of the activity and sensitivity curves can be modified and complex response characteristics (including stochastic resonance phenomena, "gain adjustment" or differential stimulus encoding) can be obtained.

INTRODUCTION

Neuronal oscillations with rhythmic spike-generation are a wide-spread feature in the central and peripheral nervous system (CNS and PNS) and can arise either from intrinsic membrane properties of individual neurons (endogenous oscillations) or from synaptic interactions within neuronal networks (network oscillations). The physiological relevance of oscillatory activity in the CNS is usually seen in context with neuronal synchronization [1–4] while

---

[1]) Supported by the Deutsche Forschungsgemeinschaft, the Kempkes Stiftung and the Friedrich-Naumann-Stiftung.

an information carrying role is generally denied. So far, only studies of sensory transduction in thermo- and electrosenstive skin afferents emphasize the signal-encoding properties of oscillatory activity [5–9].

Noise, in the nervous system, can originate from many different sources (e.g. synaptic and membrane noise) and is so far considered physiologically relevant mainly in context with stochastic resonance (SR) [10–13]. However, there are experimental and model data which indicate that noise can be of particular value for signal encoding in a much broader stimulus range [6,14] especially when combined with oscillatory mechanisms of impulse generation.

THE "NOISY OSCILLATOR"

Oscillating membrane potentials can generate rhythmic neuronal discharges which either consist of impulse groups (bursts) or single spikes. However, here we will focus on another type of rhythmically generated patterns which can only be understood when the requirement of noise is considered (Fig.1, upper trace).

Impulse Patterns

The typical impulse pattern of noisy oscillators arise from membrane potential oscillations which operate close to the threshold of spike-generation ("subthreshold oscillations") where it depends on stochastic components and whether the actual oscillation cycle triggers an impulse or not. In recent years, such oscillations with "skippings" have been recorded in diverse cortical and subcortical neurons [3,15,16]. Corresponding impulse patterns are also known from extracellular recordings from peripheral thermo- and electrosensitive skin afferents [5,6]. Because of the small size of the peripheral sensory nerve endings, intracellular membrane potential recordings have not been possible so far;

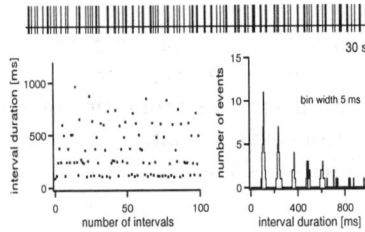

FIGURE 1. Impulse sequence (upper trace) recorded from a peripheral cold receptor of the cat skin with corresponding plot of single interspike-intervals (ID, left) and interval-histogram (right).

however, the multimodal interval distributions of the impulse sequences (Fig. 1, lower traces) clearly indicate that the impulse activity originates from oscillations with noise [5,8,17]. Moreover, the stimulus-response characteristics of these receptors elucidate that the temporal patterns essentially contribute to the sensory encoding mechanisms. Such impulse patterns can generally be modulated in two different ways: changing the oscillation frequency (OF) or the spiking probability (SP). Both parameters can be calculated from the interval distributions [8] with the mean firing frequency $F = OF \cdot SP$.

The Computer Model

To allow a focus on the basic principles of signal encoding our model consists of only a minimal set of ionic conductances. The algorithms are similar to those of recently described simulations of burst activity according to peripheral cold receptor discharges [7] with the additional restriction that here we exclusively consider the range of "skippings", i.e. the activity range where noise is an ultimate component of the encoding process. This, for example, corresponds to signal transduction in shark electroreceptors [6] and seems to be also the situation in many cortical and subcortical neurons [3,15,16].

The membrane potential V is calculated from the differential equation

$$C_M \frac{dV}{dt} = f(V) = -g_l(V - V_l) - I_d - I_r - I_{sd} - I_{sr} + I_{ext} \quad (1)$$

where $C_M = 2nF$ represents the membrane capacitance and $g_l = 0.1\mu S$ the leak conductance. The passive time constant thus is given by $t_M = C_M/g_l = 20ms$. $V_l = -60mV$ is the leak potential.

Noise g_w is calculated according to the Box-Mueller algorithm [18]:

$$g_w = \sqrt{-4\,d\,dt \cdot \ln(a)} \cdot \cos(2\pi b) \quad (2)$$

a,b are random numbers (0 to 1, b in mV)

Noise intensity is adjusted by the dimensionless parameter d and is implemented as described in Fox et al. (1988) [18]:

$$V_{t+dt} = V_t + f(V)dt + g_w \quad (3)$$

I_{ext} is for additional application of an external current (e.g. synaptic current). The remaining terms describe voltage dependent de - and repolarizing (d,r) currents. Two of them I_d and I_r, are for the generation of action potentials according to the Hodgkin-Huxley currents. Only two additional, more slowly (s) activating currents, I_{sd} and I_{sr}, are required to account for the generation of "subthreshold" oscillations. For simplification any current inactivation is neglected.

The voltage dependent currents are modelled as:

$$I_i = g_i a_i (V - V_i) \tag{4}$$

with V_i the reversal potential and g_i the maximum conductance.

Activation of the fast depolarizing current I_d is instantaneous ($a_d = a_{d\infty}$). The activation variable a_i of the voltage dependent currents, I_{sd} and I_r, is described by the differential equation

$$\frac{da_i}{dt} = \frac{\phi(a_{i\infty} - a_i)}{\tau_i} \tag{5}$$

with ϕ as a temperature dependent scaling factor:

$$\phi = 3.0^{(T-T_o)/10} \tag{6}$$

The steady-state activation variable $a_{i\infty}$ follows a sigmoidal curve:

$$a_{i\infty} = \frac{1}{1 + e^{(-s_i(V-V_{0i}))}} \tag{7}$$

with s_i the steepness and V_{0i} the half-activation value. The slowly repolarizing current I_{sr} is directly coupled to the slowly depolarizing current I_{sd} by

$$\frac{da_{sr}}{dt} = \frac{\psi(-\eta I_{sd} - k a_{sr})}{\tau_{sr}} \tag{8}$$

The numerical values of the parameters are:
$g_{sd} = 0.25$, $g_{sr} = 0.4$, $g_d = 1.5$, $g_r = 2.0$ (in μS); $\tau_{sd} = 10$, $\tau_{sr} = 20$, $\tau_r = 2$ (in ms); $\eta = 0.0012$, $k = 0.17$; $V_{sd} = V_d = 50$, $V_{sr} = V_r = -90$, $V_l = -60$ (in mV); $s_d = 0.09$, $s_r = 0.25$, $s_{sd} = 0.25$; $T_0 = 25^\circ C$.

Signal Encoding

The deterministic model without noise is either completely silent (no spikes) or exhibits tonic firing (Fig. 2, upper traces) depending on the actual parameter values which in this case is the level of current injection. With the addition of noise, the situation changes considerably: occassional spikes are induced in the range of previously subthreshold oscillations and skippings occur in otherwise regular impulse sequences (Fig. 2, lower traces).

In this way, noise linearizes the step-like transfer characteristic of the deterministic model and thereby allows gradual encoding of external currents which might correspond physiologically to synaptic input. The firing rate mainly reflects the spiking probability because current injection only has minor effects on the oscillation frequency but mainly changes the oscillation amplitude. This leads to an approximately sigmoidal relation between firing rate and current injection with maximum steepness, i.e. sensitivity (or "gain"), in a mid-range around 0.5 spiking probability. Increasing noise intensity reduces the slope of the sigmoidal transfer function and thereby extends the range of variable spiking probability which means that also SR effects can be obtained.

Current injection is one of the more simple examples of stimulus encoding in noisy oscillators although different transfer characteristics can be obtained. Depending on the current input, such a neuron can be tuned from maximum sensitivity in a mid-activation range down to a range of stochastic resonance and finally to complete silence and insensitivity. Towards the other extreme of maximum activation the sensitivity decreases as well which here, however, means that the neuron can obtain rather stable pacemaker properties.

Still more complex response characteristics can be obtained when the oscillation frequency is changed which is done with modifications of the ionic kinetics (time constants). We herefore use a physiologically plausible, temperature-like scaling factor [9]. The examples in Fig. 3 emphasize a remarkable effect of such modifications. Although the oscillation frequency is considerably increased from left to right, the number of spikes, i.e. the firing rate, remains almost the same. The reason is that increasing oscillation frequencies in turn reduce the spiking probability because of the shortened oscillation cycles. Assuming

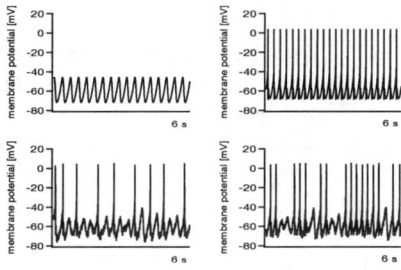

FIGURE 2. Computer simulated voltage traces under deterministic conditions (upper plots) and with addition of noise (d=0.5, lower plots). Left traces: $I_{ext} = -0.5$ in nA, right traces: $I_{ext} = 0$. Temperature scaling: $T = 20°C$ in all recordings.

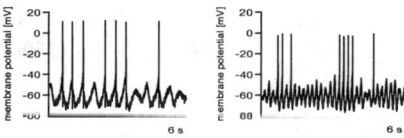

FIGURE 3. Modifications of ionic kinetics (time constants) change oscillation frequency and spiking probability in noisy simulations. Temperature scaling: $T = 15°C$ (left), $T = 25°C$ (right).

the most simple but physiologically realistic case [5] that both parameters are linearily related to a given stimulus (S) a parabolic frequency curve is obtained according to $OF(S) \cdot SP(S) = F(S^2)$.

CONCLUSIONS

Without going into the details, it is easy to see that complex reponse characteristics can be obtained when the modulatory effects of noise intensity and, especially, of different oscillation parameters (frequency, amplitude, base-line) are combined. So far, peripheral electroreceptors of sharks are the best illustration of "noisy oscillators" with functionally relevant encoding properties [6]. However, corresponding mechanisms can also be particularly significant in information processing in the CNS where a variety of neurotransmitters, hormones and physical/chemical stimuli interfere with the intrinsic membrane properties of the neurons.

REFERENCES

1. Eckhorn, R., Bauer, R., Jordan, W., Brosch, M., Kruse, W., Munk, M., and Reitboeck, H.J., *Biol. Cybern.* **60**, 121-130 (1988).
2. Engel, A.K., Kreiter, A.K., König, P., and Singer, W., *Proc. Natl. Acad. Sci. USA.* **88** , 6048-6052 (1991).
3. Gutfreund, Y., Yarom, Y., and Segev, I., *J. Physiol.* **483.3** , 621-640 (1995).
4. McCormick, D.A., and Feeser, H.R., *Neuroscience* **39(1)**, 103-113 (1990).
5. Braun, H.A., Bade, H., and Hensel, H., *Pflügers Arch.* **386**, 1-9 (1980).
6. Braun, H.A., Wissing, H., Schäfer, K., and Hirsch, M.C., *Nature* **367**, 270-273 (1994).
7. Braun, H.A., Huber, M., Dewald, M., Schäfer, K., and Voigt, K., *Int. J. Bifurcation and Chaos,* in press (1997).
8. Schäfer, K., Braun, H.A., and Isenberg, C., *J. Gen. Physiol.* **88**, 557-576 (1986).
9. Longtin, A., and Hinzer, K., *Neural Computation* **8**, 215-255 (1995).
10. Douglass, J., Wilkens, L., Pantazelou, E., and Moss, F., *Nature* **365**, 337-340 (1993).
11. Bulsara, A.R., and Gammaitoni, L., *Physics Today* **March**, 39-45 (1996).
12. Collins, J.J., Chow, C.C., and Imhoff, T.T., *Nature* **376**, 236-238 (1995).
13. Levin, J.E., and Miller, J.P., *Nature* **380**, 165-168 (1996).
14. Chialvo, D.R., Longtin, A., and Müller-Gerking, J., *Phys. Rev. E*, in press (1997).
15. Lampl, I., and Yarom, Y., *J. Neurophysiol.* **70**, 2181-2186 (1993).
16. Pare, D., Pape, H.C., Dong, J., *J. Neurophysiol.* **74**, 1179-1191 (1995).
17. Longtin, A., Bulsara, A., and Moss, F., *Phys. Rev. Lett.* **67**, 656-659 (1991).
18. Fox R.F., Gatland, I.R., Roy, R., and Vemuri, G., *Phys. Rev. A* **38**, 5938-5940 (1988).

Travelling waves in a circular array of integrate–and–fire neurons

P. C. Bressloff and S. Coombes

Nonlinear and Complex Systems Group
Department of Mathematical Sciences, Loughborough University
Loughborough, Leics. LE11 3TU U. K.

Abstract. We investigate travelling waves in a circular array of pulse–coupled integrate–and–fire oscillators with distributed delays and non–local excitatory connections. The model is analysed in the weak coupling regime where averaging leads to an effective phase–coupled model with distributed phase–shifts.

INTEGRATE–AND–FIRE MODEL

Consider a circular array of $N = 2M + 1$ pulse–coupled integrate-and-fire (IF) neurons labelled $n = 0, \pm 1, ..., \pm M$. Let $U_n(t)$ denote the somatic membrane potential of the nth neuron at time t. Suppose that $U_n(t)$ satisfies the set of coupled equations

$$\frac{dU_n(t)}{dt} = f(U_n(t)) + \epsilon \sum_{m=-M}^{M} W(m) \int_0^\infty P(\tau) E_{n+m}(t-\tau) d\tau \quad (1)$$

supplemented by the reset condition $U_n(t^+) = 0$ whenever $U_n(t) = 1$. Here $E_m(t)$ represents the spike train transmitted from the mth neuron at time t and $P(\tau)$ represents a distribution of delays. (Periodic boundary conditions are incorporated by setting $E_{n-M}(t) \equiv E_{n+M+1}(t)$, and similarly for other site–dependent variables). The weight function $W(m)$ is assumed to be (a) a positive symmetric function of m, $W(-m) = W(m)$, and (b) to be a monotonically decreasing function of $|m|$. For the moment, we shall take f to be a linear function $f(U_n) = -U_n + I$ for some constant bias I. The system of equations (1) has $\mathbf{D}_N$ as its symmetry group (cyclic permutations and reversal of the ordering of sites around the array).

The distribution $P(\tau)$ can be used to incorporate a number of important aspects of neural processing including axonal transmission delays, α–functions associated with synaptic processing [1], and dendritic processing [2]. We shall focus on the last case. For simplicity, suppose that the dendrites are represented by a semi–infinite uniform cable $0 \leq \xi < \infty$ with the soma at $\xi = 0$. Let $J(\xi)$ denote a distribution of axo–dendritic connections from a pre–synaptic to a post–synaptic neuron. Using standard cable theory [3], it can be shown that diffusion along the cable generates an effective distribution of delays $P(\tau) = \int_0^\infty J(\xi) G(\xi, \tau) d\xi$ where $G(\xi, t) = (1/\sqrt{4\pi D t}) \exp(-\xi^2/(4Dt) - t/\tau_s)$ is the fundamental solution of the cable equation, D is the diffusion constant and τ_s is the membrane leakage time constant of the cable. For convenience, we set $D = \tau_s = 1$ so that ξ is measured in terms of electronic distance, which has typical values in the range $1-10$ cm. Typical values for τ_s are $5-20$ msec. For simplicity, we are assuming that the distribution $P(\tau)$ is independent of the separation $|n - m|$ between neurons. (A more biologically realistic model would need to take into account correlations between delays and the relative positions of neurons in the array [2,4]).

We shall restrict our attention to phase–locked solutions in which every neuron fires periodically with the same frequency $\omega = 2\pi/T$ that must be determined self–consistently. The state of each neuron can then be characterized by a constant phase ϕ_n. Neglecting the shape of an individual action potential, the resulting spike train is $E_n(t) = \sum_{j=-\infty}^{\infty} \delta(t - jT + \phi_n T/2\pi)$. Generalizing the analysis of Ref. [5], one can integrate equation (1) over the interval $[-T\phi_n/2\pi, T - T\phi_n/2\pi]$ and use the reset condition $U_n(T - T\phi_n/2\pi) = 1$ to obtain the set of equations

$$1 = (1 - e^{-T})I + e^{-T} \sum_m W(m) \int_0^T e^{t'} P_T(t' + T(\phi_{n+m} - \phi_n)/2\pi) dt' \quad (2)$$

where $P_T(t) = \sum_{n=0}^{\infty} P(t+nT)$ for $0 < t < T$ and $P_T(t)$ is extended outside this range by making it a periodic function of t. Equation (2) determines the phases ϕ_n (up to an arbitrary phase–shift) and the period T. Note that equation (2) is invariant under the action of the group $\mathbf{D}_N \times \mathbf{S}^1$ with $\mathbf{S}^1$ representing constant phase–shifts. One class of *maximally–symmetric* solution is $\phi_n = n\beta$ with $\beta = 2\pi n_b/N$, $n_b = 0, 1, ..., N-1$. This corresponds to a travelling wave solution $U_n = U(\omega t + n\beta)$ where U is some periodic waveform and T satisfies equation (2) with $\phi_{n+m} - \phi_n = m\beta$. (The state $\beta = 0$ is the synchronous or in–phase state). Previous work has shown that even for two coupled IF neurons, there is a rich bifurcation structure in the space of phase–locked solutions as a function of some appropriate parameter such as the dendritic location ξ_0 (when $P(\tau) = G(\tau, \xi_0)$) [6]. We are currently using group theoretic methods that exploit the underlying symmetry of equation (2) to investigate analogous results for circular arrays where the nature of the coupling $W(m)$ also comes into play. However, analysing the linear stability of such solutions in terms of perturbations of the firing–times is difficult due to the presence of delays. Therefore, we shall consider an alternative, phase–coupled model that is valid in the weak coupling regime.

PHASE–COUPLED MODEL

Suppose that in the absence of any coupling, $\epsilon = 0$, each neuron fires with the same period T_0 where $T_0 = \int_0^1 dU/f(U)$ and we no longer restrict f to be linear. Following, Ref. [5], we introduce the nonlinear transform $u_n(t) \to \phi_n(t)$ according to

$$(\text{mod } 2\pi) \quad \phi_n(t) + \omega_0 t \equiv \Psi(U_n(t)) = \omega \int_0^{U_n(t)} \frac{du'}{f(u')} \tag{3}$$

where $\omega_0 = 2\pi/T_0$. Under such a transformation equation (1) becomes

$$\frac{d\phi_n(t)}{dt} = \epsilon F(\phi_n + \omega_0 t) \sum_m W(n-m) \int_0^\infty P(\tau) E_m(t-\tau) d\tau \tag{4}$$

where $F(z) = \omega_0/[f \circ \Psi^{-1}(z)]$, $F(z + 2\pi j) = F(z)$, $j \in \mathbb{Z}$. The function F may be interpreted as the instantaneous phase–coupling response function of the system. When $\epsilon = 0$, the phase variable $\phi_n(t)$ is constant and the firing times of the nth neuron are $t = (j - \phi_n/2\pi)T_0$. Neglecting the shape of an individual action potential, the resulting spike train is $E_n(t) = \sum_j \delta(t - jT_0 + \phi_n T_0/2\pi)$.

Now suppose that the oscillators are weakly coupled (ϵ small). To a first approximation, each oscillator still fires with period T_0 but now the phases $\phi_n(t)$ slowly drift according to equation (4). Under the assumption that ϵ is small, we can average equation (4) over a single period. Using the fact that both F and E_n are periodic functions, we obtain the phase equation

$$\frac{d\phi_n}{dt} = \sum_m W(m) \int_0^\infty P(\tau) F(\phi_n - \phi_{n+m} + \omega_0 \tau) d\tau \tag{5}$$

where the factor ϵ has been absorbed into $W(m)$. Equation (5) shows that delays in the propagation of signals between pulse–coupled neurons reduce to phase shifts in the corresponding phase–coupled model. Note that if $f(U) = -U + I$ then $T_0 = \ln[I/(I-1)]$ and $F(\phi) = \exp(T_0 \phi)/[IT_0]$ for $0 \leq \phi < 1$. One can then show that fixed point solutions (in a rotating frame) of equation (5) are fixed point solutions of equation (2) to first order in ϵ. Equation (5) is also invariant under the symmetry group $\mathbf{D}_N \times \mathbf{S}^1$. In the following we shall assume that $F(\phi) = -\sin \phi$, which is known to be a reasonable first approximation when f takes an experimentally determined form such as a cubic [7].

As in the pulse–coupled model, we shall consider travelling wave solutions of equation (5) $\overline{\phi}_n(t) = n\beta + \Omega t$. Substitution of such a solution into equation (5) with $F(\phi) = -\sin \phi$ shows that the frequency Ω must satisfy the dispersion relation $\Omega = -\Delta_s(\omega_0) H(\beta)$ where

$$H(p) = \sum_m W(m) \cos(mp), \quad \Delta(\omega_0) = \int_0^\infty P(\tau) \exp(i\omega_0 \tau) d\tau \tag{6}$$

and $\Delta_c(\omega_0) = \operatorname{Re} \Delta(\omega_0)$, $\Delta_s(\omega_0) = \operatorname{Im} \Delta(\omega_0)$.

LINEAR STABILITY ANALYSIS

In order to analyse the local stability of the travelling wave solutions, we linearize equation (5) by setting $\phi_n(t) = \overline{\phi}_n(t) + \theta_n(t)$ and expanding to first-order in θ. The resulting linear equation has solutions of the form $\theta_n(t) = e^{\lambda_p t + inp}$ where

$$\text{Re } \lambda_p = \frac{1}{2}\Delta_c(\omega_0)\widehat{H}(p,\beta), \quad \widehat{H}(p,\beta) = [H(p+\beta) + H(p-\beta) - 2H(\beta)] \quad (7)$$

$$\text{Im } \lambda_p = \frac{1}{2}\Delta_s(\omega_0)\left[H(p-\beta) - H(p+\beta)\right] \quad (8)$$

A travelling wave solution will be stable provided that $\text{Re}\lambda_p < 0$ for all $p \neq 0$. (The neutrally stable mode $\lambda_0 = 0$ corresponds to constant phase shifts). Note that in the special case $\Delta_c(\omega_0) = 0$ we have $\text{Re } \lambda_p = 0$ for all p, β so that all travelling wave solutions are marginally stable.

Let us first consider the stability of the synchronous state $\beta = 0$, for which λ_p is real for all p. Since $W(m)$ decreases monotonically with $|m|$ it follows that $\max_p H(p) = H(0)$, and hence that the synchronous state is stable (unstable) if $\Delta_c(\omega_0) > 0$ ($\Delta_c(\omega_0) < 0$). The condition $\Delta_c(\omega_0) = 0$ determines a bifurcation point where a degenerate real eigenvalue crosses the imaginary axis ($\lambda_p = 0$ for all p). As a particular example, consider the effects of dendritic processing with a distribution of axo–dendritic connections of the form $J(\xi) = \delta(\xi - \xi_0)$. In other words, every synapse is located at the same distance ξ_0 from the soma. Equations (4)–(6) then hold with $P(\tau) = G(\xi_0, \tau)$. The function $\Delta(\omega_0)$ can be determined explicitly using a Fourier representation of the fundamental solution $G(\xi, \tau)$ and performing a contour integral. The result is

$$\Delta(\omega_0) = \frac{1}{2r}\exp\left(-r\xi_0\cos(\theta/2) + i\left[r\xi_0\sin(\theta/2) + \theta/2\right]\right) \quad (9)$$

where $r^2 = \sqrt{1 + \omega_0^2}$, $\theta = \tan^{-1}(\omega_0)$, $0 \leq \theta \leq \pi/2$. Hence $\Delta_c(\omega_0)$ is positive if $\cos\left(r|\xi_0|\sin(\theta/2) + \theta/2\right) > 0$ and is negative otherwise. We deduce from the above analysis that as the distance $|\xi_0|$ of the synapse from the soma increases from zero, it reaches a critical value $\xi_{0c} = \pm(\pi - \theta)/(2r\sin(\theta/2))$, beyond which the synchronous state becomes unstable. Increasing ξ_0 further produces alternating bands of stability and instability of the synchronous state, see figure 1(a). (These regions of stability/instability would be reversed in the case of inhibitory weights). Note that ξ_0 characterizes the effective delay due to diffusion along the dendrites.

In contrast to the synchronous state, the stability of a travelling wave solution will depend on the range of interactions as specified by the function $W(m)$. As an illustrative example, consider the step interaction function $W(m) = W_0\Theta(L - |m|)$ where L specifies the range of interactions and $\Theta(x) = 1$ if $x \geq 0$ and is zero otherwise. For this choice of $W(m)$, the function $H(p)$ of equation (7) becomes $H(p) = W_0 \sin([2L+1]p/2)/\sin(p/2)$. A

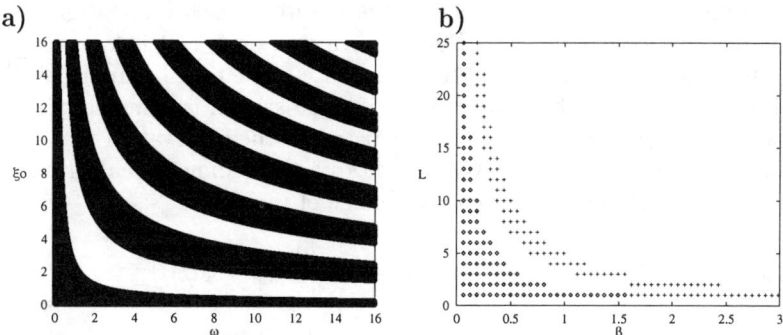

FIGURE 1. (a) Stability diagram for the synchronous state in the ξ_0–ω_0 plane. White (black) regions correspond to instability (stability). (b) Stability of travelling wave solutions with wavenumber β as a function of the interaction length L. $\diamond$'s (+'s) indicate states that are stable when $\Delta_c(\omega_0) > 0$ ($\Delta_c(\omega_0) < 0$).

travelling wave with wave number $\beta \neq 0$ will be stable provided that either (i) $\widehat{H}(p, \beta) < 0$ for all $p \neq 0$ and $\Delta_c(\omega_0) > 0$ or (ii) $\widehat{H}(p, \beta) > 0$ for all $p \neq 0$ and $\Delta_c(\omega_0) < 0$. The stability results for the step interaction function are shown in figure 1(b) for a range of values of the interaction length L, $0 \leq L \leq M/2$. (We take $M = 50$). For each L, the wavenumbers β that satisfy stability condition (i) or (ii) are indicated. We see that there are two stability bands. The first (labelled by $\diamond$'s) consists of travelling wave states that are stable when $\Delta_c(\omega_0) > 0$. This band becomes thinner as L increases from zero until only the synchronous state remains stable. Note that the so-called splay state ($\beta = 1$), where the phases are uniformly spaced around the circle, maintains stability for the largest value of L. The second band (labelled by +'s) represents travelling wave states that are stable when $\Delta_c(\omega_0) < 0$, that is, when the synchronous state is unstable. We conclude that there are two mechanisms whereby a synchronous state can be destabilized: (a) if $\Delta_c(\omega_0) > 0$ then a sufficiently large perturbation is needed to induce a transition into the basin of attraction of a stable travelling wave satisfying (i), (b) $\Delta_c(\omega_0)$ becomes negative leading to the formation of a stable travelling wave satisfying (ii). The picture $L \leq M/2$ should be contrasted with the case of all–to–all coupling ($L = M$), where $H(p) = W_0 N \delta_{p,0}$ so that $\widehat{H}(p, \beta) = W_0 N [\delta_{p,\beta} + \delta_{p,2\pi-\beta}]$ for $\beta \neq 0$ and $\widehat{H}(p, 0) = -2W_0 N$ for all $p \neq 0$. One then finds that all travelling wave solutions are stable ($\Delta_c(\omega_0) < 0$) or unstable ($\Delta_c(\omega_0) > 0$) in 2 eigendirections and marginally stable in the remaining $N - 2$ eigendirections. This basic result extends to more general solutions as shown by Watanabe and Strogatz [8].

We note from equations (7) and (8) that every travelling wave solution becomes marginally stable at the bifurcation point $\Delta_c(\omega_0) = 0$ and M pairs of complex conjugate roots (λ_p and λ_{-p}) cross the imaginary axis simultaneously. An interesting question concerns the nature of such bifurcations; for example, whether or not multiple Hopf bifurcations occur? It is known that Hopf bifurcations do not occur in the special case of all–to–all coupling since, as established by Watanabe and Strogatz [8], the system is completely integrable when $\Delta_c(\omega_0) = 0$. Their analysis is based on the observation that each trajectory of the system is actually confined to a three–dimensional subspace $(\Theta(t), \Psi(t), \gamma(t))$. This follows from the change of variables

$$\tan\left[\frac{1}{2}(\phi_n(t) - \Theta(t))\right] = \sqrt{\frac{1+\gamma(t)}{1-\gamma(t)}} \tan\left[\frac{1}{2}(\psi_n - \Psi(t))\right] \qquad (10)$$

where the ψ_n are constants and $0 \leq \gamma(t) < 1$. It can further be shown that $\Theta(t)$ is passively driven by $\gamma(t), \Psi(t)$ and the dynamics of the latter two variables is characterized by the existence of a Lyapunov function $\mathcal{H}$ such that $d\mathcal{H}/dt = R(t)^2 \Delta_c(\omega_0)$ where $R(t)$ is an order parameter that measures the degree of coherence of the system, $R(t) \exp[i\overline{\phi}(t)] = \frac{1}{N} \sum_{n=1}^{N} \exp[i\phi_n(t)]$. (Note that Watanabe and Strogatz [8] only consider a single phase–shift τ for which $\Delta_c(\omega_0) = \cos\omega_0\tau$. However, their analysis carries over to the case of distributed phase–shifts on using our more general expression for $\Delta_c(\omega_0)$). The following results thus hold for almost all initial conditions: (i) If $\Delta_c(\omega_0) > 0$ then $\mathcal{H}(t) \to \infty$, $\gamma(t), R(t) \to 1$ and the system converges to the synchronous state. (ii) If $\Delta_c(\omega_0) < 0$ then $\mathcal{H}(t), \gamma(t), R(t) \to 0$ and the system converges to an $(N-3)$-dimensional manifold of incoherent states. Such a manifold consists of constant states ϕ_n satisfying $\sum_n \exp[i\phi_n] = 0$. There are $(N-2)$ neutrally stable directions around each such incoherent state. (iii) If $\Delta_c(\omega_0) = 0$ then the system is completely integrable with $\mathcal{H}$ a conserved quantity. Trajectories run along the contours of $\mathcal{H}$ and correspond to periodic motion in (Ψ, γ) space. In the full phase space, the motion is quasiperiodic on 2–tori.

REFERENCES

1. Jack, J. J. B., Noble, D. and Tsien, R. W., *Electric Current Flow in Excitable Cells*, Clarendon Press, Oxford (1975).
2. Bressloff, P. C. and Coombes, S.,*Phys. Rev. Lett* **78**, 4665–4668 (1997).
3. Rall, W. In *Neural Theory and Modelling*, R. Reiss (editor), Stanford University Press, Stanford (1964).
4. Crook, S. M., Ermentrout, G. B., Vanier M. C. and Bower, J. M., *J. Comp. Neurosci.* **4**, 161–172 (1997).
5. Van Vreeswijk, C., Abbott, L. F. and Ermentrout, G. B., *J. Comp. Neurosci.* **1**, 313–321 (1994).
6. Coombes, S. and Lord, G. J., *Phys. Rev. E* **55**, 2104R–2107R (1997).
7. Abbott, L. F. and Kepler, T. B. In *Statistical Mechanics of Neural Networks*, L. Garrido (editor) pp. 5–18, Springer–Verlag (1990).
8. Watanabe, S. and Strogatz, S. H., *Physica D* **74**, 197–253 (1994).

Subthreshold Coincidence Detection

André Longtin and Jacques M. Laniel

Département de Physique
Université d'Ottawa, 150 Louis Pasteur, Ottawa, Ont., Canada K1N 6N5

Abstract. Many neurons in the central nervous system are thought to perform coincidence detection between two or more inputs. These inputs are individually subthreshold for action potential generation. However, in the presence of even a small background noise, the neuron can fire in the absence of such inputs. We give preliminary results on the effect of noise intensity and input amplitude on coincidences between action potentials and single or temporally coincident double inputs. The fractions of single and double coincidences can be either monotonic or nonmonotonic functions of noise intensity and input amplitude.

MOTIVATION

The problem of coincidence detection by neurons has received much renewed attention, in cells ranging from the sensory periphery up to the cortex. The main questions revolve around the mechanism for the detection of coincidences between two or more subthreshold excitatory postsynaptic potentials (EPSP's) [1,2]. This problem has been previously addressed in the context of simple noiseless integrate-and-fire neurons (see e.g. [3]). Coincidence detection is thought of as a "multiplication" property, as opposed to the more usual "summation" property where neurons integrate synaptic inputs and fire when the voltage reaches a threshold. A multiplier neuron, on the other hand, has a mean firing rate proportional to the product of the mean firing rates in (usually) two synaptic input channels (see [1,4,5] for reviews). One of the differences between a summator and a multiplier is the integration time window [4].

[3] discusses a simple coincidence detector based on integrate-and-fire dynamics without noise. The neuron fires when two input excitations, each subthreshold, are separated by at most a time Δ. For exponentially decaying pulses such as those used below, this window is related to the initial height h and rate of decay τ of the excitation [3] $\Delta = \tau \ln[h/(\theta - h)]$ where θ is the distance between the rest and threshold voltages. For two inputs streams of excitations of (constant) mean rates ν_1 and ν_2, one obtains the coincidence

rate $\nu_c = 2\Delta\nu_1\nu_2$: the mean firing rate in the detector is proportional to the product of the mean rates at the two input synapses. Similar expressions are found in the literature on coincidence detection in particle physics [6]. More sophisticated studies [2] of the biophysical details of active dendritic processing in cortical neurons further predicts that coincidence detection may underly neural computation on the sub-millisecond time scale.

There is, to our knowledge, little discussion in the literature of the effect of background noise in the cell. While this noise, of synaptic origin or other, may be small in certain coincidence detectors closer to the sensory periphery, it is usually large in cortical cells due, in particular, to the large number of inputs these cells receive. Even a small background noise can then produce a finite probability that the neuron will fire spontaneously. For example, the noisiness produced by hundreds of inputs to a cortical cell will influence the way the cell detects coincidences between any two or more such inputs, in particular between two important or "priviliged" inputs (such as those with stronger synaptic weight).

Further, there will be false coincidences ("false alarms"), not only for detection of two inputs, but even of one input and no inputs. This will be true also for suprathreshold EPSP's which can be deleted by a negative-going noise fluctuation. Action potentials can occur because a single EPSP gets kicked up to threshold by the noise. They can also occur when two temporally coincident EPSP's, which would not ordinarily reach threshold, do so with a noisy fluctuation. Recent related studies based on noisy dynamical models have focussed on the usefulness of noise for e.g. detecting single constant input signals [7] or for sharpening the timing properties of neuron networks detecting a single common EPSP [8].

We present preliminary results on these effects and their dependencies on noise intensity and input amplitude. For simplicity, we have singled out two input channels of events which the cell seeks to coincide. All other inputs and membrane fluctuations are considered to generate the background noise on which detection is to occur.

MODEL SYSTEM

We investigate the problem of coincidence detection of two randomly distributed streams of EPSP input events in the presence of a background neuronal noise. The model neuronal system is the Fitzhugh-Nagumo system operated in the excitable regime [9], which is more realistic than that studied in [3]. The fast variable is driven by three random processes: two independent renewal point processes as well as an Ornstein-Uhlenbeck (OU) process with small correlation time. The dynamics are thus governed by the following coupled differential equations:

$$\epsilon \frac{dv}{dt} = v(v-a)(1-v) - w + \eta(t) + P_1(t) + P_2(t) \qquad (1)$$

$$\frac{dw}{dt} = v - dw - b \qquad (2)$$

$$\frac{d\eta}{dt} = -\lambda\eta + \lambda\xi(t) \qquad (3)$$

The variable v is the fast voltage-like variable, while w is a recovery variable. Also, $\xi(t)$ is a zero-mean Gaussian white noise, which is lowpass filtered to produce the Gaussian Ornstein-Uhlenbeck-type noise denoted by η. The autocorrelation of the white noise is $\langle \xi(t)\xi(s)\rangle = 2D\delta(t-s)$. We will call D the intensity of the white noise. The OU correlation time is $t_c = \lambda^{-1}$. The stochastic processes P_1 and P_2 consist of simple exponentially decaying pulses:

$$p(t) = h\exp[-(t-t_i)/\tau] \qquad (4)$$

where τ is the rate of decay of the pulse, t_i the onset time of the pulse, and h the height of the pulse at time t_i. These pulses are meant to simulate excitatory (since $h > 0$) postsynaptic potentials (EPSP's) caused by action potentials from neurons which connect onto this model neuron. The interval $I_i = t_i - t_{i-1}$ is a random variable with a gamma-type distribution:

$$P(I) = \frac{\lambda e^{-\lambda I}(\lambda I)^{s-1}}{\Gamma(s)} \qquad (5)$$

where Γ is the Gamma function. The mean interval is then given by $\langle I \rangle = s/\lambda$, while the interval variance is $\sigma_I^2 = s/\lambda^2$. In our work, P_1 and P_2 have the same statistics.

We have studied the occurrence of coincidences between an action potential (AP) and a single EPSP, as well as between an AP and two temporally-coincident EPSP's. This was done for various intensities D of the OU process as well as for various EPSP amplitudes h (all EPSP's in both streams have the same height). This was done by numerical integration of the above model equations as in [9]. At each time step, the values of the processes $P_1(t)$ and $P_2(t)$ are evaluated, as both involve one of many superimposed exponentials. To avoid stalling the simulation by accumulated EPSP's at one synapse or the other, the effect of an EPSP was neglected once its value decayed to $10^{-7}h$. For all simulations, action potentials were not counted if they occurred within 0.4 sec of the preceding AP (the refractory period in scaled units). Numerical implementation produces EPSP's whose heights vary randomly within a small interval, since event times determined by the gamma distribution do not coincide exactly with multiples of the integration time step.

RESULTS

In the absence of P_1 and P_2 inputs, the FHN model fires spontaneously at a rate that is well-fitted by a Kramers-type expression (not shown; see e.g. [10]) $\langle R \rangle \propto \exp(1/D)$. Thus, among other complications, and in contrast to a deterministic model such as that in [3], there are action potentials due to the background noise alone. We have investigated the effect of EPSP amplitude and noise intensity on the occurrence of single (SC) and double (DC) coincidences between action potentials and EPSP's, with a detection criterion determined as follows.

Detection Criterion

We have numerically determined the distribution of times between an action potential T_{AP} and the onset time T_{P_1} of the closest P_1 event before the action potential. For these simulations, we have set $P_2 = 0$. The background noise amplitude is set to $D = 10^{-6}$. The height h of the EPSP's is just at the deterministic threshold, in order that almost all action potentials be caused by P_1 events (rather than the background). We have checked that this distribution does not vary significantly for other values of D and h. The distribution of intervals $T_{AP} - T_{P_1}$ is shown in Fig.1. It is peaked around 0.09 sec, and is also of gamma-type as the intervals between P_1 events (although over much smaller intervals). In trying to pinpoint causal relationships between P_1 events and action potentials, we say that a "single coincidence" has occurred if $T_{AP} - T_{P_1} \in (0.035, 0.23)$, i.e. within the lower and upper limits of the distribution in Fig.1. Likewise, we say that a "double coincidence" has occurred if both $T_{AP} - T_{P_1}$ and $T_{AP} - T_{P_2}$ fall within the interval $(0.035, 0.23)$, where T_{AP} refers to the time of the **same** action potential. The event rates in each stream are low enough that DC's due to events in the same stream are very rare.

Coincidences versus EPSP Height

We have investigated the effect of EPSP height h on the number of SC's and DC's for a given noise intensity $D = 10^{-5}$. In a real neuron, this height varies stochastically from one event to the next; it is kept fixed in our simulations, although the numerical implementation makes them vary a bit (see previous section). Here, both P_1 and P_2 are nonzero. An action potential can either be due to an SC, and DC, or neither, but not both. Figure 2a shows the SC and DC numbers as fractions of the total number of action potentials obtained at a given value of h. We note that the SC fraction first decreases with increasing h while the DC fraction increases. We see that the DC fraction goes through a maximum approximately at $h = 0.018$, which is half the value of h that

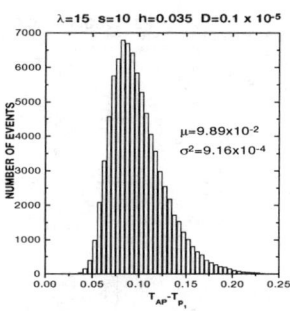

FIGURE 1. Distribution of intervals between the time of occurrence T_{AP} of an action potential and the time of onset T_{P_1} of the closest EPSP preceeding it. The intensity of the white noise is $D = 0.1 \times 10^{-5}$. The mean time between EPSP's is $\langle I \rangle = s/\lambda = 0.66$. The mean and variance are shown on the plot.

gives action potentials even in the absence of noise $D = 0$. In other words, the fraction of DC's goes down when a DC begins to be a suprathreshold event. After this point, the SC's become increasingly important to the detriment of the DC's. Note that the total fraction SC+DC is approximately constant for $h < 0.05$. At the highest values of h shown, where the number of SC events begins decreasing, the physiological relevance of the results becomes dubious since EPSP's then very strongly perturb the FHN model.

Coincidences versus Noise Intensity

We have also investigated the effect of noise intensity on SC and DC for a given EPSP height $h = 0.01$ (i.e. an EPSP is subthreshold without noise). Again, P_1 and P_2 are nonzero. Figure 2b shows that the fraction of SC's goes through a maximum around $D = 2.5 \times 10^{-6}$. The fraction of DC's decreases sigmoidally with increasing D. However, the absolute number of these latter double coincidences goes through a maximum with increasing D (not shown). The behaviors of the fractions are qualitatively the same for higher values of h. However, when one EPSP is suprathreshold, the fraction of SC's simply decreases sigmoidally with increasing D; this is expected since at low noise all the action potentials are due to single coincidences.

CONCLUSION

We have shown preliminary results of an analysis of coincidence detection in an excitable neuron governed by the Fitzhugh-Nagumo equations driven by two Gamma-distributed point processes and an Ornstein-Uhlenbeck process. We have computed the distribution of times between an action potential and

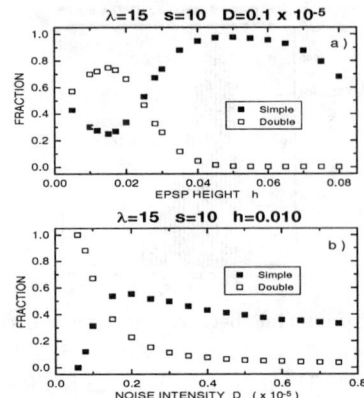

FIGURE 2. Fraction of the total number of action potentials corresponding to single coincidences (closed squares) and double coincidences (open squares), plotted as a function of a) the height h of an EPSP with $D = 0.1 \times 10^{-5}$, and b) the background noise intensity D with $h = 0.01$. The mean interval between either $P-1$ or P_2 events is 0.66. The detection criterion is based on the distribution in Fig.1.

the onset of an EPSP. This distribution is then used to determine whether action potentials are coincident with an EPSP. We have found that the number of single and double coincidences behave differently as a function of EPSP height and background noise intensity. Our results point to the complexity of possible computations that noisy "multiplying" neurons can perform, depending on the level of noise and other parameters. Further results (not shown) indicate that the detection criterion may have to be refined in order to decrease the number of false alarms and obtain a more accurate picture of the SC and DC detection processes.

REFERENCES

1. Koch, C., and Poggio, T., *Single Neuron Computation*, San Diego: Academic Press, 1992, ch. 12, pp.315-345.
2. Softky, W., *Neuroscience* **58**, 15 (1994).
3. Srinivasan, M.V., and Bernard, G.D., *Biol. Cybern.* **21**, 227 (1976).
4. Konig, P., Engel, A.K., and Singer, W., *Trends Neurosci.* **19**, 130 (1996).
5. Bugmann, G., *Network* **2**, 489 (1991).
6. Segrè, E., *Nuclei and Particles*, New York: Benjamin, 1964, ch. 3, pp.110-113.
7. Stemmler, M., *Network Comput. Neur. Syst.* **7**, 687 (1996).
8. Pei, X., Wilkens, L., and Moss, F., *Phys. Rev. Lett.* **77**, 4679 (1996).
9. Longtin, A., *J. Stat. Phys.* **70**, 309 (1993).
10. Collins, J.J., Chow, C., and Imhoff, T., *Phys. Rev. E* **R52**, 3321 (1995).

Similarity regime in the brain activity

E. Novikov*, A. Novikov*, D. Shannahoff-Khalsa*,
B. Schwartz† and J. Wright*

*Institute for Nonlinear Science, University of California, San Diego,
La Jolla, CA 92093 - 0402
†Scripps Research Institute, La Jolla

Abstract. The spectral analysis of multi-channel magnetoencephalographic data from a group of adults is performed. This analysis revealed a local similarity regime in brain activity, which is analogous to similarity regimes in other systems with strong interaction of many degrees of freedom (turbulence etc., see E. A. Novikov, Phys. Rev. E **50**, R3303 (1994) and references therein).

The study of systems with strong interactions of many degrees of freedom is one of the most important subjects in physics. Examples of these systems are turbulent flows of ordinary fluid and plasma, the global structure of the universe and the unified theory of fields (superstrings). Turbulence is more accessible (analytically, experimentally, and numerically) and its study can provide guidelines for the analysis of other systems. The nature of these systems and the mechanism of interaction can be quite different. However, in many of these systems we may expect characteristic cascade processes and a regime of scale similarity (see [1] and references therein).

The present work is a test of scale-similar activity in the brain. The human brain consists of about hundred billion neurons. Each neuron has up to ten thousand connections. Neurons are assembled in a hierarchical structure - from a small group of neurons to larger groups, etc. A flux of activity in this hierarchy can produce a similarity regime.

Data for this study was obtained from the Scripps Research Institute (La Jolla, CA), using a dual probe 37-channel "SQUID" magnetoencephalograph (the term "SQUID" is an acronym for a Superconducting QUantum Interference Device, see [2] for details). We did an analysis of the measurements of the normal component of the magnetic field on the brain surface from 70 channels in the range from 0.1 Hz to 115 Hz. The distance of each sensor from the skull is between 2cm and 2.5cm. The magnetometer was in a magnetically shielded room and the power of the environmental noise was, at least, one order of magnitude lower than that of the neural signals. Power-line interference (highly correlated and spectrally localized) was eliminated from the data.

The data presented here is from a healthy female (age 39) lying on her right side, while resting with her eyes closed. The data was recorded over 30 min with a sampling frequency of 231.5 Hz. The power spectra were calculated by the FFT method, averaging 50 intervals with 8192 points each.

The power spectrum of individual channels has peaks corresponding to typical brain rhythms [fig.1(a,b)]. However, when we took the spectrum of the difference between the signals of two channels, the peaks practically disappear in many cases and we obtained a similarity regime (power law) for, at least, two decades of frequencies [fig.1c]. The similarity parameters (α, β) and the error ε were determined by minimizing the mean square deviation

$$Q = <[\log F(\omega) - \alpha - \beta \log \omega]^2>, \varepsilon = \frac{Q^{1/2}}{<|\log F(\omega)|>} \quad (1)$$

where $F(\omega)$ is the power spectrum and brackets $<>$ indicate averaging over the range of frequencies ω. The range in this presentation was from 0.4 Hz to 40 Hz, but analogous results also hold for the ranges $0.2 - 20$ and $0.5 - 50$ Hz. The parameters $(\alpha, \beta, \varepsilon)$ were calculated for the spectra of 70 channels and for the spectra of signal differences for all 2415 pairs of channels. Let us denote $\{\}$ for averaging over all pairs of channels. The averaged parameters and rms deviations for the range 0.4-40 Hz are: $\{\alpha\} = 23.11, \sigma(\alpha) = 0.856, \{\beta\} = -0.980, \sigma(\beta) - 0.158, \{\varepsilon\} = 0.022, \sigma(\varepsilon) = 0.0060$, where $\sigma(\gamma) \equiv \{(\gamma - \{\gamma\})^2\}^{1/2}$ is the rms deviation for a characteristic γ, which can be α, β or ε. For the ranges $0.2 - 20$ Hz and 0.5-50 Hz the averaged errors are : $\{\varepsilon\} = 0.015$ and $\{\varepsilon\} = 0.025$ correspondingly.

The three-dimensional distances r between the positions of the sensors (channels) were from 2cm to 21cm. All pairs of channels were ordered with the increase of the distance between probes. We denote $\{\}_n$ the averaging over the first n pairs of channels. Fig.2 shows the global tendency of averaged error $\{\varepsilon\}_n$ to increase with the distance, but for some of the pairs of relatively close channels we still observed peaks in the spectra of the signal difference. Our interpretation of this data is that scale similarity is generally local, but there are some preferable regions in the brain (probably, functionally connected), for which scale similarity is more pronounced. The locations of some neighboring pairs of channels with relatively large ε may indicate the boundaries between such regions. We plan to use the parameters of scale similarity $(\alpha, \beta, \varepsilon)$ for a mapping of the brain and connect this mapping with anatomy and physiology. Similar data was taken from 18 other subjects and the analysis of the data from several subjects revealed the same phenomena of local scale similarity.

The similarity regime with even smaller errors was observed for the higher order moments of the signal difference [fig.3(a,b), the power spectrum was

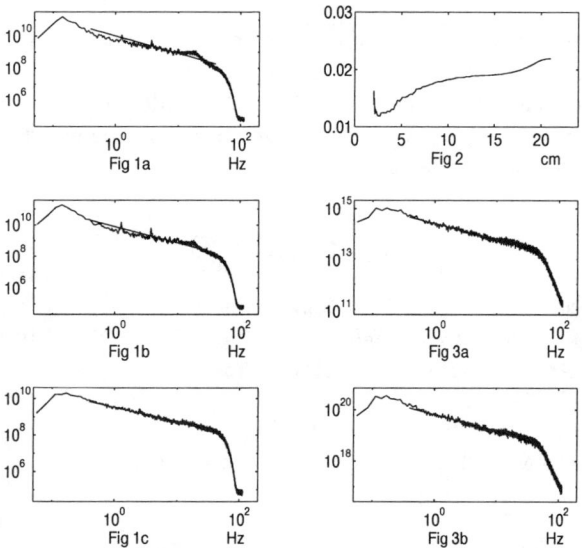

FIGURE 1. (a,b,c) Power spectra of second order $F(\omega)$: a) R16 (channel 16 in the right hemisphere), $\alpha = 23.1, \beta = -1.09, \varepsilon = 0.0217$; b) R32, $\alpha = 22.8, \beta = -1.03, \varepsilon = 0.0173$; c) R16 - R32 (signal difference), $r = 2.4cm$, $\alpha = 21.9, \beta = -0.783, \varepsilon = 0.00809$. **FIGURE 2.** Averaged error $\{\varepsilon\}_n$ as function of distance between channels r (see text). **FIGURE 3.** (a,b) Power spectra of fourth and sixth order for signal difference R16 - R32 : a) fourth order, $\alpha = 33.2, \beta = -0.628, \varepsilon = 0.00499$; b) sixth order, $\alpha = 45.6, \beta = -0.633, \varepsilon = 0.00360$.

taken from the square and cube of the signal difference]. This suggests the use of the infinitely divisible distributions [3,1] for the modeling of brain activity.

To the best of our knowledge, the similarity regime for such local characteristics of brain activity has not been previously observed. Brain rhythms are apparently synchronized (at least in functionally connected areas) and play a role similar to coherent structures in turbulent flow. The signal difference between channels is analogous to the velocity increment between points in turbulence in the sense that both quantities correspond to local structure and reveal the similarity regime (see [4,5] and references therein).

From these results we can conclude that there is a local similarity regime in brain activity. We plan to study the similarity regime with a larger group of subjects and in greater detail, including additional channels and a wider range of frequencies. Corresponding distributions and similarity exponents may be important characteristics of brain activity, that reflect a hierarchical processing of information. They may serve as a diagnostic tool for certain mental disorders (this work is in progress and the results will be presented elsewhere).

Let us stress that the local similarity regime observed in this work (in more than two decades of frequencies), in our opinion, places spontaneous brain activity in the general framework of the scale-similar phenomena for systems with strong interactions. It also provides new parameters for non-invasive experimental studies of the brain.

REFERENCES

1. E. A. Novikov, Phys. Rev. E **50**, R3303 (1994).
2. S. Saito (ed.), *Magnetoencephalography* (Raven Press, N.Y. , 1990)
3. W. Feller, *An Introduction to Probability Theory and its Applications* (Wiley, N.Y. 1991)
4. E. A. Novikov, Phys. Rev. A **46**, R6147 (1992)
5. G. Pedrizzetti & E. A. Novikov, J. Fluid Mech. **280**, 69 (1994)

Transition to Subthreshold Activity with the use of Phase Shifting in a Model Thalamic Network

Elizabeth Thomas and Thierry Grisar

Institut Leon Fredericq
University of Liège, Place Delcours 17
Bat L1, 4000 Liège, Belgium

Abstract. Absence epilepsy involves a state of low frequency synchronous oscillations by the involved neuronal networks. These oscillations may be either above or subthreshold. In this investigation, we studied the methods which could be utilized to transform the threshold activity of neurons in the network to a subthreshold state. A model thalamic network was constructed using the Hodgkin Huxley framework. Subthreshold activity was achieved by the application of stimuli to the network which caused phase shifts in the oscillatory activity of selected neurons in the network. In some instances the stimulus was a periodic pulse train of low frequency to the reticular thalamic neurons of the network while in others, it was a constant hyperpolarizing current applied to the thalamocortical neurons.

INTRODUCTION

Absence epilepsy is a pediatric epilepsy characterized by the appearance of low frequency (<10 Hz), high amplitude oscillations in the EEG of the patient during a seizure. Through the use of animal models, it has been found that the neuronal circuit which is most implicated in this problem is the thalamocortical circuit (3, 4). The isolated thalamus has been observed to generate low frequency synchronous oscillations (4). Due to its potential for pacemaking activities, many attempts to understand epilepsy have focused on the thalamus. In this report, we will focus on the question of how to make the oscillations of the thalamus subthreshold. In the subthreshold range, the thalamic neurons fail to reach the critical amplitude necessary to emit a driving signal that would entrain other coupled neurons (particularly cortical) into epileptiform activity.

The effects of both single as well as periodic stimuli on oscillatory activity in various excitable tissue have been studied (7, 8). A stimulus was frequently found to either shorten or lengthen the period of the perturbed cycle. The exact outcome of the perturbing stimulus was found to depend on factors such as the phase at which the stimulus was introduced, the amplitude of the stimulus and whether or not the stimulus was excitatory or inhibitory. The purpose of this study will be to make use of such phase shifts in neuronal oscillators in order to make the oscillations in our model subthreshold.

The model consisted of two reticular thalamic (RE) neurons and one thalamocortical (TC) neuron (figure 1). This circuit contains all the connections that have been found critical for the generation of low frequency oscillations in the thalamus (1, 4). As the name implies, the thalamocortical (TC) neurons project to the cortex and therefore provide a rhythmic driving force to the cortex during a seizure. The connection of the RE neurons on the other hand are only intra-thalamic. The question of whether subthreshold activity had been attained was therefore addressed to the TC neurons but not the RE neurons of the model.

THE MODEL

The model network architecture is illustrated in figure 1. The activity of the neurons was described using the Hodgkin Huxley framework. Neurons in this framework are modeled as RC circuits. The equations and parameters necessary to describe the oscillatory activity of thalamic networks have been previously developed (2). For a neuron j in the network, the change in membrane potential V_j was described as follows

$$C\frac{dV_j}{dt} = -I_{int} - I_{syn} + I_{app} \tag{1}$$

where C is the capacitance of the neuronal membrane, I_{int} is the sum of currents intrinsic to the neuron, I_{syn} (synaptic current) is the total current due to input from other coupled neurons and I_{app} is the current injected by an external source.

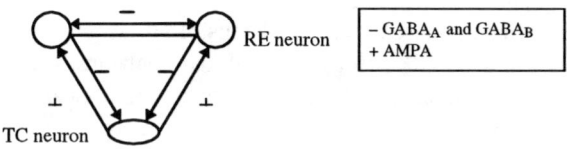

FIGURE 1. Circuit diagram of network (see Method section for explanation of $GABA_A$, $GABA_B$ and AMPA).

A current that tends to increase V_j is called depolarizing or excitatory while a current that decreases V_j is hyperpolarizing or inhibitory. An applied current $I_{app} > 0$ is therefore depolarizing while $I_{app} < 0$ is hyperpolarizing.

The contribution of any individual intrinsic current i_{int} to I_{int} can be described by the following equation

$$i_{int} = g(V_j, t)(V_j - E_{int}) \tag{2}$$

where $g(V_j, t)$ is the conductance function for i_{int}, and E_{int} is called the reversal potential of the current i_{int}. The direction of i_{int} switches when $E_{int} = V_j$. The intrinsic currents for the model TC neuron were the low threshold calcium, the hyperpolarization activated mixed cationic current, the fast sodium, fast potassium and the leak current. The intrinsic currents for the RE neurons were the low threshold calcium current, the fast sodium, fast potassium and leak current. The conductance function $g(V_j, t)$ for the model calcium currents were such that a depolarizing applied current $I_{app} > 0$, could lead unexpectedly to the hyperpolarization of the neuron (2).

The membrane potential of a neuron can also change as the result of input from all the coupled neurons (I_{syn}). The synaptic current i_{syn} due to the input from a neuron k to the neuron j can be expressed as:

$$i_{syn} = s_k w_{jk} r_j(t)(V_j - E_{syn}) \tag{3}$$

s_k is a switch function that ensures that a signal from neuron k is only sent to the connecting neurons if $V_k > V_{thresh}$. The threshold potential V_{thresh} for the release of a signal was 0 mV. If the activity of a neuron became subthreshold, it was no longer able to send out signals to the adjacent neurons. Input from the RE neurons have both a GABA$_A$ and GABA$_B$ component. The reversal potentials for these synaptic currents are $E_{GABAA} = -80$ mV and $E_{GABAB} = -103$ mV respectively. They therefore usually act to hyperpolarize the postsynaptic neuron (neuron receiving the signal). Input from the TC neurons are mediated by the AMPA synapse with $E_{AMPA} = 0$ mV and usually act to depolarize the postsynaptic neuron.

The phase as reported in this paper has a value between 0 and 1. It was computed as the ratio δ/T_o. Where δ is the time between the previous marker event (the neuron reaching threshold) and the stimulus. The normal period of the neuronal oscillatory activity without any perturbation is T_o (figure 2).

The success of an experiment in each case was judged by the number of cycles in which the TC activity became subthreshold. The fraction of activity above threshold was the number of cycles in which the TC neuron activity was above threshold compared to the total number of cycles.

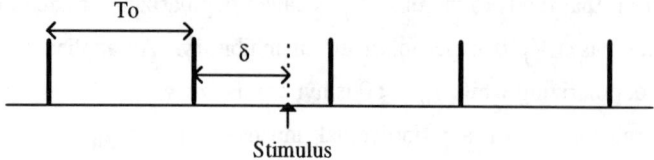

FIGURE 2. The vertical black bars represent neuronal firing once threshold in reached. The phase at which the stimulus is introduced is described as δ/T_o. Where δ is the duration between the marker event (neuron reaching threshold) and the introduction of stimulus I_{app}; T_o is the unperturbed cycle length.

EPILEPTIFORM ACTIVITY

An initial pulse of 1 ms applied to the TC neuron of the model network resulted in the autorhythmic activity of the network (figure 3a). Such low frequency autorhythmic activity in the TC neurons of the thalamocortical network plays a crucial role in maintaining epileptic activity. The frequency of the neuronal oscillations in the model was about 5 Hz.

The intrinsic calcium current I_T in the RE neurons, has been found to be an important control parameter in the transition of the system from an oscillatory to non-oscillatory state (5). In some experimental situations, low frequency oscillations in the thalamic system have been observed to come about as the result of a distributed mechanism in which the oscillations of the TC and RE neurons were interdependent (6). In other experimental preparations, the network has been found to function more like a system of coupled independent oscillators (4). In our model, we found that one of the factors that determined whether the system functioned in a distributed manner or whether it functioned as a system of coupled independent oscillators, was the conductance of the intrinsic current I_T in the RE neurons (unpublished observation). Since both mechanisms have been observed experimentally, we carried out our tests on both these oscillating systems.

SUBTHRESHOLD ACTIVITY

Phase shifting with a periodic stimulus

A periodic, excitatory stimulus applied to the RE neurons of the network were found capable of completely suppressing threshold activity in the TC neuron. Figure 3b demonstrates the completely subthreshold activity of the TC neuron as the result of the periodic excitation I_{app} applied to the RE neurons every 180 ms beginning at the phase 0.9. The amplitude of each pulse was 5 µA/cm^2 lasting for 1 ms. The period of the TC neuron activity in the subthreshold range was about 180 ms. This is slightly lower than the period of the unperturbed cycle which is 200 ms. The lowered periodicity was due to the stimulus in the RE neuron which slightly advanced the firing of the RE neuron and consequently the firing of the TC neuron. The properties of the applied current I_{app} were varied in order to ensure that subthreshold activity could be obtained for a range of stimulus parameters. The pulse amplitude was varied over the range 2-5 µA/cm^2. The pulse for each of these amplitudes was introduced at the phases 0.9, 0.925 and 0.95 with the periods 180, 185 and 190 ms respectively. The fact that significant suppression of TC threshold activity was able to take place over this range of amplitude and phase values indicates that such procedures may be successful in the experimental situation. Similar success was obtained with both the distributed system as well as the system of coupled independent oscillators.

Phase shifting with constant hyperpolarization

We next applied periodic hyperpolarizing pulses (I_{app}<0) to the TC neurons in order to achieve subthreshold activity through an increase of the TC neuron cycle

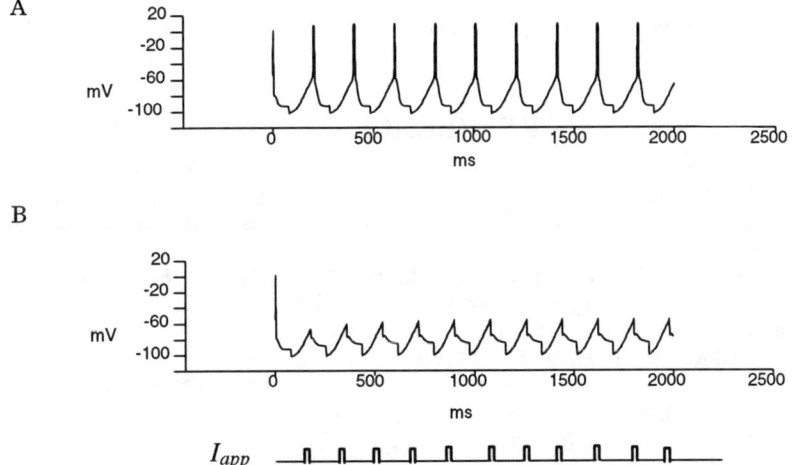

FIGURE 3. A) Activity of the TC neuron without any perturbing inputs. B) Subthreshold activity of the TC neuron following the application of periodic depolarizing pulses I_{app} to the RE neurons of the network.

TABLE 1. Results from tests in which a constant hyperpolarizing current was applied to the TC neurons. The fraction of TC neuron activity above threshold is displayed for each amplitude of current utilized in separate tests.

I_{app} ($\mu A/cm^2$)	Fraction of TC activity above threshold
-0.5	1
-1.0	0
-1.5	0
-2.0	0

length. We found however, that the use of a periodic hyperpolarizing input did not result in the stable entrainment of the TC neurons and significant subthreshold activity was not obtained. A constant low amplitude hyperpolarizing current to the TC neuron however, was able to accomplish this task. Table 1 displays the amplitudes of hyperpolarizing currents delivered to the TC neurons and the fraction of TC activity above threshold. The critical mechanism for the success of this method once again, was shifting the firing of the TC neuron with respect to the RE neuron so that the RE neuron consistently fired ahead. Unlike the case with delivering excitatory pulses to the RE neuron, the period of network oscillations had increased to 220 ms.

ACKNOWLEDGEMENTS

This work has been funded by the FNRS and Leon Fredericq Foundation of Belgium.

REFERENCES

1. Crunelli, V., J.P. Turner, S.R. Williams, A. Guyon, N. Leresche N., *Idiopathic Generalized Epilepsies: Clinical, Experimental and Genetic Aspects*. John Libbey and Company Ltd., 1994, pp. 201-213.
2. Lytton W. and E. Thomas. Modeling thalamocortical oscillations. In *Cererebral Cortex. Volume 13. Models of Cortical Circuitry*, New York, Plenum Press, 1997. (in press).
3. Marescaux, C., M. Vergnes and A. Depaulis. *J. Neural Transm. [Suppl]* **35** 37 69 (1992).
4. Steriade, M. and R.R. Llinas. *Physiological Review* **68** 649-742 (1988).
5. Thomas E. and Grisar T. *Abstracts of the Society for Neuroscience* **22** 822.22 (1996).
6. von Krosigk, M., T. Ball, D.A. McCormick. *Science* **261** 361-364 (1993).
7. Winfree A. T. *Science* **197** 761-763 (1977).
8. Winfree A.T. When time breaks down: The three dimensional dynamics of electrochemical waves and cardiac arrythmias. Princeton: Princeton University Press (1987).

Mass separation by ratchets

B. Lindner*, L. Schimansky-Geier*, P. Reimann[†], and P. Hänggi[†]

*Humboldt-University at Berlin, Invalidenstr.110, D-10115 Berlin, Germany
[†]University of Augsburg, Memminger Str. 6, D-86135 Augsburg, Germany

Abstract. The fluctuation-induced transport of massive Brownian particles is investigated. Analytic approximations for their current in periodic 'ratchet'-potential driven additively by the Ornstein-Uhlenbeck noise are in qualitative agreement with the result of numeric simulations. The ability to separate particles with different masses in situations with a constant bias is discussed.

BROWNIAN PARTICLES IN RECTIFIERS

We consider the one-dimensional motion of a Brownian particle with coordinate $x = x(t)$, mass m, and viscous friction η,

$$m\ddot{x} = -\eta\dot{x} - V'(x) + F + y(t) + \sqrt{2D}\,\xi(t) , \qquad (1)$$

in an asymmetric periodic potential (ratchet) $V(x)$ of period L. F stands for an additional 'load force'. Both last expressions in (1) are noisy source terms. Thermal fluctuations are modelled by the Gaussian white noise $\xi(t)$. Its intensity D is given due to an Einstein relation by $D = \eta k_B T \equiv \eta/\beta$ and, hence, the particle performs Brownian motion in a thermal bath with temperature T. Furthermore, we introduce the action of external fluctuating forces $y(t)$ on the particle which exposes the system out of equilibrium.

We assume $y(t)$ to be an Ornstein-Uhlenbeck-process with zero average and correlation time τ. The intensity $\int \langle y(t)y(0)\rangle dt$ is labelled by Q and is independent of the correlation time τ. It is convenient to express Q in units of the thermal noise strength D introducing R by $Q = RD$.

Overdamped situations $\eta \to \infty$ of eq.(1) with different types of $y(t)$ was the topic of a lot of studies [1,2], recently, pointing in their application to biological systems. A rich material for several physical situations of driving, either by periodic or by stochastic forces y, was investigated in detail (for review see [2]). Also the fluctuation induced transport was verified experimentally [3].

Less well elaborated is the case where inertial effects of the particles come into play. Only in [4] a complex behaviour of the current for a periodic external force without thermal noise was reported. The inclusion of inertial effects, however, would generalize the effect of stochastic ratchets and achieves importance in technical applications, especially if particles should be separated by ratchets.

The quantity of foremost interest is the steady state particle current $\langle \dot{x} \rangle$ [1]- [4]. For zero load $F = 0$ in two limits, in the white noise limit $\tau \to 0$ as well as in the zero amplitude limit $\tau \to \infty$, an equilibrium-like situation is approached. The stationary distribution achieves a canonical like shape $P^{st} = P(H)$ with H being the Hamilton-function of the particle. The mean current vanishes in both limits. In the intermediate regime $0 < \tau < \infty$ the external force $y(t)$ violates the detailed balance in the systems and the non equilibrium situation will exhibit a non-vanishing current $\langle \dot{x} \rangle$ [2].

In the following sections two different approximation schemes are presented and compared with results of numerical simulations. Specifically, we will address the dependence of the current $\langle \dot{x} \rangle$ upon the correlation time τ and the particle mass m, assuming thereby all other parameters are kept fixed. In numerical evaluations we use the special potential: $V(x) = -[\sin(2\pi x) + 0.25 \sin(4\pi x)]/(2\pi)$.

UNIFIED COLORED NOISE APPROXIMATION

The unified colored noise approximation (UCNA) has originally been developed for overdamped stochastic dynamics driven by OUP. Later refinements and generalisations have been elaborated [5]. It has proved to yield good approximations over wide parameter regimes in different situations [6] and was applied already to fluctuation induced transport [7].

The objective in the UCNA is to find an approximate Markovian description of the generally intractable non-Markovian dynamics (1) [6]. First a non-linear coordinate transformation to (approximately) decoupling stochastic variables is performed. In a second step, a separation of time scales for those new variables is established, thus admitting the adiabatic elimination of the 'fast' ones.

Adapting this general line to (1) we find expressions for small correlation times τ and, simultaneously, for a strongly overdamped dynamics $m/\eta \to 0$. Within these restrictions, the following approximate Langevin equation (in Stratonovich interpretation) as Markovian approximation of (1) is derived:

$$\eta g(x) \dot{x} = -V'(x) + F + \sqrt{2D(1+R)}\, \xi(t) \qquad (2)$$

where the state- and mass-dependent dressing $g(x)$ of the friction reads

$$g(x) = 1 + \frac{d}{dx} \frac{\tau R[V'(x) - F]}{(1+R)(\eta + \frac{m}{\tau}) + \tau V''(x)}. \qquad (3)$$

The steady state probability current J in (2) follows by means of a standard calculation [8]. With $\langle \dot{x} \rangle = JL$ we arrive at

$$\langle \dot{x} \rangle = \frac{L(1+R)\left[1 - e^{\beta \Phi(L)}\right]}{\beta \eta \int\limits_0^L dx\, g(x)\, e^{-\beta \Phi(x)} \int\limits_x^{x+L} dy\, g(y)\, e^{\beta \Phi(y)}}. \quad (4)$$

Here we introduced an effective potential

$$\Phi(x) = \int_0^x \frac{g(y)}{1+R}\left[V'(y) - F\right] dy. \quad (5)$$

In the white noise limit $\tau \to 0$, the current $\langle \dot{x} \rangle_{\tau=0}$ predicted by (4) is independent of m and coincides with the exactly solvable case $F \neq 0, \tau = 0, m = 0$ [8]. If $F = 0$, the current is generically non-zero if $0 < \tau < \infty$ for non-symmetric potentials $V(x)$. The asymptotic behaviour of (4) for small τ and zero load $F = 0$ is obtained as

$$\langle \dot{x} \rangle = -\frac{\tilde{\tau}^2 L R}{\eta^3 A (1+R)^2} \int_0^L V'(y) V''(y)^2 \, dy \quad,\quad \tilde{\tau} = \frac{\tau}{1 + m/\eta\tau} \quad (6)$$

$$A = \int\limits_0^L dx \int\limits_x^{x+L} dy\, e^{\beta[V(y) - V(x) + (x-y)F]/(1+R)}. \quad (7)$$

Thus a τ^2 decay for moderately small τ is predicted, crossing over to a τ^4 decay for extremely small τ. The m-dependence of the UCNA result (4) can be completely absorbed into the renormalised correlation time $\tilde{\tau}$, for this reason the maximum of $\langle \dot{x} \rangle$ is independent of m.

To verify the approximative results we performed numerical simulations of the Langevin equation (1) using the algorithm of Fox [9]. For each set of parameter values we integrated the stochastic dynamics over 10^7 time steps $\Delta t = 10^{-2}$. This was repeated 20 times to obtain the average current and its estimated accuracy. Since the relative numerical error increases both with decreasing η and decreasing τ we could not reach the deep asymptotic regime, assumed in our derivation of the UCNA result (4). The comparison for moderately large η and moderate-to-small τ of the UCNA with the simulations is depicted in Fig. 1 (left) for different m-values. The maximum of the UCNA shifts in τ with increasing mass while the simultaneous decreasing of the maximum is not predicted.

PATH INTEGRAL APPROACH

In this section we aim the calculation of the steady state current by help of path-integrals. As in quantum mechanics the re-formulation of stochastic dynamics yields a compact representation [10,11]. In practise, however, a further analytical evaluation of the resulting expressions is possible for weak noise only, i.e. for situations where the (effective) potential barriers between adjacent local minima are large compared to the strength of the fluctuations.

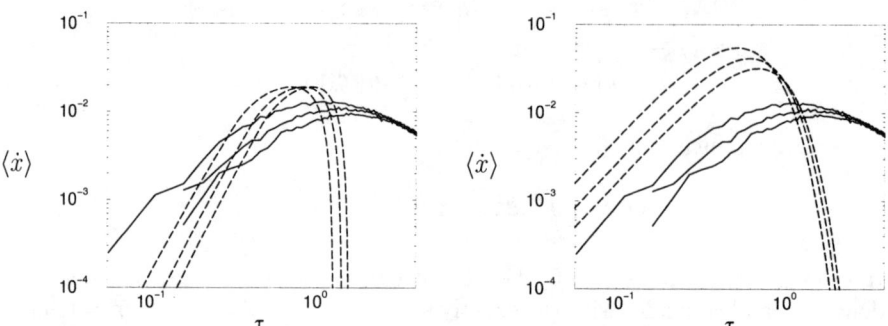

FIGURE 1. Particle current vs. correlation time. Left figure: UCNA (dashed) and numerical simulations (solid). From left to right: $m = 0.5, 1, 1.5$. ($\eta = 2$, $D = \eta/\beta = 0.1$, $Q = RD = 0.5$, $F = 0$). Right figure: Path-integrals (dashed) and simulations (solid).

Within this restriction the current can be approximated by

$$\langle \dot{x} \rangle = L[k_+ - k_-], \qquad (8)$$

where $k_+(k_-)$ are the hopping rates to the right (left) between neighbouring local minima.

For small temperatures (large β) these rates approach an Arrhenius-like dependence $k_\pm = \zeta_\pm \exp \beta \Delta \Phi_\pm$ where $\Delta \Phi_\pm$ are temperature-independent 'effective' potential barriers. The $\zeta_\pm$ are prefactors with a much weaker temperature dependence.

In the case of small τ [12] we obtain

$$\Delta \Phi_\pm(\tau) = \frac{V(x^\#) - V(x_\pm) + (x_\pm - x^\#)F}{1+R} + \tau^2 \frac{\eta R}{(1+R)^2} \int_{-\infty}^{\infty} \ddot{q}_\pm^2(t)\, dt, \qquad (9)$$

where $x^\#$ is one of the local maxima of $V(x) - xF$ and x_+ and $x_- = x_+ - L$ its neighbouring local minima to the right and left, respectively. The functions $q_\pm(t)$ are the trajectories found from $m\ddot{q}_\pm(t) = -\eta \dot{q}_\pm(t) - V'(q_\pm(t)) + F$ with boundary conditions $q_\pm(t = -\infty) = x^\#$ and $q_\pm(t = \infty) = x_\pm$.

Concerning $\zeta_\pm$ we restrict ourselves to the 0-th order approximation $\zeta(\tau) \simeq \zeta(\tau = 0)$ with the effective temperature $T(1+R)$. Closer inspection involving detailed-balance arguments as well as explicit perturbation calculations [13] have shown that the identity $\zeta_+(\tau = 0) = \zeta_-(\tau = 0)$ should hold true in the spatial diffusion regime whenever the concept of an escape rate makes sense. We thus infer that (with $\Delta \Phi_\pm^{(1)}$ being the second term in (9))

$$\langle \dot{x} \rangle = B[e^{-\beta \tau^2 \Delta \Phi_+^{(1)}} - e^{-\beta \{\tau^2 \Delta \Phi_-^{(1)} + LF/(1+R)\}}] \qquad (10)$$

and where $B = L k_+(\tau = 0)$. We will utilise here our observation from the previous section that the current $\langle \dot{x} \rangle_{\tau=0}$ is apparently almost m-independent.

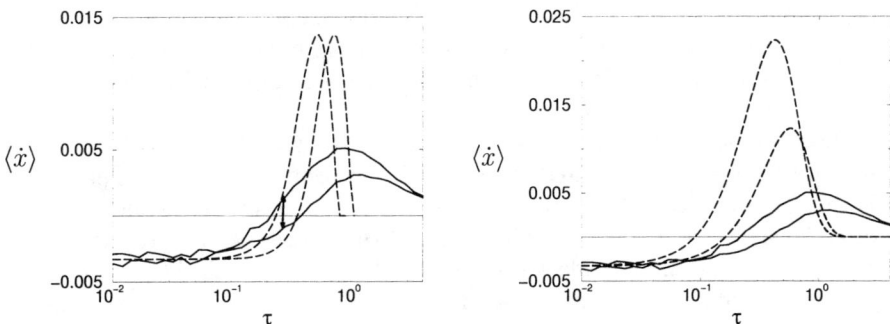

FIGURE 2. Same as in Fig.1 but for $F = -0.01$, $Q = 0.25$, $m = 0.5$ (curve, that reaches first the maximum), $m = 1.5$. The arrow indicates a τ-value allowing for mass separation.

So, the same m-independence is inherited by B and by setting $m = 0$ we obtain the approximative prefactor

$$B = L(1+R)/\beta \eta A \tag{11}$$

and A is due to (7). For zero load $F = 0$ this yields the leading order small-τ behaviour

$$\langle \dot{x} \rangle = -\frac{\tau^2 L R}{A(1+R)} \int_{-\infty}^{\infty} [\ddot{q}_+(t)^2 - \ddot{q}_-(t)^2] \, dt. \tag{12}$$

The comparison of the path integral prediction (10), (11) with numerical simulations is shown in Fig. 1. The agreement is rather satisfactory up to about $\tau = 0.5$. The shift and the decreasing of the maximum are well described. In particular, the asymptotics (12) seems to agree better with the numerics than that from the UCNA approach (7).

MASS SEPARATION

Below we obtained a displacement of the maximal current for increasing values of the mass m. This shift can be used with the purpose to separate mesoscopic particles with different masses. Beside a size-depending separation due to different friction constants of the particles, the separation by mass is a new and independent possibility, as easy to survey in the case of particles of different masses but of the same size.

With zero load $F = 0$ flux reversals do not occur in the considered case. Otherwise, adding a constant force against the preferred direction of the ratchet a flux reversal will be exhibited in a finite region of τ. Beginning from a value τ_1 until a second value τ_2 the noise-induced current overcompensates the action of the (small) load force. Both values, τ_1 and τ_2 depends on the mass of the particle. Hence for two different values of the mass non-overlapping regions of the flux reversals are possible which yields the separation of the two species of particles.

Results of simulations and of the estimations from the UCNA and the path-integral approach are depicted in Fig. 2. For a specific range of τ the particles have different signs of velocity, and hence flow in the average in different directions. An increase of the differences of the masses would strengthen the speed of separation and enlarge the region of possible correlation times.

In conclusion we have investigated inertial effects of Brownian particles in sawtooth-like potential. In the region of moderate to strong damping and for small correlation times of the considered external forces we found numerically a shift of the maximal flux with increasing mass towards larger τ. The observation was confirmed by two approximation schemes. This dependence of the flux on the mass can be exploited for separation of particles with different inertial properties.

REFERENCES

1. Magnasco, M. O., *Phys. Rev. Lett.* **71**, 1477 (1993).
2. Doering, C. R., *Il Nuovo Cim.* **17 D**, 685 (1995); Hänggi, P. and Bartussek, R., in: Parisi, J., Müller, S. C., and Zimmermann, W. (eds.) *Nonlinear Physics of Complex Systems-Current Status and Future Trends*, Lectur Notes on Physics, vol. 476, Springer, Berlin, 1996, pp. 294-308 .
3. Rousselet, J., Salome, L., Ajdari, A., and Prost, J.: *Nature* **370**, 446 (1994); Faucheux, L. P., Bourdieu, L. S., Kaplan, P. D., and Libchaber, A. J., *Phys. Rev. Lett.* **74**, 1504 (1995).
4. Jung, P., Kissner, J. G., and Hänggi, P., *Phys. Rev. Lett.* **76**, 3436 (1996).
5. Jung, P. and Hänggi, P., *Phys.Rev. A* **35**, 4464 (1987).
6. Hänggi, P. and Jung, P., *Adv. Chem. Phys.* **89**, 239 (1995).
7. Bartussek, R., Hänggi, P., Lindner, B., and Schimansky-Geier, L., *Physica D* in press.
8. Stratonovich, R. L., *Topics in the Theory of Random Noise*, Vol. 2, New York: Gordon and Breach, 1967; Büttiker, M., *Z. Phys. B* **68**, 161 (1987).
9. Fox, R. F., Gatland, I. R. , Roy, R., and Vemuri, G., *Phys. Rev. A* **38**, 5938 (1988); Lindner, B., Masther thesis, Humboldt University at Berlin, 1996.
10. Graham, R. and Tel, T., *Phys. Rev. A* **31**, 1109 (1985); Hänggi, P., *Z. Phys. B* **75**, 275 (1989); Wio, H. S. et al. *Phys. Rev. A* **40**, 7312 (1989).
11. Eichcomb, S. B. J. and McKane, A. J., *Phys. Rev. E* **51**, 2974 (1995).
12. Reimann, R., *Phys. Rev. E* **52**, 1579 (1995); Rattray, K. M. and McKane, A. J., *J. Phys. A* **24**, 1215 (1991).
13. Pollak, E. and Talkner, P., *Phys. Rev. E* **47**, 922 (1993).

Stability and Chaos in an Inertial Two Neuron System

Diek W. Wheeler[1] and W. C. Schieve

Ilya Prigogine Center for Studies in Statistical Mechanics and Complex Systems
and
Physics Department, The University of Texas, Austin, TX 78712

Abstract.
Inertia is added to a continuous-time, Hopfield [1] effective-neuron system. We explore the effects on the stability of the fixed points of the system. A two neuron system with one or two inertial terms added is shown to exhibit chaos. The chaos is confirmed by Lyapunov exponents, power spectra, and phase space plots.

INTRODUCTION

To simplify the study of large, Hopfield neural networks [1], Schieve, et al., [2] developed the notion of <u>effective</u> neurons. Their concept uses Haken's slaving principle [3,4], which describes the adiabatic elimination of fast-relaxing slave variables in favor of slow-varying master variables in solutions to the long-time dynamics of a system.

The basic effective-neuron equation is derived via Haken's slaving principle [3,4] from the deterministic Hopfield model [1],

$$\eta_i \dot{U}_i = -K_i U_i + \sum_{\substack{j=1 \\ j \neq i}}^{N} J_{ij} \tanh(U_j), \qquad (1)$$

where U is the neuron potential, η is the resistance term, K is related to the capacitance, and the J_{ij} terms are the connection strengths between neurons. The main distinction between an effective neuron and a regular Hopfield neuron is the presence of self-connection terms which arise naturally from the derivation.

The object here is to show how the positioning and behavior of the fixed points of the system are affected when an inertial/inductance term is added

---

[1] D.W.W. received partial support from Robert A. Welch Foundation Grant F-0365.

to the effective-neuron equation. The inertia may generate chaos in neural systems [5–8].

Some biological support exists for an inductance term in a neural equation [9–13]. However, in biological neural circuitry, the inductance is often phenomenological rather than actual.

TWO NEURONS WITH ONE INERTIAL TERM

For two effective neurons, only one of which has an added inertial term, the system is described by three, first-order differential equations,

$$\dot{U}_1 = U_2, \tag{2}$$

$$\dot{U}_2 = -\frac{\eta_1}{M_1} U_2 - \frac{K_1}{M_1} U_1 + \frac{J_{11}}{M_1} \tanh(U_1) + \frac{J_{13}}{M_1} \tanh(U_3), \tag{3}$$

$$\dot{U}_3 = -\frac{K_3}{\eta_3} U_3 + \frac{J_{31}}{\eta_3} \tanh(U_1) + \frac{J_{33}}{\eta_3} \tanh(U_3). \tag{4}$$

The fixed points are found by setting the first time derivatives to zero,

$$0 = U_2, \tag{5}$$
$$0 = -K_1 U_1 + J_{11} \tanh(U_1) + J_{13} \tanh(U_3), \tag{6}$$
$$0 = -K_3 U_3 + J_{31} \tanh(U_1) + J_{33} \tanh(U_3). \tag{7}$$

All of the fixed points occur at $U_2 = 0$. Coordinates U_1 and U_3 must be found numerically. It is evident that the locations of the fixed points are <u>independent</u> of the inertia.

Although the coordinates of the fixed points are unchanged by the inertial term, their behavior is modified. To determine the stability of the fixed points, the characteristic equation,

$$\begin{vmatrix} -\lambda & 1 & 0 \\ -\frac{K_1}{M_1} + \frac{J_{11}}{M_1}\text{sech}^2(U_1) & -\frac{\eta_1}{M_1} - \lambda & \frac{J_{13}}{M_1}\text{sech}^2(U_3) \\ \frac{J_{31}}{\eta_3}\text{sech}^2(U_1) & 0 & -\frac{K_3}{\eta_3} + \frac{J_{33}}{\eta_3}\text{sech}^2(U_3) - \lambda \end{vmatrix} = 0, \tag{8}$$

is evaluated. This results in

$$0 = M_1 \eta_3 \lambda^3 + [\eta_1 \eta_3 + M_1(K_3 - J_{33}\text{sech}^2(U_3))]\lambda^2 +$$
$$[\eta_1(K_3 - J_{33}\text{sech}^2(U_3)) + \eta_3(K_1 - J_{11}\text{sech}^2(U_1))]\lambda +$$
$$[(K_1 - J_{11}\text{sech}^2(U_1))(K_3 - J_{33}\text{sech}^2(U_3)) - J_{13}J_{31}\text{sech}^2(U_1)\text{sech}^2(U_3)], \tag{9}$$

which is solved numerically for the roots. We now assume $\eta_1 = \eta_3 = K_1 = K_3 = 1$ in equation (9) to reduce the parameter search space.

Chaos might be found when the origin is unstable, and the only fixed point. The value of M_1 affects the behavior of the origin, but not the number of fixed points. Through numerical search, a connection matrix

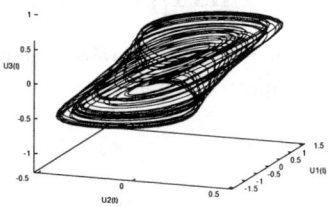

FIGURE 1. Phase space plot for $M_1 = 2.5$. After a delay of 1000 seconds, the plot was iterated for another 1000 seconds.

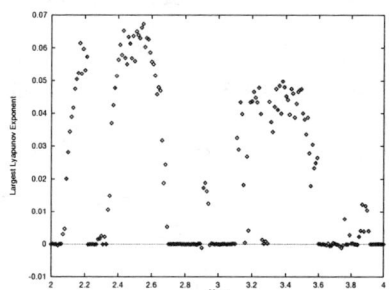

FIGURE 2. Largest Lyapunov Exponent vs. Mass.

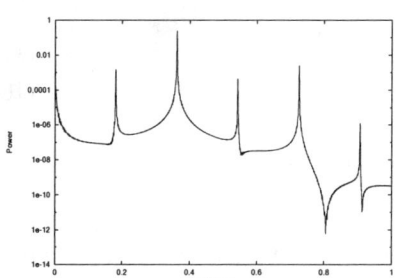

FIGURE 3. Power Spectral Density for $M_1 = 2.0$. The power peaks are relatively sharp.

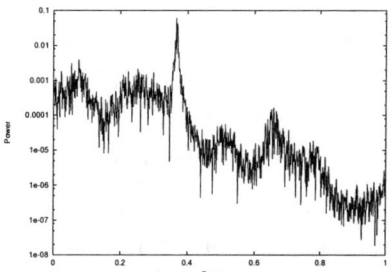

FIGURE 4. Power Spectral Density for $M_1 = 2.5$. The baseline and power peaks exhibit signs of broadband noise.

$$J = \begin{vmatrix} 0.43 & 1.50 \\ -0.25 & 1.44 \end{vmatrix} \tag{10}$$

was found to yield a strange attractor for $M_1 = 2.5$ (Figure 1). Further exploration led to the discovery that by changing the value of M_1, one could pass in and out of chaotic regimes. For this particular connection matrix, chaos is only found between the values of $M_1 = 2.0$ and $M_1 = 4.0$. The chaotic regimes appear only in bands between these mass values as depicted in Figure 2. For all points outside the graph, the Lyapunov exponents are all zero. The Lyapunov exponents were calculated using data sets generated by a fourth-order Runge-Kutta algorithm and code provided by Alan Wolf.

A comparison of power spectra for U_1 (Figures 3-4) lends support to the existence of chaos in this system. The power spectrum for $M_1 = 2.0$ in Figure 3 only shows spikes. The spectrum in Figure 4 for $M_1 = 2.5$ exhibits signs of broadband noise from spectral coupling of the system's nonlinearities. Further investigation of the dynamics for varying the parameter values is continuing.

TWO NEURONS WITH TWO INERTIAL TERMS

Four, first-order differential equations are required for a system of two effective neurons when each neuron has an added inertial term,

$$\dot{U}_1 = U_2, \tag{11}$$

$$\dot{U}_2 = -\frac{\eta_1}{M_1}U_2 - \frac{K_1}{M_1}U_1 + \frac{J_{11}}{M_1}\tanh(U_1) + \frac{J_{13}}{M_1}\tanh(U_3), \tag{12}$$

$$\dot{U}_3 = U_4, \tag{13}$$

$$\dot{U}_4 = -\frac{\eta_3}{M_3}U_4 - \frac{K_3}{M_3}U_3 + \frac{J_{31}}{M_3}\tanh(U_1) + \frac{J_{33}}{M_3}\tanh(U_3), \tag{14}$$

The first time derivatives are set to zero,

$$0 = U_2, \tag{15}$$
$$0 = -K_1 U_1 + J_{11}\tanh(U_1) + J_{13}\tanh(U_3), \tag{16}$$
$$0 = U_4, \tag{17}$$
$$0 = -K_3 U_3 + J_{31}\tanh(U_1) + J_{33}\tanh(U_3). \tag{18}$$

All of the fixed points occur at $U_2 = U_4 = 0$. Again, coordinates U_1 and U_3 can only be found numerically. As in the case with a single added inertial term, the inertia has no affect on the positions of the fixed points.

The stability of the fixed points is found by evaluating the characteristic equation,

$$\begin{vmatrix} -\lambda & 1 & 0 & 0 \\ -\frac{K_1}{M_1} + \frac{J_{11}}{M_1}\text{sech}^2(U_1) & -\frac{\eta_1}{M_1} - \lambda & \frac{J_{13}}{M_1}\text{sech}^2(U_3) & 0 \\ 0 & 0 & -\lambda & 1 \\ \frac{J_{31}}{M_3}\text{sech}^2(U_1) & 0 & -\frac{K_3}{M_3} + \frac{J_{33}}{M_3}\text{sech}^2(U_3) & -\frac{\eta_3}{M_3} - \lambda \end{vmatrix} = 0. \tag{19}$$

This results in

$$0 = M_1 M_3 \lambda^4 + [M_1 \eta_3 + \eta_1 M_3]\lambda^3 +$$
$$[\eta_1 \eta_3 + M_1(K_3 - J_{33}\text{sech}^2(U_3)) + M_3(K_1 - J_{11}\text{sech}^2(U_1))]\lambda^2 +$$
$$[\eta_1(K_3 - J_{33}\text{sech}^2(U_3)) + \eta_3(K_1 - J_{11}\text{sech}^2(U_1))]\lambda +$$
$$[(K_1 - J_{11}\text{sech}^2(U_1))(K_3 - J_{33}\text{sech}^2(U_3)) - J_{13}J_{31}\text{sech}^2(U_1)\text{sech}^2(U_3)]. \tag{20}$$

Once again to ease the evaluation of the above equation, we will set $\eta_1 = \eta_3 = K_1 = K_3 = 1$.

As before, the origin is the only fixed point where chaos is found. The values of M_1 and M_3 will alter the behavior of the origin, but not the number of fixed points. The same connection matrix as was used in the case for a single added inertial term was found to yield a strange attractor for $M_1 = 3.5$ and $M_3 = 1.0$. Results show that by changing the value of M_1 and holding M_3

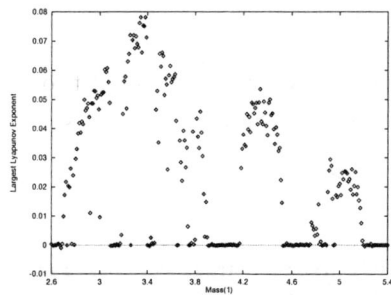
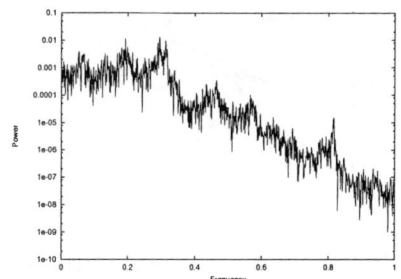

FIGURE 5. Largest Lyapunov Exponent vs. Mass. $M_3 = 1.0$ while M_1 is varied with respect to the largest Lyapunov exponent.

FIGURE 6. Power Spectral Density for $M_1 = 3.5$ and $M_3 = 1.0$. The spectrum shows obvious signs of broadband noise.

fixed, one can explore the chaotic regimes of the system. Chaos appears only between the values of $M_1 = 2.6$ and $M_1 = 5.4$ for this particular connection matrix. As in the single inertial term case, the chaotic regimes appear in bands between the mass values depicted in Figure 5. The areas outside of the graph all have Lyapunov exponents of zero.

A plot of the power spectrum for U_1 at $M_1 = 3.5$ and $M_3 = 1.0$ (Figure 6) shows added evidence for chaos in this two effective-neuron system. The spectrum is greatly affected by broadband noise in both the baseline and the power peaks.

There has been some other work on small neural systems [5,15–18], but to the best of our knowledge, the model presented here describes the lowest dimensional, autonomous, continuous-time, neural system that exhibits chaos.

CONTROL

We have begun to explore the effects of control on the neural system with a single added inertial term. In particular, we are examining entrainment with arbitrary goal dynamics as a means of controlling its behavior [19,20]. Entrainment control works through the gradual overlaying of the goal dynamics onto the normal system dynamics. We selected entrainment control because it allows us to direct the system from a chaotic attractor to a stable limit cycle, or to any other dynamical system we choose, as demonstrated in Figures 7-8. We are currently trying to find further evidence for a theoretical limit on the parameters needed for full control of a system with arbitrary goal dynamics.

REFERENCES

1. Hopfield, J. J., *Proc. Natl. Acad. Sci. USA* **81**, 3088 (1984).

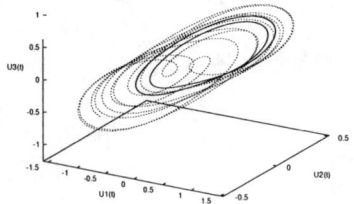 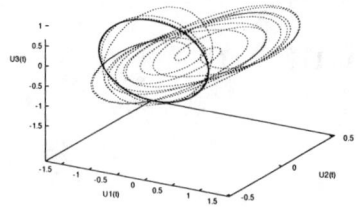

FIGURE 7. Entrainment control from a chaotic attractor into a stable limit cycle with two periods.

FIGURE 8. Entrainment control from a chaotic attractor into a two-dimensional sinusoid.

2. Schieve, W. C., Bulsara, A. R., and Davis, G. M., *Phys. Rev. A* **43**, 2613 (1991).
3. Haken, H., *Synergetics*, Berlin; New York: Springer-Verlag, 1977.
4. Haken, H., *Advanced Synergetics*, Berlin; New York: Springer-Verlag, 1983.
5. Babcock, K. L., and Westervelt, R. M., *Phys. D* **428**, 305 (1987).
6. Tani, J., and Fujita, M., *IEICE Trans. Fundamentals* **E75-A**, 578 (1992).
7. Tani, J., *Electron. Comm. Japan 3* **75**, 62 (1992).
8. Tani, J., *IEEE Trans. Syst. Man Cybern.* **26**, 421 (1996).
9. Ashmore, J. F., and Attwell, D., *Proc. R. Soc. London, Ser. B* **226**, 325 (1985).
10. Angelaki, D. E, and Correia, M. J., *Biol. Cybern.* **65**, 1 (1991).
11. Mauro, A., Conti, F., Dodge, F., and Schor, R., *J. Gen. Physl.* **55**, 497 (1970).
12. Puil, E., Gimbarzevsky, B., and Spigelman, I., *J. Neurphysl.* **59**, 77 (1988).
13. Koch, C., *Biol. Cybern.* **50**, 15 (1984).
14. Rader, S., Wheeler, D. W., Schieve, W. C., and Das, P., *Z. Naturforsch. A* **50**, 718 (1995).
15. Das, P., Schieve, W., and Zeng, Z., *Phys. Lett. A* **161**, 60 (1991).
16. Das II, P. K., and Schieve, W. C., *Phys. D* **88**, 14 (1995).
17. Matsuda, H., and Uchiyama, A., *IEICE Trans. Fundamentals* **E76-A**, 1544 (1993).
18. Marcus, C. M., Waugh, F. R., and Westervelt, R. M., *Phys. D* **51**, 234 (1991).
19. Jackson, E. A., and Grosu, I., *Phys. D* **85**, 1 (1995).
20. Chen, C.-C., *Phys. Lett. A* **213**, 148 (1996).

Random and Chaotic Oscillations in a Model of Childhood Epidemics Caused by Seasonal Variations of the Contact Rate

Polina S. Landa [*] and Alexei A. Zaikin [*†]

[*] *Department of Physics, Lomonosov Moscow State University, 119899 Moscow, Russia*
[†] *Humboldt-Universität zu Berlin, Invalidenstraße 110, 10115 Berlin, Germany*

Abstract. We consider an alternative approach to a standard model for seasonal oscillations of childhood infections by introducing a noisy variation of control parameter instead of periodic one. Chaotic and noise-induced oscillations are compared and the problem of distinguishing between these two kinds of oscillations is addressed. In a certain range of the action frequencies the synchronization of noise-induced oscillations takes place in the sense that the mean frequency of the oscillations becomes close to the action frequency. We demonstrate also the effect of synchronization between two interacting populations.

INTRODUCTION

The nature of irregularity is a challenging problem in ecology [1], and in epidemiology of host populations in particular. An epidemiological, or so-called SEIR model for seasonal oscillations of childhood infections, such as chickenpox, measles, mumps and rubella, under the influence of contact rate fluctuations, was suggested by Dietz [2]. Later this model was intensively

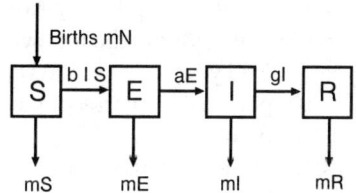

FIGURE 1. The scheme of the compartmental SEIR model.

studied (see, e.g. [3,4]) and it was shown that periodic variations of the contact rate result in chaotic oscillations of childhood infections. The model involves four components: (1) Susceptibles (S); (2) Exposed but not yet infective (E); (3) Infective (I); (4) Recovered and immune (R). Mutual relations between these components are illustrated schematically in Fig. 1.

The model has the following form:

$$\dot{S} = m(1-S) - b(t)SI,$$
$$\dot{E} = b(t)SI - (m+a)E, \qquad (1)$$
$$\dot{I} = aE - (m+g)I,$$
$$\dot{R} = gI - mR, \qquad (2)$$

where $1/m$ is the average expectancy time, $1/a$ is the average latency period, $1/g$ is the average infection period, b is the contact rate (the average number of susceptibles contacted yearly with infective). The total number of children N is normalized to 1. The assumptions of the derivation can be found in [4].

In experimental data the strong seasonality in outbreaks of epidemics has been found. London and Yorke [5] addressed the problem, whether there is an underlying seasonal variation in the contact rate. Usually the seasonal variations are taken into account in the form: $b(t) = b_0(1 + b_1 f(t))$, where $f(t)$ is a function describing the shape of the contact rate variation. If the parameter b varies with time then the variables S, E and I oscillate around a stable singular point with coordinates ($S_0 = (m+a)(m+g)/ab_0$, $E_0 = m/(m+a) - m(m+g)/ab_0$, $I_0 = am/((m+a)(m+g)) - m/b_0$).

PERIODIC AND RANDOM VARIATION OF THE CONTACT RATE

In [3] it was assumed that owing to seasonal variations of environmental conditions the contact rate b depends periodically on time with the period equal to one year, viz., $f(t) = \cos(2\pi t)$ (time is measured in years).

In this case either periodic or chaotic oscillations of S, E, and I appear. The transition from periodic to chaotic oscillations as the parameter b_1 increases occurs via the sequence of period-doubling bifurcations. Chaotic oscillations were observed by Olsen and Schaffer [3] for the following values of the parameters: $m = 0.02\,\text{year}^{-1}$, $a = 35.84\,\text{year}^{-1}$, $g = 100\,\text{year}^{-1}$, $b_0 = 1800\,\text{year}^{-1}$, $b_1 = 0.28$ (Fig. 2 a). These parameters correspond to estimates made for childhood diseases in first world countries.

From a physical standpoint, random variation of the contact rate is more justified than periodic. For the case when $f(t) = \xi(t)$ is a colored (band limited white) noise with the center frequency $\omega = 2\pi$, the results of numerical simulation of Eqs. (1) with the integration step 10^{-4} are shown in Fig. 2a for the same values of the parameters as for periodic variation of contact rate (b_1

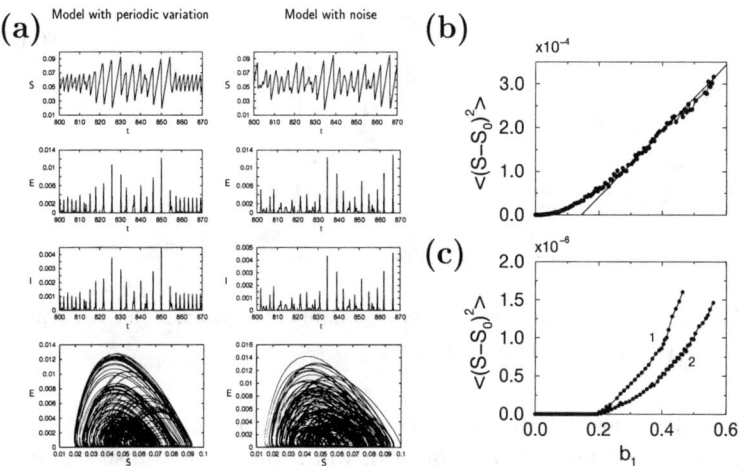

FIGURE 2. (a) Chaotic ($f(t) = \cos(2\pi t)$) and noise-induced ($f(t) = \xi(t)$) oscillations. Solutions and their projection on the (E, S) plane. (b) the dependence of the variance of S on the parameter b_1, (c) the same dependence, if the additive noise is absent. The term $-\alpha E'^3$ is added to the equation for E' to avoid instability. (curve 1: $\alpha = 20 \cdot 10^6$, curve 2: $\alpha = 30 \cdot 10^6$).

is chosen so that the variance of $S(t)$ would be approximately the same as for $f(t) = \cos(2\pi t)$. As the parameter b_1 increases the variance of $S(t) - S_0$ increases too (see Fig. 2b).

In a certain range of b_1 the dependence of $\overline{(S - S_0)^2}$ on b_1 can be approximated by a straight line. The value of b_1 for which this straight line intersects the abscissa's axis can be taken for the point of the noise-induced phase transition (in the sense, discussed in [6]). It seems reasonable to say that this phase transition differs in the mechanism of its appearance from that considered in [6] for a pendulum with a randomly vibrating suspension axis. The difference lies in the fact that this phase transition is caused primarily by additive noise but not multiplicative. To prove this let us rewrite model equations (1) for new variables $S' = S - S_0$, $E' = E - E_0$, $I' = I - I_0$ to detach additive noise:

$$\dot{S}' = -mS' - b(S'I' + S'I_0 + S_0I') - b_0b_1S_0I_0\xi(t),$$
$$\dot{E}' = b(S'I' + S'I_0 + S_0I') - (m+a)E' + b_0b_1S_0I_0\xi(t), \quad (3)$$
$$\dot{I}' = aE' - (m+g)I'$$

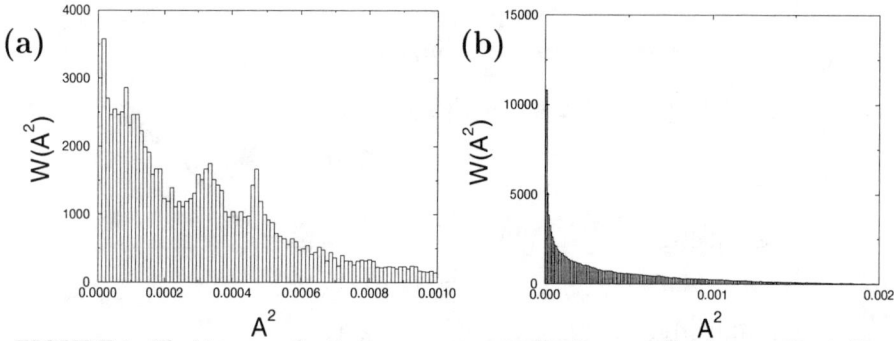

FIGURE 3. The histogram for the instantaneous amplitude squared in the case of periodic (a) and random (b) variation of the contact rate.

If in Eqs. (3) we artificially eliminate additive noise (the term $b_0 b_1 S_0 I_0 \xi(t)$), we obtain that the threshold value of b_1 for which the phase transition appears is larger than for the original equations (1)(see. Fig. 3b).

Comparing noise-induced and chaotic oscillations shown in Figs. 2a we can conclude that the oscillations are at least superficially similar.

DISTINGUISHING CHAOTIC AND NOISE-INDUCED OSCILLATIONS

Despite the similarity, distinctions between these two kinds of oscillations exist. First and foremost the power spectra of the oscillations excited are different. In the case of harmonic variation of the contact rate the power spectrum contains a discrete line at the frequency of this variation. In addition, in the spectrum there are peaks at subharmonics of the variation frequency. In the case of random vibration of the contact rate the power spectrum contains no discrete lines. The second essential difference manifests itself in the correlation dimension of the attractor reconstructed from time series. In distinction to noise-induced oscillations the correlation dimension estimated in simulations for the harmonic variation of the contact rate saturates as the embedding space dimension increases.

In the case of random variation of the contact rate the calculated correlation dimension increase monotonically with the increase in the embedding space dimension. The latter indicates that the correlation dimension of the attractor are infinite or very large. Such a dissimilarity of the result obtained for noise-induced pendulum's oscillations [6], for which the correlation dimension was found to be finite, is attributable to the influence of additive noise.

Finally, the noise-induced and chaotic oscillations can be distinguished by use of the Rytov-Dimentberg criterion, initially proposed in [8,9] to solve the problem of distinguishing between noise passed through a linear narrow-band

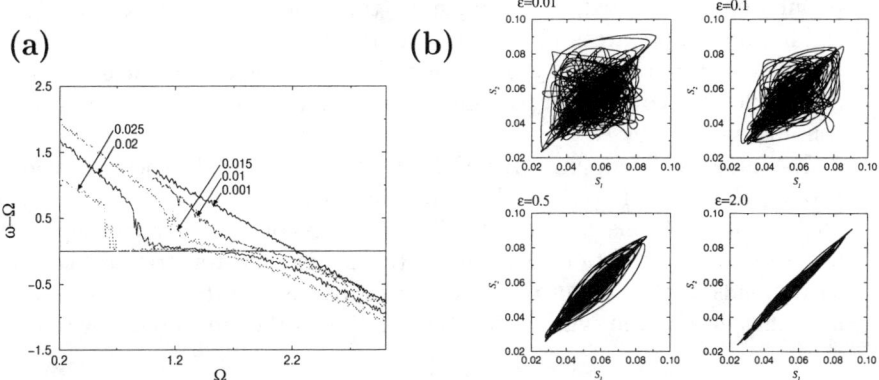

FIGURE 4. (a) The dependence of the difference $\omega - \Omega$ on the frequency of the external force Ω for different values of ε (shown in the fig.). (b) The illustration of the synchronization between two identical systems in the phase space $(S_1; S_2)$. The different amplitude of the coupling is shown in the figure.

filter and periodic but noisy self-oscillations. According to this criterion, the probability distributions for the process itself and for the instantaneous amplitude squared are monotonic in the case of noise-induced oscillations, whereas for chaotic oscillations these distributions have to have peaks. The distinction in the probability distribution for the instantaneous amplitude squared is illustrated in Fig. 3 a,b. The instantaneous amplitude was calculated by means of the Hilbert transform (see [7]).

SYNCHRONIZATION OF NOISE-INDUCED OSCILLATIONS

Another interesting question is the interaction of two neighbouring populations, for example the synchronization. To study it we first consider the simplest case of the synchronization by the external force. We add the term $\varepsilon \cos(\Omega t)$ to r.h.s of the eq. (1) for S, which can model, for example, the periodic influence of economical cycles on the number of susceptible children. Here ε, Ω are the amplitude and frequency of the external force. We have found that in a certain range of the force frequencies, a synchronization of oscillations takes place in the sense that the mean frequency of noise-induced oscillations ω is approximately equal to the action frequency, if ε is large enough (Fig. 4 a). Such interpretation of the problem of the synchronization and calculation technique was proposed in [7,10] and used in [11].

It is interesting to note that frequency locking can be also treated as the possibility to control the frequency of oscillations by the harmonic action.

Another kind of synchronization is the mutual synchronization between two populations, interacted, for example, by travel of groups of possibly infected children in remote areas. To study it we consider two populations, described by eqs. (1) each, coupled by the term $\varepsilon(S_{2,1} - S_{1,2})$, added to the equation for S. Indexes 1, 2 correspond to different subpopulations, and the amplitude of coupling is given by ε.

The increase of coupling strength leads to the almost full synchronization of two populations. This can be illustrated by Fig. 4b, where the solution of coupled model equations is shown for different values of ε. For large enough ε the projection of the solution to the plane (S_1, S_2) is almost a straight line, what corresponds to the case of the synchronization between two populations. We note that despite the synchronization, solution of the model remains random.

SUMMARY

To summarize, we have studied the behaviour of the model with random variation of the contact rate and compared chaotic and noise-induced oscillations. We have shown that the model undergoes phase transition (in the sense discussed in [6]). The Rytov-Dimentberg criterion is found to be valid for distinguishing chaotic and noise-induced oscillations in the model considered. In a certain range of the action frequencies, a synchronization of noise-induced oscillations by the external force takes place. We demonstrate also the effect of the almost full synchronization between noise-induced oscillations in two populations.

Acknowledgments: We are grateful to J. Kurths, M. Rosenblum, and L. Schimansky-Geier for valuable discussions. A.Z. acknowledges support from DFG (grants number SCHI 354/5-1, 436 RUS 113/334/0 R).

REFERENCES

1. R. Pool, Science **243**, 25 (1989).
2. K. Dietz, Lect. Notes Biomath. **11**, 1 (1976).
3. L.F. Olsen and W.M. Schaffer, Science **249**, 499 (1990).
4. R. Engbert and F.R. Drepper, Chaos, Solitons & Fractals **4**, 1147 (1994).
5. W.P. London and J.A. Yorke, Am. J. Epidem. **98**, 453 (1973).
6. P.S. Landa and A.A. Zaikin, Phys. Rev. E **54**, 3535 (1976).
7. M. Rosenblum, A. Pikovsky, and J. Kurths, Phys.Rev.Lett. **76**, 1804 (1996).
8. S.M.Rytov, *Introduction to Statistical Radiophysics*, (Nauka, Moscow, 1966).
9. M.F.Dimentberg, *Nonlinear Stochastic Problems of Mechanical Oscillations*, (Nauka, Moscow, 1980) (in Russian).
10. A.S. Pikovsky, M.G. Rosenblum, G.V. Osipov, and J. Kurths, Physica D, **104**, 219 (1997).
11. P.S. Landa, A.A. Zaikin, M.G. Rosenblum, and J. Kurths, Phys. Rev. E **56**, 1465 (1997).

SIGNAL AND IMAGE PROCESSING

Chaotic Encoding of Information without Synchronization

V.B.Ryabov, P.V.Usik, and D.M.Vavriv

Institute of Radio Astronomy, 4 Krasnoznamennaya St., 310002 Kharkov, Ukraine

Abstract. A method of secure communication based on the inversion of chaotic dynamical systems is analyzed. The approach does not require chaotic synchronization of two or several oscillators to be present. The influence of noise in the communication channel and parameter mismatch on the quality of decoded signals is discussed. Both autonomous Rössler and non autonomous pendulum oscillators are used as examples.

INTRODUCTION

Considerable progress has been made recently in achieving the goal of secure communication by means of chaotic synchronization. In the framework of this approach, the property of mutual synchronization [1] of two or more identical nonlinear oscillators is used for decoding of an information bearing signal previously masked by mixing it with a chaotic oscillation. Depending on the particular realization of the synchronization scheme, the signal is decoded either by subtracting the chaotic component generated by the synchronized oscillator in the receiver or via auto synchronization [2]. It is known that the practical implementation of any scheme utilizing the synchronization property has several drawbacks originating from the finiteness of synchronization time, and the absence of stability of the synchronous regime as well. Another approach [3] uses the concept of the inverse system which does not require a synchronization to be used in some special cases of coder-decoder design. However, as it is claimed in [3], the practical implementation of such a scheme is extremely susceptible to noise in a communication channel because of at least three operations of differentiation necessary for the decoding of a chaotically masked signal.

PROPOSED METHOD

In the present communication we show that the chaotic masking with subsequent decoding can be obtained without the utilization of a synchronization technique and with the one differentiation operation only. The principal idea of the proposed approach is based on the invertibility of non linear equations demonstrating chaotic dynamics [3]. For reducing the number of necessary differentiations in the receiver end we use a non autonomous chaotic dynamical system as a generator of a masking signal and two communication channels instead of a traditionally used single one. Our approach can be demonstrated, for example, as follows. Consider a nonlinear oscillator of the type

$$\frac{d^2x}{dt^2} + f\left(x, \frac{dx}{dt}\right) = F(t) \tag{1}$$

where $F(t)$ is an external force exciting a chaotic oscillation in the oscillator (1). Given the particular shape of a nonlinear function $f\left(x, \frac{dx}{dt}\right)$ and a chaotic solution $x_{ch}(t)$, one can find a straightforward way to solve the inverse problem of reconstructing the exciting signal (external force) from the chaotic response of the system. Indeed

$$F(t) \equiv \frac{d^2 x_{ch}(t)}{dt^2} + f\left(x_{ch}(t), \frac{dx_{ch}(t)}{dt}\right). \tag{2}$$

So defined procedure turns out to be a sufficiently robust one, in the sense that small uncertainties in specifying the function f() in the reconstruction process result in small differences between the reconstructed signal $\tilde{F}(t)$ and the original one $F(t)$. Another important property of the proposed technique is its asymptotic stability, i.e., the absence of error accumulation effect with time. In the decoding procedure the errors are not accumulated with time, as could be expected in any chaotic system with exponential divergence of any nearby trajectories, mainly due to the local character of signal transformation (2) at any time moment. So defined decoding process is in essence just a kind of nonlinear filtration, which preserves one-to-one correspondence between input and output signals.

As for the problem of secure communication, the described invertibility property can be exploited in the following way. Let $F(t)$ be the sum of a harmonic signal and a small addition $i(t)$ carrying information

$$F(t) = A\cos(\omega t) + i(t)$$

If $i(t) \equiv 0$, chaotic oscillations are excited in system (1). Assume that the perturbation $i(t)$ is sufficiently small and does not lead to the destruction of the chaotic motion. Then, if the oscillator (1) is in a chaotic regime, the masking is performed in a way similar to the one reported in [2]. In such a method of encoding the signal undergoes complicated nonlinear transformation which can not be expressed in terms of an additive or multiplicative mixture of the signal and the masking chaotic oscillation. It should be noted, that the frequency of the external force does not required to be in the frequency band occupied by the information carrying signal, as the most intensive part of the chaotic oscillation $x_{ch}(t)$ is as a rule located in the lower frequency band compared to ω (e.g., if the chaotic attractor is formed via a cascade of period doubling bifurcations).

To decode the chaos + signal mixture, i.e. separate the $i(t)$ component, it is necessary to design a receiver capable of inverting Eq. (1) in accordance with Eq. (2). In other words, the received signal has to be twice differentiated and combined with its copy passed through a nonlinear element with the response $f\left(x, \frac{dx}{dt}\right)$ that can be readily realized with conventional electronic circuits. As a result, the original exciting oscillation $F(t)$ is recovered which is an additive mixture of the harmonic and information signals. If the frequency of the harmonic component is chosen outside the frequency band of the information signal, it can be easily suppressed by means of conventional filtering, thus leading to the recovery of the signal $i(t)$. The utilization of the non autonomous dynamical system thus enables one to reduce by one the number of necessary differentiations in the receiver. Further reduction can be accomplished through increasing the number of communication channels, i.e., by transmitting, for example, x and $\frac{dx}{dt}$, with subsequent single differentiation.

EXAMPLE OF A PENDULUM

We demonstrate the feasibility of the proposed ideas on the example of pendulum equation with harmonic excitation. For information bearing signal we used various possibilities, which included a sinusoid, amplitude modulated oscillations, wide band sound records, and highly dimensional chaotic signals obtained from the well known Mackey-Glass equation

$$\frac{dx}{dt} = \frac{a x(t-\tau)}{1 + [x(t-\tau)]^{10}} - b x(t)$$

at $a = 0.2$, $b = 0.1$, $\tau = 100$. The results obtained for different types of the signals are qualitatively similar, so we present below only the ones for the chaotic signals which are known to be the most difficult to recover in conventional methods using synchronization.

We start with the pendulum equation

$$\frac{d^2 x}{dt^2} + \alpha \frac{dx}{dt} + \omega^2 \sin(x) = A \sin(\omega t + \varphi_0) + i(t) \qquad (3)$$

where x is generalized phase of the pendulum, ω, its natural frequency, α, the dissipation parameter, A and ω are the amplitude and phase of the external force, $i(t)$, information bearing signal.

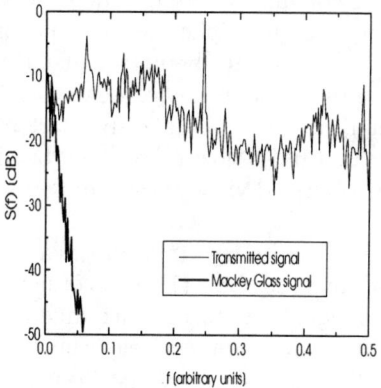 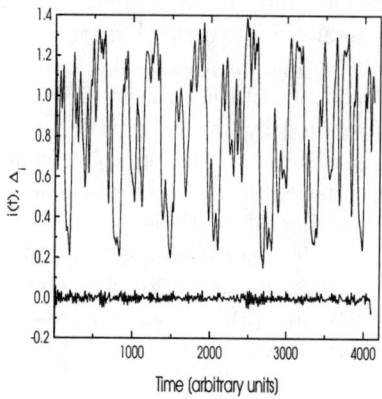

FIGURE 1. Spectra of two signals: (i) transmitted mixture of chaotic oscillations in pendulum oscillator (masking) with information bearing chaotic signal from Mackey-Glass equation (thin line); (ii) the signal of Mackey-Glass equation alone (thick line).

FIGURE 2. Original and recovered signals at zero noise level and identical parameter values in the transmitter and receiver (two upper lines). The difference between them (lower line).

Pendulum equation is a convenient object for our purpose, sinse it is a well documented and easily controlled model. In addition, chaotic regimes of the pendulum appear to be stable with respect to comparatively strong perturbations in the form of intense driving signal $i(t)$. In ideal situation the signal $i(t)$ can be restored exactly in the receiver as

$$i(t) = \frac{d^2 x^*(t)}{dt^2} + \alpha \frac{dx^*(t)}{dt} + \omega^2 \sin(x^*(t)) - A\sin(\Omega t + \varphi_0)$$

where $x^*(t)$ is an arbitrary solution of (3) transmitted through the communication channel. However, in any implementation we encounter the problem of extreme sensitivity of the decoding process upon frequency Ω mismatch between the transmitter and receiver systems. As can be easily seen, arbitrarily small error in specifying the frequency Ω and the initial phase φ_0 leads to accumulation of the error in the decoded signal with time. Thus, if to apply the proposed procedure directly, it appears necessary to transmit one more signal in order to synchronize the sources of harmonic oscillations at both ends. We have found another solution to this problem which consists in the utilization of a harmonic excitation with the frequency Ω, higher than the frequency band of the information bearing signal, and subsequent low-

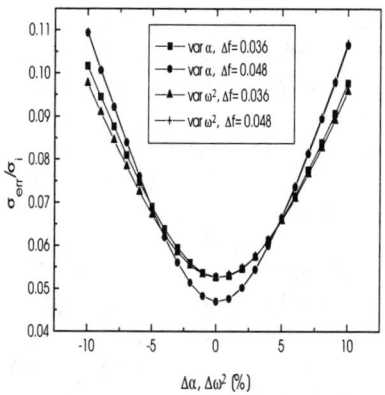 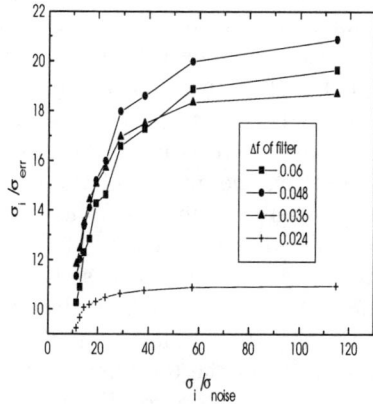

FIGURE 3. The normalized error in the recovered signal vs. parameter mismatch for different bandpasses of the low-pass filter at the output of the receiver.

FIGURE 4. The normalized error in the recovered signal vs. normalized noise intensity in the communication channel for different bandpasses of the low- pass filter.

pass filtering at the output of the receiver. Note that even with such "high frequency" excitation of the pendulum, the spectrum of the induced chaotic signal has intensive components in the low frequency domain, and the goal of chaotic masking is successfully achieved. An example of power spectrum of the transmitted signal $\frac{dx^*(t)}{dt}$ is shown in Fig. 1, together with spectrum of $i(t)$. As we already mentioned, in our computer experiments we used two communication channels for transmitting both $\frac{dx^*(t)}{dt}$ and $\sin(x^*(t))$, mainly for the purpose of noise reduction.

The result of the decoding and filtration is depicted in Fig. 2, where both the initial signal $i(t)$ at the input of the transmitter and the recovered one $\tilde{i}(t)$ at the output of the receiver are shown together with their difference. Our analysis shows that in the absence of noise in the communication channel the magnitude of error is mainly defined by the quality of the utilized procedure of low-pass filtering of the harmonic component (driving force).

As the quality of chaotic masking is defined by the selectivity property of the receiver with respect to the variation of control parameters, we have performed the analysis of the influence of non identity of control parameter values in the encoding and decoding systems on the quality of the recovered signal. In Fig. 3 we plot the magnitude of the normalized error $error = \sqrt{D\left(i(t) - \tilde{i}(t)\right)/D(i)}$ where $D(\cdot)$ denotes dispersion, vs. detuning in the parameters of dissipation and natural frequency. It is evident that comparatively small deviations in parameter values are not crucial for the quality of the decoded signal. For example, the mismatch in the natural frequency about 10% results in about 10% error value. Larger deviations lead to complete deterioration of the recovered signal that is an important property for the secure communication purpose (selectivity with respect to parameter values).

We also have studied the influence of additive noise in the transmission channel on the quality of the recovered signal. The results are presented in Fig.4 where the inverse value of normalized error is plotted vs. the inverse normalized noise intensity at different values of the bandpass of the low-pass filter. We conclude about sufficient robustness of the proposed scheme to the effect of noise if the bandwidth of the signal is much less than the bandwidth of the chaotic masking oscillation. As one can see, the signal to noise ratio at the output of the decoding device is not much greater that at its input and strongly depends upon the bandpass of the filter used. It should be noted that the effect of noise is considerably reduced due to filtering and is not very crucial as could be expected from the operation of differentiation.

DISCUSSION

We performed the analysis of feasibility of the encoder-decoder scheme based on the notion of an inverse system and chaotic masking. It has been demonstrated that this concept, which does not require synchronization of transmitter and receiver oscillators, can be effectively implemented if to use two communication channels and a non autonomous chaotic system as encoder. Intrinsic properties of the proposed scheme require the narrow bandwidth of the information bearing signal compared to the bandwidth of the masking oscillation. This restriction is motivated first by the demand of the robustness of the algorithm to the noise in communication channel, and, second, by the necessity to remove the driving sinusoidal oscillation from the decoded signal.

The proposed method of chaotic encoding-decoding may be also used with autonomous systems, like Rössler oscillator. However its practical implementation encounters several difficulties originating from the amplification of noise in differentiating circuits of the receiver. Nevertheless, as our calculations show, the output noise can be substantially reduced by using multichannel communication schemes and smoothing filters in the differentiation operators.

REFERENCES

1. Pecora, L.M., Caroll, T.L. *Phys. Rev. A* **44**, 2374 (1991).
2. Parlitz, U., Kocarev, L., Stojanovsky, T., and Preckel. H., *Phys. Rev. E* **53**, 4351, (1996).
3. Feldmann, U., Hasler, M., and Schwarz, W., *Int. J. of Circuit Theory and Applications*, **24**, 551 (1996).

Enlightenment in Shadows

Isla Gilmour and Leonard A. Smith

Mathematical Institute, University of Oxford
Oxford, OX1 3LB, U.K.

Abstract. Numerical weather forecasting has functioned both as one of the major inspirations for the development of the theory of nonlinear dynamical systems, and as one of its leading applications. While ensemble forecasts used by operational forecast centres both in the US and the EC provide the best operational estimates of the reliability of a given day's forecast, many open questions regarding the construction and evaluation of the ensembles remain. The concepts of shadowing are illustrated and applied to evaluate ensembles for the thermally driven rotating fluid annulus. Low-dimensional dynamical systems are obvious test-beds for proposed improvements, yet the question arises of whether the simplicity that one often observes in very high-dimensional weather models (with millions of apparent degrees of freedom) fails 'even in or only in' low-dimensional chaotic systems; this is addressed and initial results on the uniformity of 'the linear range' presented for the annulus.

Forecasting nonlinear phenomena is a driving force of the applications of nonlinear dynamical systems as we head into the next millennium; the forecasting of physical phenomena in general, and the Earth's atmosphere in particular, has been perhaps the major single force in the formation and perpetuation of nonlinear dynamics this millennium. The role of atmospheric dynamics at the turn of this century has many parallels with that of celestial mechanics at the turn of the last: providing a slightly too difficult problem.

We consider the prediction problem given an imperfect model, an uncertain initial condition and an incomplete understanding of observational noise, as opposed to assuming a perfect model with uncertain parameters or inexact arithmetic. Two physical systems are considered, the thermally driven rotating fluid annulus and the Earth's atmosphere [3]. The annulus has the advantage of being somewhat simpler physically than the atmosphere, admitting rather simpler forecast models, and enabling analysis of data sets with a longer duration (measured in characteristic times of the system); nevertheless both are infinite dimensional (fluid) systems for which no perfect model exists.

Implications of imperfect models and uncertain observations are discussed, illustrating the concepts of shadowing for the annulus using a radial basis function (RBF) model [5]. Moving from $\mathbf{R}^5$ to $\mathbf{R}^{10^6}$, the question arises of whether the simplicity observed in high dimensional systems fails 'even in or only in' (EIOOI) low dimensional chaotic systems. A viable test of internal consistency is proposed, given the operational constraints in the weather forecasting scenario, and results presented for the annulus.

Dynamics of Uncertainty

Given a series of observations and some knowledge of the observational uncertainty, we may define a perfect model as one which is able to generate a solution which differs from the observations in a manner consistent with the uncertainty. In general, imperfect models will differ from the perfect model scenario in more fundamental ways than simply having inexact (estimated) parameters[1]. Often the structure of the model is wrong in the sense that the true dynamics cannot be represented over 'long' periods of time by *any* set of model parameters (*e.g.* the model consists of a truncated Taylor expansion of the perfect model), also the state-space of the model may differ from that of the system. How should the 'best' parameters in such models be defined?

In practice, the lack of sufficient observations to completely specify a model state, the knowledge that the available observations are uncertain, and discrepancies between the model state-space and that of the physical system make it inadvisable to initiate a forecast from an initial condition based solely on current observations. An *analysis*, $A(t, \tau)$, for time t is formed by combining observations until time τ using the model dynamics; thus the initial analysis $A(0,0)$ serves as the best estimate of the state corresponding to the initial system state given past observations for this model. Knowledge of the uncertainty distribution is represented by a 'ball' around the analysis (an isopleth of the corresponding probability distribution function). If the model is perfect then (i) there exists a model state within the t=0 ball which represents the system state and (ii) evolution of this model state under the model will yield a trajectory which passes through the uncertainty balls at t=1, 2, 3, ... This model state need not be unique (fig. 1a), and will not be if the model is also hyperbolic. An alternative schematic, more familiar to low dimensional dynamical systems, is given in fig. 1b; the subset of consistent initial conditions is determined by considering the intersection of forward and backward projections (under the model) of past and future uncertainty bounds respectively, with that at $t = 0$.[2]

The model is never perfect. If the system is uniformly hyperbolic and the model is good then there will exist a system trajectory which stays close to, or

---

[1] The underlying physical system may have dynamics which can not be represented by a system of ODEs (or PDEs), or may not be closed in the thermodynamic sense.
[2] The consistent subsets of fig. 1a are the intersection of the states within the uncertainty at $t = 0$ and the backward projections (onto $t = 0$) of uncertainties at times $t = 1, t = 1, 2, \ldots$

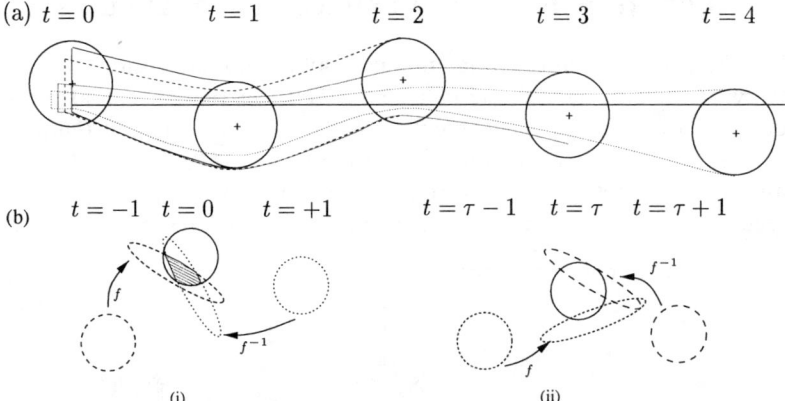

FIGURE 1. (a) Perfect model scenario: the 'true' trajectory (horizontal line), analyses (+) and corresponding uncertainty (circles). Subsets of initial conditions consistent with observations at $t = 0, 1$ (solid line), $t = 0, 1, 2$ (long dashes),... demonstrate the collapse of the subset of consistent initial conditions to a non-empty subset containing the 'true' initial state as more future observations are obtained. (b) Schematic reflecting the existence of a shadowing trajectory from projection of uncertainty distributions. (i) Imperfect model; an ι-shadowing trajectory exists, passing through the shaded region. (ii) No ι-shadowing trajectory exists as there is no intersection between projections of past and future uncertainties and the current uncertainty.

ϵ-shadows the model trajectory. (The Anosov-Bowen lemma states that every 'good' model can be ϵ-shadowed [1].) But the question of interest here is finding model trajectories sufficiently near the observed system trajectory to be consistent with the observational uncertainty distribution. In general, we can make precise statements about the dynamics of the model (we know its functional form), but not about the system itself (we know only the observations). And inasmuch as the Hénon map is not hyperbolic, it is doubtful that current weather models are. In reality, there may be a time τ_ι, the ι-shadowing time, at which *all* initial conditions consistent with the observational uncertainty at $t = 0$ are inconsistent with some observation at time $\tau \in (0, \tau_\iota]$ (see fig. 1bii). Note that this time, τ_ι, is a function of the initial state of the system, the initial analysis and associated uncertainty, and the particular model.

Variational data assimilation schemes may needlessly degrade the analysis at $t < \tau$ and $t > \tau$ due to the model error. By explicit use of the distribution of measurement uncertainty and the acceptance of systematic (if unknown) macroscopic model error, ι-shadowing is able to cut a trajectory at a location like fig. 1bii. This suggests regions for model improvement and provides a measure of optimality, the distribution of τ_ι, which may be used to contrast models with similar average forecast errors. Unlike a root mean square error cost function, ι-shadowing times will not penalise realistic model sensitivity.

Low–dimensional dynamics in the Annulus

The thermally driven rotating annulus is an infinite dimensional laboratory analogy of the mid–latitude circulation systems in the Earth's atmosphere. Fluid is held between concentric cylinders each held at constant temperature (the inner one being cooler), and the entire apparatus (including the temperature probe) rotates at a fixed rate Ω. For large Ω, the flow appears spatially irregular and local co-rotating temperature measurements are chaotic [4,5,7].

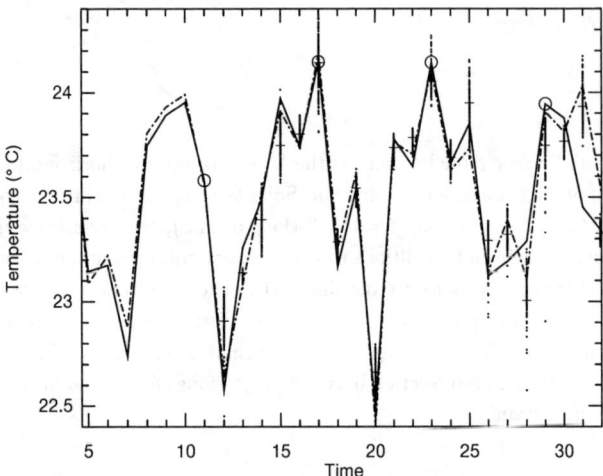

FIGURE 2. An observed temperature time series (solid) from the annulus and an ι-shadowing trajectory (dot-dashed) from a RBF model which stays 'close' to the observed trajectory for 30 time steps. Four 6–step ensemble predictions are shown: an ensemble of 128 points, normally distributed within 0.075 degrees of (each component of) the initial observation, is initiated at times 11, 17, 23 and 29 (circles) and iterated under the model to give a distribution after 1,2,3,4,5 and 6 steps (dots), the mean of which is denoted (+).

Figure 2 shows ensemble predictions for a RBF model of the annulus; in $\mathbf{R}^5$ the distributions of the ensembles show regions of rapid error growth, return of skill, and model error (*e.g.* at times 12, 13 and 31 respectively). At time 31 the scenario illustrated in fig. 1bii occurs; no shadowing trajectory exists. Even in low dimensional systems, models are imperfect.

High–dimensional Atmospheric Dynamics

The computational complexity of high-dimensional numerical weather prediction (NWP) models poses an operational restriction limiting real time ensembles to about 2^6 members. This suggests the use of constraints so that the "most relevant" perturbations are investigated. A topic of much discussion is the optimal choice of constraining sub-space, which is often defined using singular vectors of the linear propagator, $\mathcal{M}$, over an optimisation time t_{opt}. *Singular vectors* (SV) [2], employed by ECMWF, are the right singular vectors of the linear propagator over $[0, t_{\text{opt}}]$; they represent the directions of the linearised system which grow most rapidly over $[0, t_{\text{opt}}]$. The *Lyapunov vectors* (LV) are the left singular vectors of the linear propagator in the limit $t_{\text{opt}} \to \infty$, over time $[-t_{\text{opt}}, 0]$ and represent the directions of the linearised system which have grown the most. If the model is perfect and the perturbation infinitesimal the *breeding vectors* (BV) [8], employed by NCEP, evolve toward the local orientation of the first global LV. Given an imperfect model and finite perturbation the BV contains information on model error; in this case there may be no LV, as no trajectory need pass through the analysis.

Ideally, the best (possible) representation of the system initial condition for a given model is included in the ensemble; such a current initial condition may be defined by trajectory which shadows farthest both from the past and into the future. Defining the perturbation from the current analysis to such an initial condition as the 'dream perturbation', the projection of the constrained vectors onto this dream perturbation provides a method for (*a posteriori*) evaluation of ensemble construction.

The treatment of nonlinearity in NWP is based upon several assumptions which may be shown to fail *even in* low ($m \sim 2^2$) dimensional dynamical systems. A prime example is the assumption of uniformity in the time scale over which the linear propagator yields a reasonable approximation to the dynamics for different initial conditions. One familiar with high ($m \sim 2^{20}$) dimensional nonlinear dynamical systems, and the Earth's atmosphere, might respond that the huge fluctuations observed in low dimensional systems happen *only in* low dimensional systems: the 'even in or only in' (EIOOI) dilemma.

The use of SV subspaces assumes that the linearised model remains a good approximation for the optimisation time, about 2 days operationally. We plan to evaluate this assumption for NWP models by exploiting the common practice of running twin perturbations in operational ensembles in order to avoid the computational cost of additional model runs. Specifically, for every initial condition in the ensemble with perturbation $\boldsymbol{\delta}$, $\boldsymbol{\delta} \in \mathbf{R}^m$, from the control, a twin initial condition with perturbation $-\boldsymbol{\delta}$ is included in the ensemble. These twin trajectories reflect the time scale over which the linear propagator might provide a reasonable approximation of the dynamics of finite perturbations, thereby providing both a direct measure of the uniformity, or otherwise, of this time scale with initial condition and a test of the internal consistency of SV ensembles. Taking the control run as the origin, consider the dynamics of each pair of twin perturbations, $\boldsymbol{\delta}^+(t)$ and $\boldsymbol{\delta}^-(t)$. In the linear approximation $\boldsymbol{\delta}^+(t) + \boldsymbol{\delta}^-(t) = \mathbf{0}$, thus we monitor the time at which the quantity $\theta(t)$ exceeds a given threshold, where

$$\theta(t) = \frac{||\boldsymbol{\delta}^+(t) + \boldsymbol{\delta}^-(t)||}{||\boldsymbol{\delta}^+(t)|| + ||\boldsymbol{\delta}^-(t)||} \tag{1}$$

and $||\cdot||$ is one of several possible metrics. Results from twin experiments run under a RBF model for the annulus show that, for uncertainties comparable to the analysis error and a 20% threshold (*i.e.* $\theta < 0.2$), linearity breaks down rapidly and non-uniformly. Choosing an optimisation time of 4 time steps, SV perturbations to 90% of analysis values show a break down of linearity within within 6 time steps; only ~20% are linear until optimisation time. Linearity breaks down non-uniformly and slightly later in the BV subspace (90% within 7 time steps).

CONCLUSIONS

While Lyapunov exponents place no *a priori* limit on the predictability of a dynamical system [6], the ι-shadowing time does place an upper bound on the predictability given the combination of a specific model of a given system and some particular (distribution of) uncertainty. Even this bound is of limited utility in practice, since it assumes infinitely large ensembles. However, ι-shadowing offers an alternative to the usual prediction error cost functions in defining the 'best' model of a system; it penalises model error while utilising realistic model sensitivity, rather than penalising both model error and sensitivity. Shadowing failure at an anomalous result yields a trajectory containing useful information over its duration and indicating regions for model improvement. An assumption of NWP ensemble formation was tested for a low-dimensional system and the applicability of these results to high-dimensional systems queried due to the EIOOI dilemma. Twin experiments are currently being run for an operational NWP model, the results of which will address an aspect of this dilemma.

REFERENCES

1. Farmer J.D., and Sidorowich J.J., *Physica D* **47**, 373–392 (1991).
2. Mureau R., Molteni F., and Palmer T.N., *Q. J. R. Meteorol. Soc.* **119**, 299–323 (1993).
3. *Predictability*, ECMWF, Shinfield Park, Reading, U.K., 1996.
4. Read P.L. *et al.* , *J. Fluid Mech.* **238**, 599–632 (1992).
5. Smith L. A., *Physica D.* **58**, 50–76 (1992).
6. Smith L. A., *Phil. Trans. R. Soc. Lond. A* **348**, 371–381 (1994).
7. Smith R.L., *J. R. Statist. Soc. B* **54(2)**, 329–352 (1992).
8. Toth Z., and Kalnay E., *Bull. Am. Meteorol. Soc.* **74(12)**, 2317–2330 (1993).

Noise induced effects and stochastic resonance in an El Niño model

Lewi Stone* and Peter I. Saparin*,†

*The Porter Super-Center for Ecological and Environmental Studies, Tel Aviv University, Ramat Aviv 69978, Tel Aviv, Israel. E-mail: lewi@lanina.tau.ac.il
†Dept. of Nonlinear Dynamics, Inst. of Physics, Potsdam University, Am Neuen Palais, 22, Pf 601553, D-14415, Potsdam, Germany.

Abstract. El Niño, the anomalous warming of eastern equatorial waters of the Pacific, occurs sporadically every 3-7 years causing severe economical and ecological damage across the globe. Here we attempt to explain El Niño's puzzling irregularity by modelling the ocean's equatorial wave dynamics as part of a nonlinear dynamical system driven by random environmental fluctuations. The theory of Stochastic Resonance (SR) proves to be a useful framework for providing insight into the dynamics of this complex climatic signal.

The possibility that noise could play an important role in climatic systems, or more generally in organizing the spatio-temporal dynamics of nonlinear physical systems, has led to the development of the theory of stochastic resonance (SR) [1–3]. Although originally designed for the study of climatic systems over 20 years ago [1], the theory of stochastic resonance (SR) has not since been explored for its environmental applications. Here we show how noise, and the SR phenomenon, may be intimately connected with the El Niño Southern Oscillation (ENSO) cycle which heats and cools the equatorial region of the East Pacific ocean every 3-7 years. ENSO's inherent irregularity is as much an undisputed fact as it is an enigma. No two ENSO cycles have evolved exactly alike – there is great variation in the amplitude of, and in the interval between, individual episodes. One prominent school of thought argues that most of the variability is a form of correlated Gaussian noise and can be explained with a simple linear stochastic model [4]. This contrasts with a more recent competing view which asserts that the irregular ENSO oscillation is a manifestation of low-dimensional deterministic chaos arising from the nonlinearity of the ocean-atmosphere system; if so, it would account for the erratic character of the ENSO phenomenon as well as its more regular deterministic features such as El Niño's locking to the seasonal cycle (El Niño usually peaks close to December) [5,6]. The model analyzed here attempts to provide a fresh interpretation and makes transparent the crucial role of noise in acting as a control on the ocean-atmosphere dynamics.

I THE MODEL

We modify a successful heuristic model of the Pacific's equatorial wave dynamics in which the key variable $h(t)$ represents the deviation of the thermocline depth (at the Pacific's eastern boundary) from its average seasonal depth. It also provides a reasonable representation of the local sea surface temperature (SST) anomaly (i.e., the deviation from the average monthly SST) (see [5]). The "anomaly" $h(t)$ can be modelled by the delay-differential equation:

$$\frac{dh}{dt} = aA\left[h(t-\tau_1)\right] - bA\left[h(t-\tau_2)\right] + c\cos(2\pi f_a t) + Dc\xi(t), \qquad (1)$$

where $A(h) = \beta \tanh(h\kappa/\beta)$ ($\beta = 1.5$ if $h \geq 0$, and $\beta = 0.3$ otherwise).

The model's behavior is controlled via the nonlinearity of $A(h)$ by the parameter κ, which governs the coupling between ocean and atmosphere. (A detailed description on the derivation of this functional form may be found in [5].) The first term in the right-hand side of (1) represents a Kelvin wave travelling eastward to the South America coast in time $\tau_1 = 1.15$ months. The second term represents a delayed Rossby wave travelling first to the western Pacific boundary where it is reflected back to the South American coast in time $\tau_2 = 5\tau_1$ (as in [6]). The periodic cosine term of amplitude c represents the interaction of the annual (i.e., seasonal) forcing. Lastly, we introduce stochastic excitation into the model by taking $\xi(t)$ as a Gaussian distributed white noise with $\langle \xi \rangle = 0$, $Var(\xi) = 1$, and with the parameter D setting the noise level. The model was integrated with a daily time step and with noise injected daily into the system. Representative parameter values of a, b and c may be found in [6] and Saparin et al. (ms.).

In the absence of noise ($D = 0$), the model's dynamics varies as a function of the coupling parameter κ, as described by the bifurcation diagram in Fig.1. The diagram was obtained by strobing the solution of Eq.1 every year (i.e., at the same period as the periodic forcing). For $0 \leq \kappa \leq 1.233$ there is a single point solution representing the annual oscillation whose frequency we denote here by f_a. A Hopf bifurcation occurs when κ is increased beyond $\kappa = 1.233$, and a second frequency (f_E) emerges giving rise to (quasiperiodic) torus dynamics. The frequency f_E defines the model's intrinsic ENSO oscillation. Note also that windows of frequency locking and chaos are also observed in the bifurcation diagram for $\kappa > 1.403$.

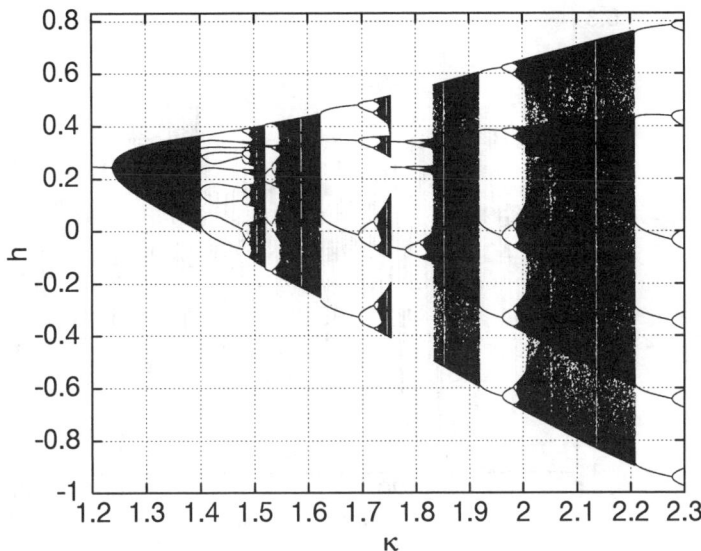

FIGURE 1. Slice of the bifurcation diagram of Eq.1 as a function of κ in the absence of noise ($D = 0$).

II NOISE INDUCED EFFECTS

The time-delay model gives rise to several interesting noise induced effects which are briefly examined below.

A. Coherence resonance (precursor to Hopf bifurcation): We first examine the delay model parameterised just below the point of Hopf bifurcation $\kappa = 1.233$ i.e., just before the ENSO oscillations f_E come into existence. Below this point, the ENSO oscillation f_E is completely absent and as a consequence with zero amplitude in the spectrum; only the annual oscillation f_a is present. We have found that the injection of a small amount of noise ($D > 0$) generates a new peak in the spectrum at the ENSO frequency f_E. This unusual effect has been explained in [7], where it has been demonstrated that the power spectrum of a system observed after a bifurcation point can nevertheless be visible even before the bifurcation actually occurs, if there is noise present. We performed a spectral analysis of model time-series for a range of different noise levels. The signal to noise ratio (SNR) of the equation's "noisy bifurcation precursor," calculated at the ENSO frequency f_E, was then determined as a function of noise level D and found to pass through a maximum, peaking at an optimal noise level. The peak in the SNR curve indicates the presence of a "coherence resonance", an important class of stochastic resonance; at low levels, noise paradoxically increases the SNR, but at higher levels the SNR deteriorates [8]. The same noisy precursor of the Hopf bifurcation has very

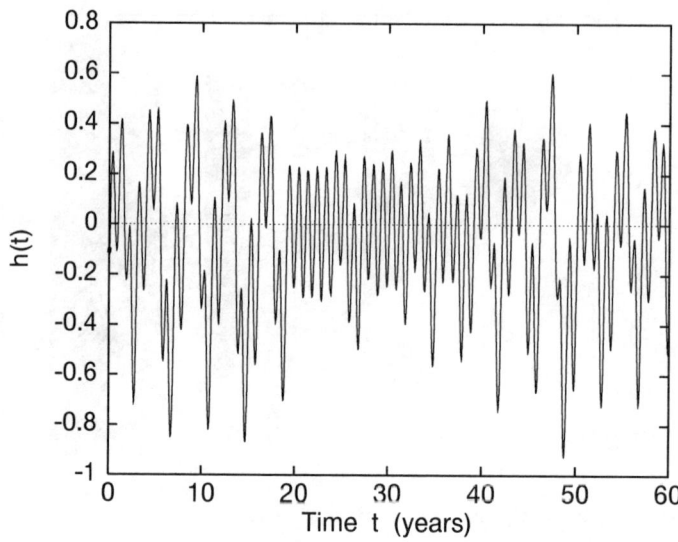

FIGURE 2. Time-series $h(t)$ generated from Eq.1 ($\kappa = 1.4, D = 0.2$)

recently been observed in more sophisticated spatial El Niño models (Blanke et al. ms.).

B1. Noise induced chaos: Eq.1 was parameterised for strictly periodic or quasiperiodic oscillations. Chaotic dynamics are then easily induced simply by perturbing the model with noise ($D > 0$), i.e., without any need for boosting the nonlinear ocean-atmosphere coupling κ. The time-series (Fig.2) has been diagnosed as noise-induced chaos from a study of the largest Lyapunov exponent as well as from its spectral characteristics. The spontaneous generation of chaos in systems parameterised for periodic or quasiperiodic dynamics arises when perturbations, due to forcing by the injection of Gaussian noise, promote random jumping between coexisting regular and chaotic attractors (of different stabilities) leaving the impression of "chaotic stochasticity".

B2. Noise induced intermittency: A particular feature of the noise-induced chaos (Fig.2) is the distinctive intermittency termed by oceanographers as "regime like behavior." This occurs when the trajectory hops erratically between the longer (≥ 3 year) chaotic ENSO oscillation (f_E), and the unstable annual cycle (f_a) that sporadically entrains the system. Perhaps most importantly, the regime like behavior has much in common with the dynamics observed in more complex ENSO models (eg., the CZ model in [9]), where regions of quiescence are often followed by bursts of activity. The intermittency may be understood in terms of the Poincaré section in Fig.3, obtained by first embedding the model trajectory ($\kappa = 1.4$) into 3-dimensional space and then slicing the resulting torus-like attractor with a plane transverse to the flow.

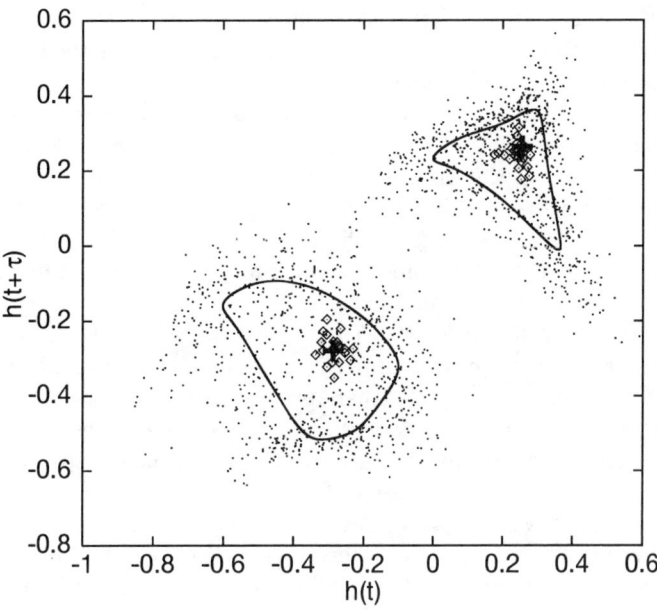

FIGURE 3. Poincaré section of model time-series obtained from Eq.1 (see text).

In the absence of noise ($D = 0$), the Poincaré section consists of two smooth closed loops (thick lines) representing the section of torus formed by quasiperiodicity in the time-series. When noise fluctuations are small, the trajectory jumps around but stays in the vicinity of these closed loops. At larger noise levels ($D = 0.2$) the trajectory is kicked further away from the loops (points marked by ".", corresponding to the time-series in Fig. 2). Occasionally the trajectory is pushed close to the singular point "+" representing the unstable annual cycle. This cycle becomes a saddle when $\kappa > 1.231$ (see the bifurcation diagram Fig. 1). Since the transverse component of the velocity is extremely small near the saddle, the trajectory may hover in the neighborhood of the annual cycle for a sustained time duration, producing a region of "quiescent" behavior in the time-series. These points are demarked by "◇" in the Poincaré section. Eventually, the trajectory escapes from the annual cycle because of saddle instability and/or noise perturbation.

B3. Coherence Resonance observed in intermittent dynamics: We examined Eq.1's intermittency effect further by plotting the probability of residing in a quiescent regime as a function of D. For each time-series, we counted the total number of period-1 orbits (i.e., of period one year), and determined their proportion in the length of the entire time-series. (We also included in this count orbits that were deemed to be "close" or well within a defined distance from the period-1 orbit, thereby allowing for noise perturbed cycles.) The resulting graph has a clear maximum at intermediate noise levels. The peak in the graph is due to the fact that sustained motion on the unstable annual cycle is difficult to induce both at low noise levels where the trajectory rarely reaches the vicinity of the saddle cycle, and for large D, where strong noise perturbations immediately kick trajectories which manage to approach the saddle cycle, far away. Only at intermediate noise levels is there an opportunity for the trajectory to reach and remain close to the saddle for lengthy periods of time. An analysis of the probability of 3–4 year orbits in the time-series results in a qualitatively similar graph which peaks at a different optimal noise intensity. These results leads to the nonintuitive conclusion that increased noise in the system may in some circumstances lead to a maximum in signal coherence i.e., a coherence resonance, (similar to that in A, above).

C. Stochastic Resonance as a Threshold Effect: The El Niño itself triggers whenever the sea-surface temperature SST anomaly exceeds a predefined positive threshold level [9]. We therefore constructed time-series of El Niño initiations or pulses at each time the smoothed time-series $h(t)$ exceeded a representative threshold level. We performed a frequency spectrum analysis of the resulting pulse trains under different noise regimes. The signal-to-noise ratio (SNR) index was calculated at the ENSO frequency f_E over a range of noise levels D. As we report elsewhere, the SNR curve has a maximum at an optimal noise level providing strong evidence for another stochastic resonance (SR) effect, as known to occur in threshold systems [2].

III DISCUSSION

Conventional El Niño models are designed specifically for simulating long time-scale ENSO dynamics (say months to years) but rarely allow for short term dynamics occurring on daily or weekly time-scales. For the El Niño model (1), such fluctuations lead to SR effects in which net order or coherency in the ENSO signal paradoxically increases with noise magnitude. In view of the ocean's ever-present short time-scale fluctuations, these SR effects suggest that environmental variability may have significant impact on the frequency, intensity, duration and synchronization of the El Niño.

REFERENCES

1. Benzi, R., Parisi, G., Sutera, A. & Vulpiani A. *Tellus* **34** 10 (1982).
2. Wiesenfeld, K. & Moss, F. *Nature* **373**, 33 (1995).
3. McNamara, B. & Wiesenfeld, K. *Phys. Rev. A* **39**, 4854 (1989).
4. Penland, C. & Sardeshmukh, P.D. *J. Climate* **7**, 1352 (1994).
5. Munnich, M., Cane, M.A. & Zebiak, S.E. *J. Atmos. Sci.* **48**, 1238 (1991).
6. Tziperman, E., Stone, L., Cane, M.A. & Jarosh, H. *Science* **264**, 72 (1994).
7. Wiesenfeld K. *J. Stat. Phys.* **38**, 1071 (1985).
8. Neiman, A., Saparin, P.I. & Stone, L. *Phys. Rev. E* **56**, 270 (1997).
9. Cane, M.A., Zebiak, S.E., & Dolan, S.C. *Nature* **321**, 827 (1986).

Processing of Vertebral CT-images and Quantification of their Structure by Measures of Complexity

Peter I. Saparin*, Wolfgang Gowin[†],
Jürgen Kurths*, Dieter Felsenberg[†]

*Dept. of Nonlinear Dynamics, Inst. of Physics, Potsdam University,
Am Neuen Palais, 22, Pf 601553, D-14415, Potsdam, Germany.
[†] Osteoporosis Research Group, Dept. of Radiology and Nuclear Medicine, University
Hospital B. Franklin, Free University, Hindenburgdamm 30, D-12200, Berlin, Germany

Abstract. We propose an algorithm which processes computed tomography images of vertebral bodies, segments them into the areas of interest and then quantifies their *structure* using approaches developed in nonlinear dynamics. Vertebral bodies are segmented from the connective and soft tissue background, and then the image of the entire vertebrae are split into the cortical and trabecular bones. At the next stage, several criteria, based on nonlinear dynamics, complexity measures and symbolic dynamics, are applied. We show that these measures indeed contribute significantly to the early diagnostics of changes in bone structure, which are specific for osteoporosis and other bone diseases.

INTRODUCTION

The architectural structure of the vertebral tissue determines the risk of fracture as well as other osseous and extra-osseous factors. It is still an open but very important question how to characterize *the structure* of the vertebral body quantitatively [1-3]. Bone mineral density (BMD) is only one quantitative technique that is useful for evaluating the density (but not the structure!) of bone mineral in a vertebra [1]. The purpose of this contribution is to show that the structure and architecture of a bone can be quantified with methods used in non-linear dynamics. We propose an algorithm which automatically separates the areas of trabecular and cortical bone of vertebral computed tomography (CT) images. We demonstrate that complexity measures [4-7] and symbolic dynamics [8] are able to evaluate the structure and the state of the bone tissue. This technique is extremely useful for the purpose of early diagnostics of changes in bone structure, which are specific for osteoporosis.

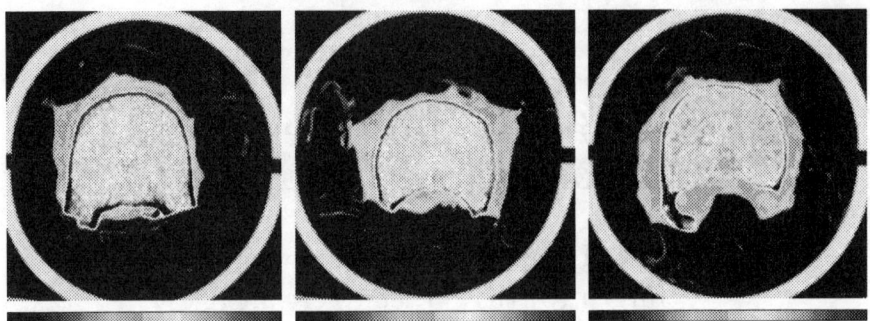

FIGURE 1. Original CT-images of 1 mm thickness axial slice of L3 vertebrae. Normal (left), and osteoporotic specimens at a weak (middle) and strong (right) grade of the disease.

I DATA

The data, axial slices of 1 mm and 10 mm thickness obtained by Quantitative Computed Tomography (QCT) from non-fractured lumbar vertebrae L3 harvested from human cadavers were acquired on a high-resolution CT-scanner (Siemens AG, Somatom Plus S, 512×512 pixels). Typical CT-images of 1 mm slice thickness are shown in Fig. 1.

II METHOD

Our techniques consists of two main stages: 1) The vertebral bodies are segmented from the connective and soft tissue background, and then the entire vertebrae are split into the cortical and trabecular bones by processing of the corresponding CT-images with a specially developed technique. 2) Structural properties of these separated areas are quantified by modern statistical, statical and dynamical approaches developed in nonlinear dynamics.

Preprocessing

At the first step, we apply threshold filtering to define the outer boundary of the bone and then define its contour using the procedure of binary contour following and an algorithm which connects separated parts of the contour (cf. [9] and references therein). An area inside this contour is used as a mask to isolate the vertebral body from the other parts of the CT-image.

At the next step, another automatic procedure splits the entire vertebrae into two areas: the trabecular bone and the cortical shell. Initially, we construct the image of edges by application of One-Pixel-Edge operator [9]. The edges generated by this operator have a width of one pixel with the result that the operator preserves the image resolution. A rough separation is based on the fact, that the magnitudes of the edge inside the cortical shell are significantly larger then the ones inside the trabecular bone. Then the mean attenuation inside this "draft" trabecular area is calculated. Now, the corrected contour of the trabecular bone is defined as a closed loop, outside of which the attenuation exceeds some threshold, based on this mean value. The area inside this contour is considered as trabecular bone, while the rest of the vertebral body is related to the cortical shell (see Fig. 2).

Assessment

Structural features of vertebra are quantified by different techniques developed in nonlinear dynamics. Statistical, statical and dynamical approaches are used to assess the spatial and architectural properties of three data sets: the entire vertebra, trabecular and compact bones.

To analyze **the images of attenuation**, as obtained from CT, we at first use well-known statistics: *moments of different order* [7] applied to the distribution of the attenuation a (or CT-numbers). We also tried to apply probability density $p(a)$ and distribution function $F(a)$, or so called coefficient of filling $c_f = 1 - F(a)$. We, however, found that such statistical approach is not sufficient to describe the changes in the bone structure, which are also obscured by a high slice-to-slice variability of these measures. Secondly, *complexity measures* [4–7,10] are introduced: one- and two-dimensional Shannon entropy, Renyi entropy of different order [4], and others [11].

Next, we use the concept of *symbolic dynamics* [8,10,11]. (cf. also [6,10] and references therein) and its generalize it to the two-dimensional case, i.e. analyze **images composed by symbols**. In order to analyze non-linear properties and a structure of a bone and also to sim-

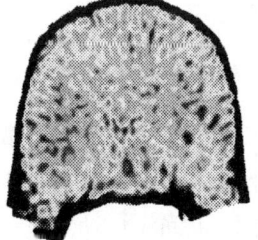 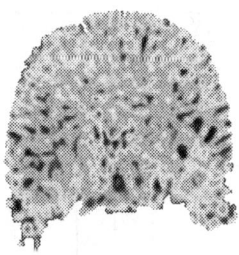 

FIGURE 2. Segmented entire vertebra, trabecular bone and cortical bone. (From axial slice shown in Fig. 1, left.) Special experiments have proved the stability of the developed segmentation algorithm for the orientation and shape of the bone, for slice thickness and the resolution of the CT-scanner, and for the wide range of healthy and osteoporotic specimens with BMD changes within the interval 21÷144 mg/ml.

plify the image, we introduce a set of *structural elements* or *an alphabet of symbols*. This is done by using a mixing of static and dynamical coding. At this point, the image of attenuations (CT-numbers) is transformed into an image composed of 5 different symbols as follows: Symbol "L", "lake", corresponds to the pixel with the attenuation close to the attenuation of water. Symbols "V" and "H", "valley" and "highland", represent the pixels with medium (larger than "lake', but less than the mean value plus the standard deviation) and high (larger than the mean plus the standard deviation) attenuation. Our experiments show that these *static* elements are not sufficient to describe the difference in the structure of healthy and pathological bones. Therefore, we additionally introduce a *dynamical* refinement of this coding procedure. We transform images of attenuation into the images of edges by application of One-Pixel-Edge operator [9] preserving the image resolution. If the edge in the current pixel (related to a difference between the pixel and its 8 neighbors) is less than some predefined limit, it is coded as described above. Additionally, symbol "I", "incline", denotes a pixel, for which the edge is larger than the limit. If the edge is higher than the threefold limit, it is coded as "C", "cliff". Typical distributions of symbols for normal and pathologic specimens are shown in Fig. 3. Entropy approach characterizes only averaged properties of probabilities ensemble and is not appropriate to assess the shape of these distribution. Therefore we propose the measure which we call "dynamical quality" of the structure: $Q_1 = \frac{p(I) \mid p(C)}{p(L)+0.001}$, $p()$ is the probability density of the corresponding symbol.

To quantify a spatial arrangement of the symbols, we introduce the notion "*blocks of symbols*" of different size (spatial analogue of one-dimensional notion "word of some length" [8,10]). Using this notion we construct several measures, based on the conditional probability, local relation between the

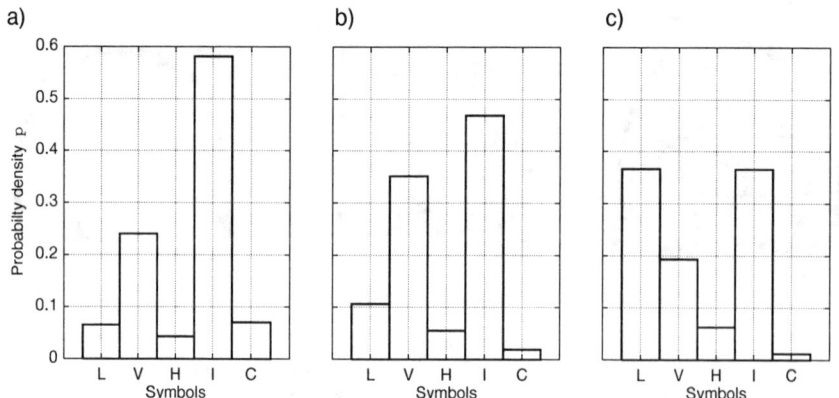

FIGURE 3. Typical distributions of symbols p within the trabecular bone: (a) normal, (b) osteopenia, or a weak grade, (c) definite osteoporosis or a strong grade of the disease.

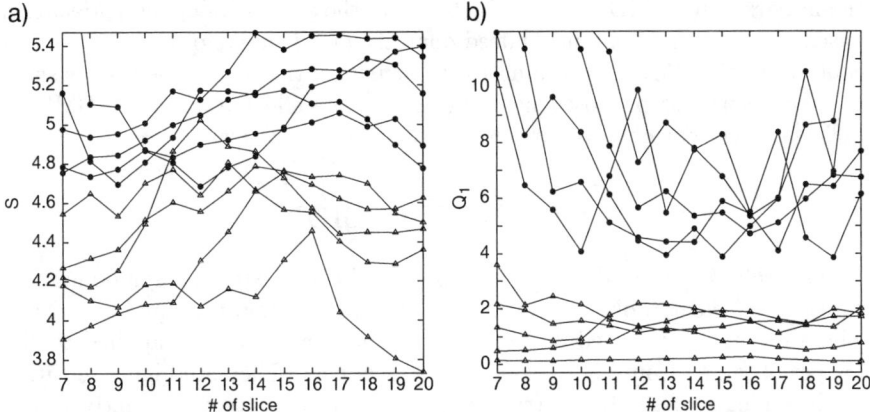

FIGURE 4. Shannon entropy S of the distribution of attenuation (a) and Dynamical quality Q_1 of distribution of symbols (b) in the trabecular bone of 5 healthy (black dots) and 5 osteoporotic (white triangles) specimens. Slices of 1 mm thickness are numbered from top to bottom of the vertebra.

probabilities of the symbols within the block, distributions of such local measures obtained from the blocks of different locations, size of block composed of the same symbols, dominated, and forbidden blocks.

III RESULTS

The behavior of complexity measures reflects the variability in structure from slice to slice both for osteoporotic and normal vertebral bodies. Healthy and pathologic structural patterns of vertebral bodies which cannot be clearly discriminated from each other by qualitative radiological reading, can be easily distinguished in several ways by entropy- and symbolic approaches (Fig. 4).

The complexity of the pathological non-fractured vertebrae always shows a lower level than the healthy bone (Fig. 4a). Some spatial structure, dynamics and corresponding elements, which are typical for the normal specimens and reflect how rich is the trabecular net, are suppressed in the case of the disease, such as "I", "inclines" and "C", "cliffs" (Fig. 3). While another symbol "L", "lake" that represents soft and marrow tissue inside the bone, becomes more and more pronounced in the case of osteoporosis. These significant relative changes in structural composition of the affected bones are captured and quantified by the symbolic measures we propose (Fig. 4b). Such differences are found for almost every slice of the pathologic and normal specimens, excluding the slices acquired near the top and the bottom of the vertebra which contain intervertebral disk information due to partial volume effects.

The behavior of the aforementioned quantities is compared with the traditional parameter BMD obtained by QCT on the same CT-scanner. Measures based on symbolic dynamics indeed quantify *the structure* of the vertebra, in contrast to the BMD that evaluates *the density* of the bone. The results show that the measures proposed here are much more effective than the BMD in differentiating normal / osteopenic / osteoporotic specimens.

IV CONCLUSION

Our preliminary results show that proposed complexity and symbolic measures are efficient tools to quantify the structure of non-fractured vertebrae. These methods are based on non-invasive measurements and applied to the images obtained from high resolution CT-scanners. They are highly sensitive to the changes of the bone structure, and can contribute significantly to the identification of the patients at early stages of metabolic bone diseases, especially for the diagnostics of osteoporosis. Our experiments show that the introduced methods can also be successfully applied to images obtained from standard CT-scanners.

ACKNOWLEDGMENTS

This work has been made possible in part by a grant from Boehringer Mannheim, Germany. The specimens were obtained from the study BIOMED 1 (contract # BMH1-CT92-0296), funded by the European Union.

REFERENCES

1. R. Marcus, in *Osteoporosis*, edited by R. Marcus, D. Feldman, and J. Kelsey, San Diego: Academic Press, 1996.
2. L. Mosekilde, *Calcif. Tissue Int.* **53**, 121 (1993).
3. R. R. Recker, *Calcif. Tissue Int.* **53**, 139 (1993).
4. J. Balatoni and A. Renyi, in *Selected Papers of A. Renyi*, Budapest: Acadeiau, 1976, Vol. 1, p. 558.
5. R. Wackerbauer, A. Witt, J. Kurths, and H. Scheingraber, *Chaos, Solitons & Fractals* **4**, 133 (1994).
6. J. Kurths, A. Voss, P. Saparin, and A. Witt, in *Information: New Questions to a Multidisciplinary Concept*, edited by K. Kornwachs and K. Jacoby, Bremerhaven: Akademie Verlag, 1995, pp. 129–137.
7. W. Gowin, P. Saparin, J. Kurths, and D. Felsenberg, *Osteologie* **6**, 11 (1997).
8. B.-L. Hao, *Physica* **D 51**, 161 (1991).
9. R. Klette and P. Zamperoni, *Handbook of image processing operators*, Chichester, UK: John Wiley & Sons, 1996.
10. J. Kurths, A.Voss, P.Saparin, A.Witt, J.Kleiner, N.Wessel, *Chaos* **5**, 88 (1995).
11. F. Spahn, U. Schwarz, and J. Kurths, *Phys. Rev. Lett* **78**, 1596 (1997).

Some applications of nonlinear diffusion to processing of dynamic evolution images

Alexey N. Goltsov* and Sergey A. Nikishov[†]

*Institute of Physics and Technology, Prechistenka Str., 13/7, Moscow, 119034, Russia
[†]Moscow Institute for Radioengineering, Electronics and Automation,
Vernadsky Prosp., 78, 117454, Moscow, Russia

Abstract. Model nonlinear diffusion equation with the most simple Landau–Ginzburg free energy functional was applied to locate boundaries between meaningful regions of low-level images. The method is oriented to processing images of objects that are a result of dynamic evolution: images of different organs and tissues obtained by radiography and NMR methods, electron microscope images of morphogenesis fields, etc. In the methods developed by us, parameters of the nonlinear diffusion model are chosen on the basis of the preliminary treatment of the images. The parameters of the Landau–Ginzburg free energy functional are extracted from the structure factor of the images. Owing to such a choice of the model parameters, the image to be processed is located in the vicinity of the steady-state of the diffusion equation. The suggested method allows one to separate distinct structures having specific space characteristics from the whole image. The method was applied to processing X-ray images of the lung.

INTRODUCTION

Notion of image processing as essentially nonlinear process is embodied in development of methods based on application of nonlinear partial differential equations [1]–[4]. Extensive use is made of nonlinear diffusion equation [2]–[4]. In this approach, the processing of an image I_o consists in solving a non-homogeneous differential equation with the initial condition I_o. The diffusion conductance is chosen in such a way that it responds to edges and noises of the image. It can be easily done by choosing the diffusion conductance as a function of the gradient of the image. It was shown [2]–[4] that this approach results not only in smoothing noises but also in sharpening boundaries. However, the use of these models involves several practical difficulties. The major difficulty is connected with the need for reliable estimation of the gradient of the useful image contaminated with noises. The region of the gradient values of interest is not usually a priori known. We believe that using additional information on image structures along with data on gradient distribution will improve the method.

In the present approach, an attempt is made to use additional information about peculiarities of spectral characteristics of the image. The method is intended mainly for processing of images of the objects that are a result of evolutional dynamics. In the living systems, such structures arise from morphogenesis (e.g., different organs and tissues). The processing of such images is of importance in X-ray, ultrasonic, and NMR diagnosis.

Other objects of this type are dissipative structures arising in a wide range of physical and chemical systems such as reaction–diffusion, self–assembled, fractal–like, and turbulent systems. One common feature of evolution of these structures is a loss of stability of the space modes in the limited range of the wave numbers. Interaction of these modes (order parameters of a system) with each other determines forms of resulting structures. This means that, for processing images of this type, one can invoke an additional information about the space mode spectra of the images.

In the proposed approach, the basic space modes defining the structures are extracted from the structure function of the image. Based on the information obtained, the parameters of nonlinear diffusion equation are chosen such that the equation solution has the space characteristics in the detected range of the wave numbers.

During the diffusion model simulation, the selected space modes associated with the structure image are enhanced. This enhancement is accompanied by sharpening of boundaries and smoothing of noises. The noise spectrum is assumed to lie higher than the image spectrum.

In the present approach, a transition from blurred image $\phi_o(\vec{r})$ with noise to a smoothing image $\phi(\vec{r}, t)$ with enhanced edges is considered as a phase transition which results in changing the nature of the field of the order parameter $\phi(\vec{r}, t)$. The function $\phi(\vec{r}, t)$ which is continuous at initial stage of the processing (hydrodynamic modes) becomes step-wise (behaves like the Ising quantity).

MODEL EQUATION

To obtain such evolution of the two–dimensional order parameter $\phi(\vec{r}, t)$, we considered the following model nonlinear diffusion equation

$$\frac{\partial \phi(\vec{r}, t)}{\partial t} = M \nabla^2 \frac{\delta \mathcal{F}[\phi(\vec{r}, t)]}{\delta \phi(\vec{r}, t)} \qquad (1)$$

with the initial condition $\phi(\vec{r}, 0) = \phi_o(\vec{r})$, here M is a mobility coefficient.

For the functional $\mathcal{F}[\phi(\vec{r}, t)]$, the simplest Landau-Ginzburg surface free energy functional

$$\mathcal{F}[\phi(\vec{r},t)] = \int \{f[\phi] + K|\nabla\phi(\vec{r},t)|^2\}dxdy \qquad (2)$$

was chosen, where K is an expansion coefficient and $f[\phi]$ is the ϕ^4 free energy functional

$$f[\phi] = -\frac{1}{2}a(\phi - \overline{\phi})^2 + \frac{1}{4}b(\phi - \overline{\phi})^4; \qquad (3)$$

$a, b, \overline{\phi}$ and K are positive model parameters.

The substitution of (2) into (1) yields

$$\frac{\partial \phi(\vec{r},t)}{\partial t} = M\nabla^2(-a\phi + b\phi^3) - 2MK\nabla^4\phi(\vec{r},t). \qquad (4)$$

In the theory of phase separation, equation (1) is known as the Cahn–Hilliard equation [5]. It is used for description of phase transition such as spinodal decomposition of binary mixtures (alloys, polymer blends and others).

The specific feature of the solution to nonlinear equation (1) can be derived from the linear approximation. The linearized treatment suggests that the initial image $\phi_o(\vec{r})$ is blurred and a deviation of $\phi_o(\vec{r})$ from $\overline{\phi}$ is small. The spatio-temporal evolution can be described by the structure function defined as

$$S(q,t) = <|\phi(q,t)|^2>. \qquad (5)$$

In the linear approximation, it takes the following form:

$$S(q,t) = S(q,0)e^{R(q)t}, \qquad (6)$$

where $R(q)$ is the "enhancement" coefficient

$$R(q) = 2Mq^2 \left(\left.\frac{\partial^2 f}{\partial \phi^2}\right|_{\phi=\overline{\phi}} + Kq^2\right).$$

The amplitude of the homogeneous fluctuations grows with time for values of the wave numbers $q < q_c$ and decreases for $q > q_c$, where

$$q_c = \left(-\frac{1}{K}\left.\frac{\partial^2 f}{\partial \phi^2}\right|_{\phi=\overline{\phi}}\right)^{1/2}. \qquad (7)$$

The amplification factor takes its maximum value at $q_m = q_c/\sqrt{2}$.

At the late stage of the phase-ordering dynamics, the nonlinear effects become important. The field of the local order parameter $\phi(\vec{r},t)$ is divided into domains of size $L(t)$, which is greater than the correlation length $\xi = 2\pi/q_m$.

The local order parameter $\phi(\vec{r}, t)$ takes the value ϕ_1 inside the domains and the value ϕ_2 outside them. At the late stage of the evolution, the structure function $S(q, t)$ is known to have the following scaling behaviour

$$S(q, t) = L(t)^d \Phi(qL(t)), \tag{8}$$

where Φ is a universal function, $L(t)$ is a time-dependent length scale of the form $L(t) \sim t^{1/3}$, d is space dimension [6].

As can be seen from the above discussion, at the initial stage of the image processing, it is possible to enhance image structures formed by space modes in the region $q < q_c$ and to weaken the fluctuations and noises for $q > q_c$. At the late stage of the processing, the sharpening of the image boundaries occurs.

This means that, before applying this model to image processing, it is necessary to carry out the preliminary treatment of the image in order to obtain the basic space modes of the image. The information derived is used for choosing the model parameters (tuning of the model).

The space characteristics of the image being processed are extracted from the structure function

$$S(q, t) = |\sum_j g(r_j) \exp(iqr_j)|^2, \tag{9}$$

which is the Fourier transformation of the space pair-correlation function

$$g(r, t) = \frac{1}{N^2} \sum_{|\vec{r'}|} \phi(\vec{r'} + \vec{r}, t) \phi(\vec{r'}, t), \tag{10}$$

where N is a linear size of the image. The analysis of the structure function $S(q, 0)$ of the initial image yields the correlation length ξ of the order parameter field $\phi(\vec{r}, 0)$. By assuming that the linearized model equation (1) is true at the initial stage of the processing, the equation (7) can be used in finding relation between the parameters a and K. These parameters are chosen from the condition that the value of the wave number q_m of the linearized equation (1) coincides with the revealed wave number $q_{im} = 2\pi/\xi$. This results in the following relation between a and K: $a = 2q_{im}^2 K$. The parameter b is chosen from the positions of minima of the potential $f[\phi]$ (3): $\phi_{1,2} = \overline{\phi} \pm (a/b)^{1/2}$.

ILLUSTRATIVE EXAMPLE

As an illustration, we present here the application of this method to processing of X-ray image of the lung tumor growing into the bronchus (see Figure 1 (a)). Here, the goal of the processing is to separate two structures (the tumor and the bronchus) that have distinct space properties. The bronchus by its

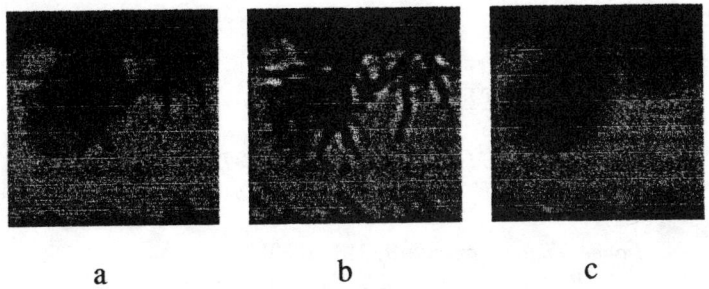

FIGURE 1. The results of extraction of different structures from X-ray image of lung cancer: (a) the initial image, (b) the bronchus structure, (c) the tumor structure.

nature is close to a dendrite–like, or a fractal–like structure. The analysis of the structure function $S(q,0)$ (9) gave the following correlation lengths $\xi_b \simeq 10$ for the bronchus and $\xi_t \simeq 50$ for the tumor structures.

The kinetic equation (4) was solved numerically by Euler's method [6] on the two–dimensional square lattice of linear size $N = 100$ with periodic boundary conditions.

The result of applying the model tuned to the processing of the bronchus structure is shown in Figure 1 (b). In the simulation process, termination of the bronchus growing into the tumor becomes clear. This is one of tumor indications in cancer diagnosis.

The result of simulation of the model tuned to the processing of the tumor is shown in Figure 1 (c). Investigation of internal structure of the tumor at its decay stage is possible in the model tuned to intermediate region of the wave numbers.

CONCLUSION

In the context of simple diffusion model, we have considered the method of using information about the space image characteristics for the image processing. It has been shown that the model equation adopted can be applied to enhancement and separation of different structures having distinct space mode spectra. However, as can be seen from the results presented, the model equation does not reproduce the dendrite–like structure of the bronchus. It seems likely that the data used are not sufficient to reproducing this structure. The model describing the evolution of such complex systems must also reproduce other characteristics of the dynamic systems such as fractal dimension, correlation dimension, and others. This information can be obtained from additional treatment of two–dimensional images and can be used for construction of some nonlinear models, which are related to real physical, chemical, and biological phenomenon.

REFERENCES

1. Osher S., and Rudin L.I., *SIAM J. Numer. Analysis* **27**, 919 (1990).
2. Perona P., and Malik J., *IEEE Trans. Pattern. Analysis Mach. Intell.* **12**, 429 (1990).
3. Nordstrom N., *Image Vision Comput.* **8**, 318 (1990).
4. Li X., and Chen T., *Pattern Recognition* **27**, 1029 (1994).
5. Cahn J.W., and Hilliard J.E., *J. Chem. Phys.* **31**, 3 (1959).
6. Oono Y., and Puri S. *Phys. Rev. A* **38**, 434 (1988).

Low Dimensional Model of Heart Rhythm Dynamics as a Tool for Diagnosing the Anaerobic Threshold

O.L. Anosov*, O.Ya. Butkovskii**, J. Kadtke[†],
Yu.A. Kravtsov[‡], V. Protopopescu[∥]

*Institute of Physiology, Charite
Humboldt-University at Berlin, Berlin, D-10117, Germany, oleg.anosov@rz.hu-berlin.de
(permanent address: Regional. Cardio Center, Vladimir, Russia)
**Vladimir State Technical University, 87, Gorki St., Vladimir, 600026, Russia,
butr@vpti.vladimir.su
[†]Institute for Pure and Applied Physical Science
University of California at San Diego, La Jolla, CA 92093-0360, USA,
jkadtke@sdphv1.ucsd.edu
[‡]Space Research Institute of RAS, 84/32 Profsoyuznaja St., Moscow, 117810, Russia,
kravtsov@asp.iki.rssi.ru.
[∥]Oak Ridge National Laboratory Oak Ridge, TN 37831-6364, USA, vvp@ornl.gov

Abstract. We report preliminary results on describing the dependence of the heart rhythm variability on the stress level by using qualitative, low dimensional models. The reconstruction of macroscopic heart models yielding cardio cycles (RR-intervals) duration was based on actual clinical data. Our results show that the coefficients of the low dimensional models are sensitive to metabolic changes. In particular, at the transition between aerobic and aerobic-anaerobic metabolism, there are pronounced extrema in the functional dependence of the coefficients on the stress level. This strong sensitivity can be used to design an easy indirect method for determining the anaerobic threshold. This method could replace costly and invasive traditional methods such as gas analysis and blood tests.

INTRODUCTION

Direct measurements of the parameters of the cardiac regulation system are rather complicated, and in many cases impossible to carry out without invasive procedures. Thus, to improve diagnosis, prevention, and treatment, it is extremely important to develop simple mathematical models that relate the "latent" processes of the cardiac regulation system to easily measured data, such as, for example, cardiac cycle duration (RR-interval).

Despite numerous publications devoted to the behavior of the cardiac regulation system, the development of accurate macroscopic models based on the description of internal biophysical and biochemical processes has met only with partial success. The principal obstacle is the so called "dimensionality curse", typical of complex biological systems. In some models, the number of equations describing the cardiac regulation system has already exceeded one hundred [1] and that is far from being exhaustive. Thus, the practical value of the present models for the routine medical practice is at best limited.

Much attention has been devoted lately to the description of cardiac activity from the point of view of nonlinear dynamics (see, for example, [2,9], [3,8]). This viewpoint allows for the possibility of describing complex behaviors using relatively simple systems. Recently, we have proposed a new procedure to construct simple (low dimensional) dynamical models from time series generated by possibly complex (high dimensional) systems [4,5,6]. We shall apply this procedure to construct a macroscopic model of the cardiac regulation system. This model will not describe physiological processes in detail, but will still reflect the qualitative processes taking place in real bio-medical instances. In particular, the reconstruction of macroscopic models for the cardiac dynamics makes it possible to evaluate the "distance" between the actual state of the cardiac regulation system and the critical points that characterize the qualitative changes in the system's behavior. In our opinion, this type of macroscopic models for cardiac dynamics may become the basis of a new diagnostic tool for the transition of cardiac rhythm from periodicity to chaos. Since ventricular fibrillation is of chaotic nature [10], the search for precursors of the transition to chaos is important for identifying patients with high risk of sudden death. In this paper we report preliminary results on low dimensional modeling of cardiac dynamics based on clinical experiments aimed at studying RR-series behavior under different stress levels.

HEART RHYTHM DYNAMICS AND STRESS LEVEL

Clinical experiments

The heart rhythm dynamics was examined during a clinical experiment aimed at studying the cardiac regulation system under various stress levels. The clinical experiments were conducted on the usual patient's (not sportsmen and without any pathology). The typical resalts for this group are presented below.

During the experiment the stress level was monitored by calibrated exertion in the "Metabolic Measurement System SensorMedics 2900" (veloergometer). Simultaneously with the registration of the RR-intervals (duration), the anal-

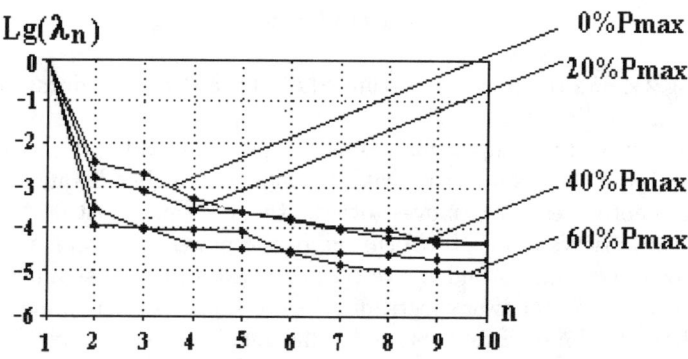

FIGURE 1. Eigenvalues of the Takens matrix for RR-series under different physical loads

ysis of the patient's exhaled gas was carried out. The patient's electrocardiogram and arterial pressure were also recorded.

The experiment had two stages. In the first stage we continuously increased the stress level and estimated the patient's maximum work capacity, P_{max}, and anaerobic threshold, AT. The anaerobic threshold AT was determined by the dependence of the discharged carbon dioxide volume (VCO2) on the consumed oxygen volume (VO2). In the second stage, we increased the stress level in steps of 0%, 20%, 40% and 60% of P_{max}. The data, registered at each step during 400 cardiocycles after the patient reached the stationary response to the load, was subjected to further analysis. Described below are the power spectrum analysis, the analysis of the RR-series dimensionality, and the results of the heart rhythm macroscopic model reconstruction obtained for various stress levels. Special attention was given to unambiguously discriminating dynamical features for two types of metabolism, namely: aerobic and aerobic-anaerobic.

Dimensionality processing

Direct methods of dimensionality evaluation require a great number of experimental points (about 10^4–10^5), which could not be obtained from the clinical experiment. Instead, we performed the qualitative analysis of RR-series dimensionality by using the eigenvalues of the Takens matrix [7] (see fig.1).

As seen in fig.1, upon increasing the load, the eigenvalue spectra level out and the contribution of the first eigenvalues decreases. Such a behavior indicates a decrease of the process dimensionality. This shows a "simplification" of the cardiac regulation system due to the stress increase.

Spectral processing

Fig.2 shows the power spectrum $S(T)$ of the RR-series duration, calculated for each loading step.

As seen in Fig.2, upon increasing the physical load and, consequently, the stress level, the respiratory peak, T_R, moves to higher frequencies, while the baroreceptor peak, T_B, moves towards the lower ones. At 60% load, (Fig.2), one can observe a structural change of the power spectrum, i.e. the baroreceptor peak vanishes and the respiratory peak increases significantly.

Such RR-series power spectrum behavior suggests the nonlinear dependence of the control mechanisms of the cardiac regulation system on the stress level.

In our experiment, the change of the RR-series power spectrum was observed in the transition from aerobic to aerobic-anaerobic metabolism, which was independently registered by the gas analyzer of the "Metabolic Measurement System SensorMedics 2900", under about 50% of the maximum load, P_{max}.

Low dimensional model of heart rhythm dynamics

The reconstruction of macroscopic model for the heart rhythm dynamics was carried out by using the minimax procedure developed by us and described in [4,5,6] The width of the time windows was $T_1 = T_2 = 40$ cardiocycles. At each loading step, we analyzed 400 cardiocycles registered after the patient reached the stationary response to the load.

Here, we sought to restore the macroscopic model which reflects the structural changes in heart rhythm dynamics that are independent of the absolute value of the patient pulse rate. To achieve this, the reconstruction was performed by using RR-series previously normalized to the mean value. During reconstruction we used models described by a differential equation system, with polynomial nonlinearities of third degree. Such a restriction was imposed in order to get a simple model, suitable for practical clinical applications.

The heart rhythm dynamics model was represented as:

$$dX/dt = Y,$$
$$dY/dt = Z,$$
$$dZ/dt = \frac{F(X,Y,Z;\mathbf{a})}{f(X,Y,Z;\mathbf{b})}. \quad (1)$$

Here, $X = (T_{RR}/<T_{RR}>) - 1$ - is the normalized variance of RR-interval duration, $<T_{RR}>$ is the mean value, $F(X,Y,Z;\mathbf{a})$ a third degree polynomial, $f(X,Y,Z;b)$ - second degree polynomial, vectors, $\mathbf{a}$ and $\mathbf{b}$ are parameter vectors.

The initial model contained 20 unknown parameters. However, during model reconstruction at maximum load step (60% of P_{max}) only four unknown parameters, namely a_1, a_2, a_3, a_4 turned out to be significant. As a result, the polynomial of the system (1) simplified to:

$$F(X,Y,Z;\mathbf{a}) = 3a_1YX^2 + 2a_2Y^2X + a_2ZX^2 + \\ + a_3Z^2Y + a_4Y^3 + 2(1+a_4)XYZ. \qquad (2)$$

Here we confine ourselves to studying the behavior of the mean values of of the parameters in (2) for various loads.

Fig.3 shows the dependence of the coefficients a_1, a_2 and a_4 of model (2) on the exertion level P/P_{max} (parameter values calculated from the experimental data are marked with circles).

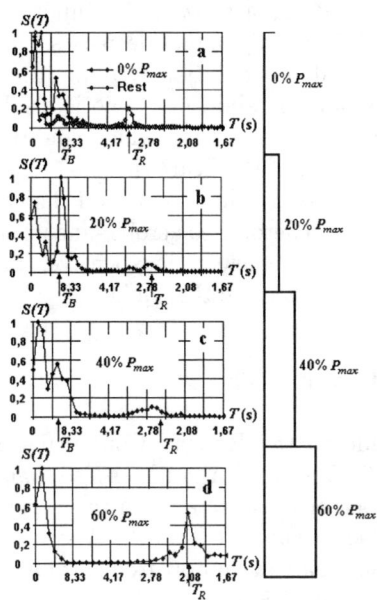

FIGURE 2. Power spectra of RR-series under different physical loads

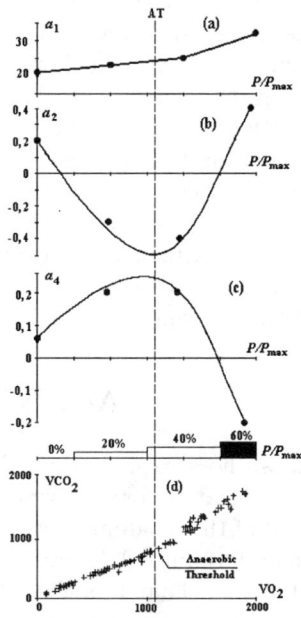

FIGURE 3. The interpolated dependences of some model (2) parameters on the physical load, (a), (b), (c), and respiratory characteristic of a patient, (d)

Fig.3 shows the dependence of the discharged carbon dioxide volume (VCO2) on the consumed oxygen volume (VO2), obtained from the exhaled gas analysis of the patient at the first stage of the experiment. The characteristic bend in this dependence corresponds to the anaerobic threshold, i.e. to the transition from aerobic to aerobic-anaerobic metabolism.

The coefficient a_1, shown in Fig.3, increases linearly under loads up to 40%, while at the anaerobic threshold it displays a marked change in slope. Coefficients a_2 and a_4, shown in Fig.3 and Fig.3, display pronounced extrema at the anaerobic threshold. These behaviors could potentially be used for developing a practical method to determine the anaerobic threshold, which would require only the numerical analysis of the RR-intervals, instead of tedious and costly gas analysis methods and blood tests.

CONCLUSION

We proposed a simple method to reconstruct and parametrize a low dimension model of the heart rhythm dynamics from clinical data. The model establishes a relationship between the features of the heart rhythm dynamics and the level of the physical stress exerted onto the human organism.

The dimensional analysis of the RR-intervals data showed that upon increasing stress level, the dimensionality of the heart dynamics decreases, thus implying the behavior simplification of the cardiac regulation system. In the transition area from aerobic to aerobic-anaerobic metabolism one notices substantial structural changes of the RR-series power spectrum which confirms the nonlinear character of the control mechanism of the heart rhythm.

Also the dependence of the model parameters on the stress level display pronounced extrema in the transition region from aerobic to aerobic-anaerobic metabolism. This effect may be used to develop efficient, inexpensive, and non-invasive diagnostic tools to determine the anaerobic threshold.

ACKNOWLEDGMENTS

The authors express their deep gratitude to the Russian Committee of Higher Education for support under grant (95-0-83-1). The authors also thank the staff of the Vladimir Regional Cardio Center for assistance in the experiments performing. O.A. gratefully acknowledges support from the Alexander von Humboldt Foundation. V.P. acknowledges partial support from the Office of Basic Energy Sciences of the US Department of Energy. J.B.K. wishes to acknowledge continuing support from the Office of Naval Research.

REFERENCES

1. Mosekilde, E. & Jensen, E. [1981] "Dynamic simulation of human ventilatory regulation", in *System Dynamics Conf.*, (Albany, New York).
2. Glass, L. & Guevara, M.R. [1983] "Bifurcations and Chaos in a Periodically Stimulated Cardiac Oscillator", *Phys.D* **70**(1-3), 89-101.
3. Goldberger, A.L. *et al.* [1995] "The fractal basis of physiology: long-range correlations in health and their breakdown with disease", in *Chaotic, Fractal and Nonlinear Signal Processing, Mystic, July 10-14, 1995. Connecticut, USA*, ed. Katz R. A. (AIP Press, Woodbury, New York).

4. Anosov, O.L. et al. [1996] "Strategy and algorithns for dynamical forecasting", in *Predictability of Complex Dynamical System* eds. Kravtsov, Yu.A. & Kadtke, J.B. (Springer, Berlin).
5. Anosov, O.L. et al. [1996] "Discriminant procedure for the inverse problem solution to the nonstationary system", in *Predictability of Complex Dynamical System* eds. Kravtsov, Yu.A. & Kadtke, J.B. (Springer, Berlin).
6. Anosov, O.L. et al. [1997] "Non linear Chaotic System Identification from Observed Time Series", *J. M3AS* **1**(7), (to be published).
7. Landa, P.S. & Rosenblum, M.G. [1989] "On a certain method of evaluation of attractor embedding dimensionality according to experiments results", *Zhurnal Teoreticheskoi Fiziki* **59**(1),13-20.
8. Anishchenko, V.S. & Smirnova, N.B. [1993] "Analysis and synthesis of dynamical systems from experimental data", *SPIE* **2098**, 137-141.
9. Mackey, M.C. & Glass, L. [1977] "Oscillation and chaos in physiological system", *Science* **197**, 287-289.
10. Witkowski, F.X. et al. [1995] "Evidance for Determinism in Ventricular Fibrillation", *PhRL* **75**(6), 1230-1233.

Multiple-Scales Analysis of Non-Stationary Biological Signals

C.-K. Peng[1]

Beth Israel Deaconess Medical Center, Harvard Medical School, Boston, MA 02215 USA
Center for Polymer Studies and Department of Physics, Boston University, Boston, MA 02215 USA

Abstract. We discuss multiple-time scale properties of neuralphysiological control mechanisms, using heart rate regulation as a model system. We find that scaling exponents can be used as prognostic indicators. Furthermore, detection of more subtle degradation of scaling properties may provide a novel early warning system in subjects with a variety of pathologies including those at high risk of sudden death.

Scale-invariant properties in biological systems have received much attention recently [1,2]. The absence of characteristic temporal (or spatial) scales may confer important biological advantages, related to adaptability of response [2,3]. Here we present some recent progress in applying scale-invariant (fractal) analysis to physiological time series. We will concentrate on the output of a model physiological systems: the human heartbeat time series under neuroautonomic control.

Clinicians often describe the normal activity of the heart as "regular sinus rhythm." But in fact cardiac interbeat intervals normally fluctuate in a complex, apparently erratic manner [2,4] (Fig. 1). This highly irregular behavior has recently motivated researchers [5,6] to apply time series analyses that derive from statistical physics, especially methods for the study of critical phenomena where fluctuations at all length (time) scales occur. These studies show that under healthy conditions, interbeat interval time series exhibit long-range power-law correlations reminiscent of physical systems near a critical point [7,8]. Furthermore, certain disease states may be accompanied by alterations in this scale-invariant (fractal) correlation property. Here we explore the potential utility of such scaling alterations in the detection of pathological states.

Our analyses are based on the beat-to-beat heart rate fluctuations of digi-

---

[1] Partially supported by NIMH (MH54081)

tized electrocardiograms recorded with an ambulatory (Holter) monitor. The time series obtained by plotting the sequential intervals between beat i and beat $i+1$, denoted by $B(i)$, typically reveals a complex type of variability (Fig. 1). The mechanism underlying such fluctuations appears to be related primarily to countervailing neuroautonomic inputs. Parasympathetic stimulation decreases the firing rate of pacemaker cells in the heart's sinus node. Sympathetic stimulation has the opposite effect. The nonlinear interaction (competition) between the two branches of the autonomic nervous system is the postulated mechanism for the type of erratic heart rate variability recorded in healthy subjects [4,9].

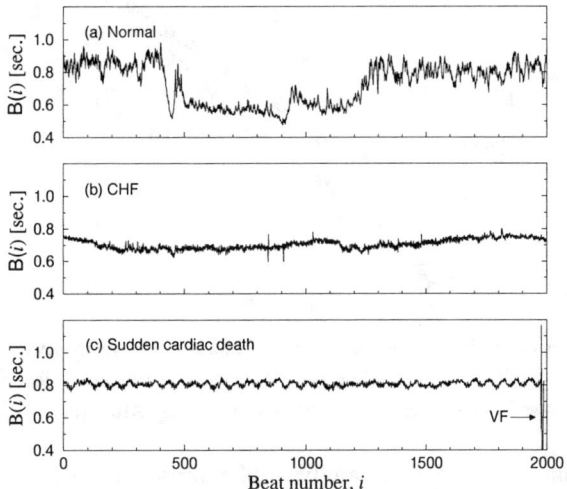

FIGURE 1. Representative complex physiological fluctuations. Cardiac interbeat interval time series of 2000 beats from (a) a healthy subject, (b) a subject with congestive heart failure (CHF) and (c) a sudden cardiac death subject with ventricular fibrillation (VF).

Detrended Fluctuation Analysis (DFA)

Difficulties of quantitatively analyzing physiological time series arise mainly from their nonstationarity and sometimes short data length. We have developed a scaling analysis, called *detrended fluctuation analysis* (DFA) [10,11], which takes these factors into account. DFA is a modified root mean square analysis of a random walk based on the following concept: A stationary time series with long-range correlations can be integrated, i.e., form an accumulated sum, to form a self-similar process. Therefore, measurement of the self-similarity scaling exponent of the integrated series can tell us the long-range correlation properties of the original time series. In short, we integrate the original time series once; then we determine the fluctuations $F(n)$ of the integrated signal around the best linear fit in a time window of size n. The slope of the line relating log $F(n)$ to log n determines the scaling exponent (self-similarity parameter) α. The DFA method has been validated on control time series that consist of long-range correlations with the superposition of a non-stationary external trend [10]. It has also been successfully applied to detect long-range correlations in highly heterogeneous DNA sequences [10,12,13], and other complex physiological signals [11,14,15].

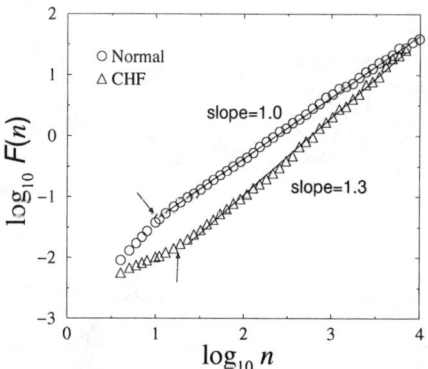

FIGURE 2. Plot of $\log F(n)$ vs $\log n$ for two 24 hours interbeat interval time series. The circles are from a healthy subject while the triangles are from a subject with congestive heart failure. Arrows indicate "crossover" points in scaling. After [11].

Figure 2 compares the DFA analysis of representative 24 hour interbeat interval time series of a healthy subject and a patient with congestive heart failure (CHF). Notice that for large time scales (asymptotic behavior), the healthy subject interbeat interval time series shows almost perfect power-law scaling over two decades ($20 \leq n \leq 10000$) with $\alpha = 1$ (i.e., $1/f$ noise) while $\alpha \approx 1.3$ (closer to Brownian noise) for the CHF patient.

We note that for short time scales, there is an apparent *crossover* exhibited for the scaling behavior of some data sets (arrows in Fig. 2). For the healthy subject, the α exponent estimated from very small n (< 10 beats) is larger than that calculated from large n (> 10 beats). This is probably due to the fact that on very short time scales (a few beats to ten beats), the physiologic interbeat interval fluctuation is dominated by the relatively smooth heartbeat oscillation associated with respiration, thus giving rise to a large α value. For longer scales, the interbeat fluctuation, reflecting the intrinsic dynamics of a complex system, approaches that of $1/f$ behavior as previously noted [5,16]. *In contrast, the CHF data set shows a very different crossover pattern* (Fig. 2). For very short time scales, the fluctuation is quite random (close to white noise, $\alpha \approx 0.5$). As the time scale becomes larger, the fluctuation becomes smoother (asymptotically approaching Brownian noise, $\alpha \approx 1.5$).

Practical Utilities

Several recent studies have demonstrated that scaling exponents (both short- and long-time scales) might be useful clinical indicators for detecting pathological dynamics. In particular, these studies found:

(1) For a groups of 12 healthy adults without clinical evidence of heart disease and a group of 15 adults with severe heart failure. The long-range exponents (for time scales $10^2 \sim 10^4$ beats) are significantly different. For the group of healthy cardiac interbeat interval time series (mean value ± S.D.): $\alpha = 1.00 \pm 0.11$. This result is consistent with previous reports of $1/f$ fluctuations in healthy heart rate (by spectral analysis) [3,16]. The pathologic group shows a significant ($p < 0.01$ by Student's t-test) deviation of the long-range correlation exponent, $\alpha = 1.24 \pm 0.22$, from normal.

(2) The above observation of a differential crossover pattern for healthy versus pathologic data motivated us to extract two parameters from each data set by fitting the scaling exponent α over two different time scales: one short, the other long. To be more precise, for each data set we calculated an exponent α_1 by making a least squares fit of $\log F(n)$ vs $\log n$ for $4 \leq n \leq 16$. Similarly, an exponent α_2 was obtained from $16 \leq n \leq 64$. Relatively short data sets are sufficient for this procedure, thereby making this technique applicable to "real world" clinical data.

We applied this quantitative fluctuation analysis to the two different groups of subjects mentioned above to measure the two scaling exponents α_1 and α_2. All data set records were divided into multiple sub-sets (each with 8192 beats ~ 2 hours) and the two exponents were calculated for each subset. For healthy subjects, we find that $\alpha_1 = 1.20 \pm 0.18$ (mean ± S.D.) and $\alpha_2 = 1.00 \pm 0.12$. For the group of congestive heart failure subjects, we find that $\alpha_1 = 0.80 \pm 0.26$ and $\alpha_2 = 1.13 \pm 0.22$, both significantly ($p < 0.0001$ for both α_1 and α_2) different from normal. Furthermore, we show in Fig. 3a that fairly good discrimination between these two groups can be achieved by using these two scaling exponents.

(3) Based on the hypothesis that there is a region of scaling behavior (in Fig. 3a) over which the normal (healthy) cardiac control operates, we have recently found another promising application of DFA in analyzing data sets from Framingham Heart Study—a prospective, population based study [17]. The primary group of interest was individuals with congestive heart failure (CHF); 28 CHF cases and 41 sex and age-matched healthy control cases were analyzed by our scaling analysis. Briefly, using Holter monitor data (approximately 2 hours) from each subject of the Framingham Study, we assigned an index (range from 0 to 1) to each individual by estimating the probability that this particular heartbeat time series was operating in the appropriate region in Fig. 3a (normal vs. pathologic). Does this measure add independent information to conventional measures? In comparison with other 10 time and frequency measures, we found that the DFA index carries prognostic informa-

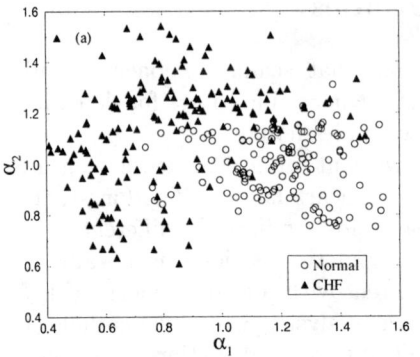
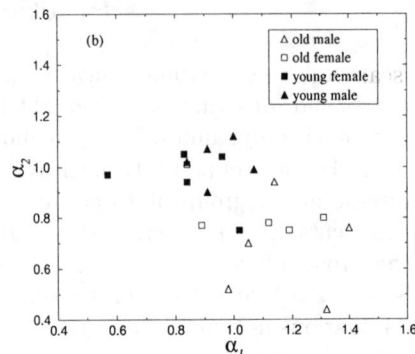

FIGURE 3. Scatter plot of scaling exponents α_1 vs α_2 for (a) the healthy subjects and subjects with congestive heart failure; (b) young and elderly subjects. Note good separation between healthy and heart disease subjects in (a), with clustering of points in two distinct "clouds." Similarly there is good separation between young and elderly subjects in (b).

tion about mortality not extractable from these traditional methods of heart rate variability analysis [17].

(4) Similar analysis was applied to study the effect of physiologic aging. Ten young (21–34 yr) and ten elderly (68–81 yr) healthy sujects underwent 2 hr of continuous supine resting ECG recording. In healthy young subjects, scaling exponent α is close to a value of 1.0. In the group of healthy elderly subjects, the interbeat interval time series also had two scaling regions. Over the short range, interbeat interval fluctuations resembled a random walk process (Brownian noise, $\alpha = 1.5$), whereas over the longer range they resembled white noise ($\alpha = 0.5$) Short-range (α_1) and long-range (α_2) exponents were significantly different in the elderly subjects compared with young (see Fig. 3b) [18].

CONCLUSIONS

It is known that biological systems contain a wide range of time scales. The scaling exponents discussed here can be thought of as a quantitative measure of how "balance" are these time scales. Therefore, subtle or intermittent degradation of scaling properties may indicate dominance (or drop-out) of certain time scales. Thus it may provide an early warning of incipient pathology. Finally, we note that to fully discribe the dynamics of these physiological systems, more sophisicated methods are needed to probe the nonlinear interaction (coupling) between those different time scales in the system.

In summary, we apply a new fluctuation analysis (modified from classical random walk analysis) to the nonstationary heartbeat time series. We show that this method can detect the presence of scaling of multiple-time scales in physiological time series. Furthermore, this method is capable of identifying crossover behavior due to differences in scaling over short versus long time scales. These findings are of interest from a physiologic viewpoint since it motivates new modeling approaches to account for the control mechanisms regulating cardiac dynamics on different time scales. From a practical point of view, quantification of these scaling exponents may have potential applications for bedside and ambulatory monitoring.

REFERENCES

1. A. Bunde, S. Havlin, eds., *Fractals in Science* (Springer-Verlag, Berlin, 1994).
2. A.L. Goldberger, D.R. Rigney and B.J. West, *Sci. Am.* **262**, 42–49 (1990).
3. A.L. Goldberger, *Lancet* **347**, 1312–1314 (1996).
4. R.I. Kitney and O. Rompelman, *The study of heart-rate variability* (Oxford University Press, Oxford, 1980).
5. C.-K. Peng, J.E. Mietus, J.M. Hausdorff, S. Havlin, H.E. Stanley and A.L. Goldberger, *Phys. Rev. Lett.* **70**, 1343–1346 (1993).
6. C.-K. Peng, S.V. Buldyrev, J.M. Hausdorff, S. Havlin, J.E. Mietus, M. Simons, H.E. Stanley and A.L. Goldberger. In: T.F. Nonnenmacher, G.A. Losa and E.R. Weibel, eds., *Fractals in biology and medicine* (Birkhaüser Verlag, Basel, 1994) 55–65.
7. H.E. Stanley, *Introduction to phase transitions and critical phenomena* (Oxford University Press, London, 1971).
8. P. Bak, C. Tang and K. Wiesenfeld, *Phys. Rev. Lett.* **59**, 381–384 (1987).
9. M.N. Levy, *Circ. Res.* **29**, 437-445 (1971).
10. C.-K. Peng, S.V. Buldyrev, S. Havlin, M. Simons, H.E. Stanley and A.L. Goldberger, *Phys. Rev. E* **49**, 1685–1689 (1994).
11. C.-K. Peng, S. Havlin, H.E. Stanley, and A.L. Goldberger, *Chaos* **5**, 82–87 (1995).
12. S.V. Buldyrev, A.L. Goldberger, S. Havlin, C.-K. Peng, H.E. Stanley and M. Simons, *Biophys. J.* **65**, 2673–2679 (1993).
13. S.M. Ossadnik, S.V. Buldyrev, A.L. Goldberger, S. Havlin, R.N. Mantegna, C.-K. Peng, M. Simons and H.E. Stanley, *Biophys. J.* **67**, 64–70 (1994).
14. J.M. Hausdorff, C.-K. Peng, Z. Ladin, J. Y. Wei and A.L. Goldberger, *J. Appl. Physiol.* **78**, 349–358 (1995).
15. J.M. Hausdorff, P. Purdon, C.-K. Peng, Z. Ladin, J.Y. Wei and A.L. Goldberger, *J. Appl. Physiol.* **80**, 1448–1457 (1996).
16. M. Kobayashi and T. Musha, *IEEE Trans. Biomed. Eng.* **29**, 456 (1982).
17. K.K.L. Ho, G.B. Moody, C.-K. Peng, J.E. Mietus, M.G. Larson, A.L. Goldberger and D. Levy, *Circulation* **96**, 842–848 (1997).
18. N. Iyengar, C.-K. Peng, R. Morin, A.L. Goldberger, and L.A. Lipsitz, *Am. J. Physiol.* **271**, 1078–1084 (1996).
19. W. B. Cannon, *Physiol. Rev.* **9**, 399 (1929).

Author Index

A

Afanasjev, V. V., 103
Allie, S., 47
Anishchenko, V. S., 267
Anosov, O. L., 359
Argoul, F., 125
Arnéode, A., 125

B

Barnett, J., 51
Bartussek, R., 243
Bauer, A., 249
Beck, C., 11
Bezrukov, S. M., 175
Birx, D., 203
Braun, H. A., 281
Bressloff, P. C., 287
Bünner, M. J., 75
Butkovskii, O. Ya., 359

C

Cao, L., 69
Čenys, A., 81
Chatterjee, N., 117
Clark, T. D., 261
Coombes, S., 287
Coombs, M. J., 203

D

Dewald, M., 281
Dialynas, T. E., 163
Diggins, J., 261
Doering, C. R., 3
Dressler, U., 237
Dykman, M. I., 157, 169

E

Ebeling, W., 151
Everitt, M., 261

F

Farmer, B. A., 187
Felsenberg, D., 347

G

Gammaitoni, L., 221
Gilmour, I., 335
Goltsov, A. N., 353
Gouesbet, G., 125
Gowin, W., 347
Grifoni, M., 139
Grisar, T., 303
Grishchenko, A. D., 30
Gupte, N., 117

H

Hammel, S., 51
Hänggi, P., 139, 243, 309
Harris, J. G., 255
Hasslacher, B., 179
Häußler, R., 243
Hibbs, A. D., 227
Hilgers, A., 11
Huber, M. T., 281

J

Jung, P., 145

K

Kadtke, J. B., 57, 109, 359
Khasilev, V., 24
Khovanov, I. A., 267
Kittel, A., 75
Kravtsov, Yu. A., 359
Kremliovsky, M. N., 57
Kurths, J., 39, 347

L

Lambert, C. J., 157
Landa, P. S., 321
Laniel, J. M., 293
Lenhart, S., 87
Le Sceller, L., 125
Letellier, C., 125
Levin, J. E., 275
Lindberg, E., 81
Lindenberg, K., 163
Lindner, B., 309
Longtin, A., 131, 293
Luchinsky, D. G., 157
Lund, C. D., 18

M

Malomed, B. A., 103
Mantegna, R. N., 197
Maple, B. M., 18
Maquet, J., 125
Marchesoni, F., 221
Mayer-Kress, G., 213
McClintock, P. V. E., 157
Mees, A., 47
Mensour, B., 131
Meyer, Th., 75
Miller, L. D., 187
Moss, F., 151
Mykolaitis, G., 81

N

Namajūnas, A., 81
Neeley, J., 255
Neiman, A., 151
Nikishov, S. A., 353
Novikov, A., 299
Novikov, E., 299

P

Parisi, J., 75
Peng, C.-K., 366
Péntek, Á., 109
Pikovsky, A. S., 39
Platt, N., 51
Prance, H., 261
Prance, R. J., 261
Protopopescu, V., 87, 359
Pyragas, K., 63

R

Rabitz, H., 169
Ralph, J. F., 261
Reimann, P., 139, 145, 309
Rosenblum, M. G., 39
Ryabov, V. B., 329

S

Santucci, S., 221
Saparin, P. I., 341, 347
Schenk zu Schweinsberg, A., 237
Schieve, W. C., 315
Schimansky-Geier, L., 151, 309
Schöllmann, V., 261
Schwartz, B., 299
Schwartz, I. B., 97
Shannahoff-Khalsa, D., 299
Smelyanskiy, V. N., 157, 169
Smith, L. A., 335
Stone, L., 341
Sulcoski, M. F., 187
Surko, C. M., 18

T

Taha, A., 203
Tamaševičius, A., 81
Thomas, E., 303
Thorwart, M., 145
Tilden, M. W., 179
Toroczkai, Z., 109
Triandaf, I., 97
Tsironis, G., 163

U

Usik, P. V., 329

AIP Conference Proceedings

	Title	L.C. Number	ISBN
No. 192	Vacuum Mechatronics (First International Workshop) (Santa Barbara, CA, 1989)	89-45905	0-88318-394-3
No. 193	Advanced Accelerator Concepts (Lake Arrowhead, CA, 1989)	89-45914	0-88318-393-5
No. 194	Quantum Fluids and Solids—1989 (Gainesville, FL, 1989)	89-81079	0-88318-395-1
No. 195	Dense Z-Pinches (Laguna Beach, CA, 1989)	89-46212	0-88318-396-X
No. 196	Heavy Quark Physics (Ithaca, NY, 1989)	89-81583	0-88318-644-6
No. 197	Drops and Bubbles (Monterey, CA, 1988)	89-46360	0-88318-392-7
No. 198	Astrophysics in Antarctica (Newark, DE, 1989)	89-46421	0-88318-398-6
No. 199	Surface Conditioning of Vacuum Systems (Los Angeles, CA, 1989)	89-82542	0-88318-756-6
No. 200	High T_c Superconducting Thin Films: Processing, Characterization, and Applications (Boston, MA, 1989)	90-80006	0-88318-759-0
No. 201	QED Structure Functions (Ann Arbor, MI, 1989)	90-80229	0-88318-671-3
No. 202	NASA Workshop on Physics From a Lunar Base (Stanford, CA, 1989)	90-55073	0-88318-646-2
No. 203	Particle Astrophysics: The NASA Cosmic Ray Program for the 1990s and Beyond (Greenbelt, MD, 1989)	90-55077	0-88318-763-9
No. 204	Aspects of Electron-Molecule Scattering and Photoionization (New Haven, CT, 1989)	90-55175	0-88318-764-7
No. 205	The Physics of Electronic and Atomic Collisions (XVI International Conference) (New York, NY, 1989)	90-53183	0-88318-390-0
No. 206	Atomic Processes in Plasmas (Gaithersburg, MD, 1989)	90-55265	0-88318-769-8
No. 207	Astrophysics from the Moon (Annapolis, MD, 1990)	90-55582	0-88318-770-1
No. 208	Current Topics in Shock Waves (Bethlehem, PA, 1989)	90-55617	0-88318-776-0
No. 209	Computing for High Luminosity and High Intensity Facilities (Santa Fe, NM, 1990)	90-55634	0-88318-786-8

Title	L.C. Number	ISBN
No. 210 Production and Neutralization of Negative Ions and Beams (Brookhaven, NY, 1990)	90-55316	0-88318-786-8
No. 211 High-Energy Astrophysics in the 21st Century (Taos, NM, 1989)	90-55644	0-88318-803-1
No. 212 Accelerator Instrumentation (Brookhaven, NY, 1989)	90-55838	0-88318-645-4
No. 213 Frontiers in Condensed Matter Theory (New York, NY, 1989)	90-6421	0-88318-771-X 0-88318-772-8 (pbk.)
No. 214 Beam Dynamics Issues of High-Luminosity Asymmetric Collider Rings (Berkeley, CA, 1990)	90-55857	0-88318-767-1
No. 215 X-Ray and Inner-Shell Processes (Knoxville, TN 1990)	90-84700	0-88318-790-6
No. 216 Spectral Line Shapes, Vol. 6 (Austin, TX 1990)	90-06278	0-88318-791-4
No. 217 Space Nuclear Power Systems (Albuquerque, NM 1991)	90-56220	0-88318-838-4
No. 218 Positron Beams for Solids and Surfaces (London, Canada 1990)	90-56407	0-88318-842-2
No. 219 Superconductivity and Its Applications (Buffalo, NY 1990)	91-55020	0-88318-835-X
No. 220 High Energy Gamma-Ray Astronomy (Ann Arbor, MI 1990)	91-70876	0-88318-812-0
No. 221 Particle Production Near Threshold (Nashville, IN 1990)	91-55134	0-88318-829-5
No. 222 After the First Three Minutes (College Park, MD 1990)	91-55214	0-88318-828-7
No. 223 Polarized Collider Workshop (University Park, PA 1990)	91-71303	0-88318-826-0
No. 224 LAMPF Workshop on (π, K) Physics (Los Alamos, NM 1990)	91-71304	0-88318-825-2
No. 225 Half Collision Resonance Phenomena in Molecules (Caracas, Venezuela 1990)	91-55210	0-88318-840-6
No. 226 The Living Cell in Four Dimensions (Gif sur Yvette, France 1990)	91-55209	0-88318-794-9
No. 227 Advanced Processing and Characterization Technologies (Clearwater, FL 1991)	91-55194	0-88318-910-0
No. 228 Anomalous Nuclear Effects in Deuterium/Solid Systems (Provo, UT 1990)	91-55245	0-88318-833-3

	Title	L.C. Number	ISBN
No. 229	Accelerator Instrumentation (Batavia, IL 1990)	91-55347	0-88318-832-1
No. 230	Nonlinear Dynamics and Particle Acceleration (Tsukuba, Japan 1990)	91-55348	0-88318-824-4
No. 231	Boron-Rich Solids (Albuquerque, NM 1990)	91-53024	0-88318-793-4
No. 232	Gamma-Ray Line Astrophysics (Paris-Saclay, France 1990)	91-55492	0-88318-875-9
No. 233	Atomic Physics 12 (Ann Arbor, MI 1990)	91-55595	088318-811-2
No. 234	Amorphous Silicon Materials and Solar Cells (Denver, CO 1991)	91-55575	088318-831-7
No. 235	Physics and Chemistry of MCT and Novel IR Detector Materials (San Francisco, CA 1990)	91-55493	0-88318-931-3
No. 236	Vacuum Design of Synchrotron Light Sources (Argonne, IL 1990)	91-55527	0-88318-873-2
No. 237	Kent M. Terwilliger Memorial Symposium (Ann Arbor, MI 1989)	91-55576	0-88318-788-4
No. 238	Capture Gamma-Ray Spectroscopy (Pacific Grove, CA 1990)	91-57923	0-88318-830-9
No. 239	Advances in Biomolecular Simulations (Obernai, France 1991)	91-58106	0-88318-940-2
No. 240	Joint Soviet-American Workshop on the Physics of Semiconductor Lasers (Leningrad, USSR 1991)	91-58537	0-88318-936-4
No. 241	Scanned Probe Microscopy (Santa Barbara, CA 1991)	91-76758	0-88318-816-3
No. 242	Strong, Weak, and Electromagnetic Interactions in Nuclei, Atoms, and Astrophysics: A Workshop in Honor of Stewart D. Bloom's Retirement (Livermore, CA 1991)	91-76876	0-88318-943-7
No. 243	Intersections Between Particle and Nuclear Physics (Tucson, AZ 1991)	91-77580	0-88318-950-X
No. 244	Radio Frequency Power in Plasmas (Charleston, SC 1991)	91-77853	0-88318-937-2
No. 245	Basic Space Science (Bangalore, India 1991)	91-78379	0-88318-951-8
No. 246	Space Nuclear Power Systems (Albuquerque, NM 1992)	91-58793	1-56396-027-3 1-56396-026-5 (pbk.)
No. 247	Global Warming: Physics and Facts (Washington, DC 1991)	91-78423	0-88318-932-1

Title	L.C. Number	ISBN
No. 248 Computer-Aided Statistical Physics (Taipei, Taiwan 1991)	91-78378	0-88318-942-9
No. 249 The Physics of Particle Accelerators (Upton, NY 1989, 1990)	92-52843	0-88318-789-2
No. 250 Towards a Unified Picture of Nuclear Dynamics (Nikko, Japan 1991)	92-70143	0-88318-951-8
No. 251 Superconductivity and its Applications (Buffalo, NY 1991)	92-52726	1-56396-016-8
No. 252 Accelerator Instrumentation (Newport News, VA 1991)	92-70356	0-88318-934-8
No. 253 High-Brightness Beams for Advanced Accelerator Applications (College Park, MD 1991)	92-52705	0-88318-947-X
No. 254 Testing the AGN Paradigm (College Park, MD 1991)	92-52780	1-56396-009-5
No. 255 Advanced Beam Dynamics Workshop on Effects of Errors in Accelerators, Their Diagnosis and Corrections (Corpus Christi, TX 1991)	92-52842	1-56396-006-0
No. 256 Slow Dynamics in Condensed Matter (Fukuoka, Japan 1991)	92-53120	0-88318-938-0
No. 257 Atomic Processes in Plasmas (Portland, ME 1991)	91-08105	0-88318-939-9
No. 258 Synchrotron Radiation and Dynamic Phenomena (Grenoble, France 1991)	92-53790	1-56396-008-7
No. 259 Future Directions in Nuclear Physics with 4π Gamma Detection Systems of the New Generation (Strasbourg, France 1991)	92-53222	0-88318-952-6
No. 260 Computational Quantum Physics (Nashville, TN 1991)	92-71777	0-88318-933-X
No. 261 Rare and Exclusive B&K Decays and Novel Flavor Factories (Santa Monica, CA 1991)	92-71873	1-56396-055-9
No. 262 Molecular Electronics—Science and Technology (St. Thomas, Virgin Islands 1991)	92-72210	1-56396-041-9
No. 263 Stress-Induced Phenomena in Metallization: First International Workshop (Ithaca, NY 1991)	92-72292	1-56396-082-6
No. 264 Particle Acceleration in Cosmic Plasmas (Newark, DE 1991)	92-73316	0-88318-948-8
No. 265 Gamma-Ray Bursts (Huntsville, AL 1991)	92-73456	1-56396-018-4
No. 266 Group Theory in Physics (Cocoyoc, Morelos, Mexico 1991)	92-73457	1-56396-101-6

Title	L.C. Number	ISBN
No. 267 Electromechanical Coupling of the Solar Atmosphere (Capri, Italy 1991)	92-82717	1-56396-110-5
No. 268 Photovoltaic Advanced Research & Development Project (Denver, CO 1992)	92-74159	1-56396-056-7
No. 269 CEBAF 1992 Summer Workshop (Newport News, VA 1992)	92-75403	1-56396-067-2
No. 270 Time Reversal—The Arthur Rich Memorial Symposium (Ann Arbor, MI 1991)	92-83852	1-56396-105-9
No. 271 Tenth Symposium Space Nuclear Power and Propulsion (Vols. I–III) (Albuquerque, NM 1993)	92-75162	1-56396-137-7 (set)
No. 272 Proceedings of the XXVI International Conference on High Energy Physics (Vols. I and II) (Dallas, TX 1992)	93-70412	1-56396-127-X (set)
No. 273 Superconductivity and Its Applications (Buffalo, NY 1992)	93-70502	1-56396-189-X
No. 274 VIth International Conference on the Physics of Highly Charged Ions (Manhattan, KS 1992)	93-70577	1-56396-102-4
No. 275 Atomic Physics 13 (Munich, Germany 1992)	93-70826	1-56396-057-5
No. 276 Very High Energy Cosmic-Ray Interactions: VIIth International Symposium (Ann Arbor, MI 1992)	93-71342	1-56396-038-9
No. 277 The World at Risk: Natural Hazards and Climate Change (Cambridge, MA 1992)	93-71333	1-56396-066-4
No. 278 Back to the Galaxy (College Park, MD 1992)	93-71543	1-56396-227-6
No. 279 Advanced Accelerator Concepts (Port Jefferson, NY 1992)	93-71773	1-56396-191-1
No. 280 Compton Gamma-Ray Observatory (St. Louis, MO 1992)	93-71830	1-56396-104-0
No. 281 Accelerator Instrumentation Fourth Annual Workshop (Berkeley, CA 1992)	93-072110	1-56396-190-3
No. 282 Quantum 1/f Noise & Other Low Frequency Fluctuations in Electronic Devices (St. Louis, MO 1992)	93-072366	1-56396-252-7
No. 283 Earth and Space Science Information Systems (Pasadena, CA 1992)	93-072360	1-56396-094-X

Title	L.C. Number	ISBN
No. 284 US-Japan Workshop on Ion Temperature Gradient-Driven Turbulent Transport (Austin, TX 1993)	93-72460	1-56396-221-7
No. 285 Noise in Physical Systems and 1/f Fluctuations (St. Louis, MO 1993)	93-72575	1-56396-270-5
No. 286 Ordering Disorder: Prospect and Retrospect in Condensed Matter Physics: Proceedings of the Indo-U.S. Workshop (Hyderabad, India 1993)	93-072549	1-56396-255-1
No. 287 Production and Neutralization of Negative Ions and Beams: Sixth International Symposium (Upton, NY 1992)	93-72821	1-56396-103-2
No. 288 Laser Ablation: Mechanismas and Applications-II: Second International Conference (Knoxville, TN 1993)	93-73040	1-56396-226-8
No. 289 Radio Frequency Power in Plasmas: Tenth Topical Conference (Boston, MA 1993)	93-72964	1-56396-264-0
No. 290 Laser Spectroscopy: XIth International Conference (Hot Springs, VA 1993)	93-73050	1-56396-262-4
No. 291 Prairie View Summer Science Academy (Prairie View, TX 1992)	93-73081	1-56396-133-4
No. 292 Stability of Particle Motion in Storage Rings (Upton, NY 1992)	93-73534	1-56396-225-X
No. 293 Polarized Ion Sources and Polarized Gas Targets (Madison, WI 1993)	93-74102	1-56396-220-9
No. 294 High-Energy Solar Phenomena: A New Era of Spacecraft Measurements (Waterville Valley, NH 1993)	93-74147	1-56396-291-8
No. 295 The Physics of Electronic and Atomic Collisions: XVIII International Conference (Aarhus, Denmark, 1993)	93-74103	1-56396-290-X
No. 296 The Chaos Paradigm: Developments an Applications in Engineering and Science (Mystic, CT 1993)	93-74146	1-56396-254-3
No. 297 Computational Accelerator Physics (Los Alamos, NM 1993)	93-74205	1-56396-222-5
No. 298 Ultrafast Reaction Dynamics and Solvent Effects (Royaumont, France 1993)	93-074354	1-56396-280-2
No. 299 Dense Z-Pinches: Third International Conference (London, 1993)	93-074569	1-56396-297-7

	Title	L.C. Number	ISBN
No. 300	Discovery of Weak Neutral Currents: The Weak Interaction Before and After (Santa Monica, CA 1993)	94-70515	1-56396-306-X
No. 301	Eleventh Symposium Space Nuclear Power and Propulsion (3 Vols.) (Albuquerque, NM 1994)	92-75162	1-56396-305-1 (set) 156396-301-9 (pbk. set)
No. 302	Lepton and Photon Interactions/ XVI International Symposium (Ithaca, NY 1993)	94-70079	1-56396-106-7
No. 303	Slow Positron Beam Techniques for Solids and Surfaces Fifth International Workshop (Jackson Hole, WY 1992)	94-71036	1-56396-267-5
No. 304	The Second Compton Symposium (College Park, MD 1993)	94-70742	1-56396-261-6
No. 305	Stress-Induced Phenomena in Metallization Second International Workshop (Austin, TX 1993)	94-70650	1-56396-251-9
No. 306	12th NREL Photovoltaic Program Review (Denver, CO 1993)	94-70748	1-56396-315-9
No. 307	Gamma-Ray Bursts Second Workshop (Huntsville, AL 1993)	94-71317	1-56396-336-1
No. 308	The Evolution of X-Ray Binaries (College Park, MD 1993)	94-76853	1-56396-329-9
No. 309	High-Pressure Science and Technology—1993 (Colorado Springs, CO 1993)	93-72821	1-56396-219-5 (set)
No. 310	Analysis of Interplanetary Dust (Houston, TX 1993)	94-71292	1-56396-341-8
No. 311	Physics of High Energy Particles in Toroidal Systems (Irvine, CA 1993)	94-72098	1-56396-364-7
No. 312	Molecules and Grains in Space (Mont Sainte-Odile, France 1993)	94-72615	1-56396-355-8
No. 313	The Soft X-Ray Cosmos ROSAT Science Symposium (College Park, MD 1993)	94-72499	1-56396-327-2
No. 314	Advances in Plasma Physics Thomas H. Stix Symposium (Princeton, NJ 1992)	94-72721	1-56396-372-8
No. 315	Orbit Correction and Analysis in Circular Accelerators (Upton, NY 1993)	94-72257	1-56396-373-6

Title	L.C. Number	ISBN
No. 316 Thirteenth International Conference on Thermoelectrics (Kansas City, Missouri 1994)	95-75634	1-56396-444-9
No. 317 Fifth Mexican School of Particles and Fields (Guanajuato, Mexico 1992)	94-72720	1-56396-378-7
No. 318 Laser Interaction and Related Plasma Phenomena 11th International Workshop (Monterey, CA 1993)	94-78097	1-56396-324-8
No. 319 Beam Instrumentation Workshop (Santa Fe, NM 1993)	94-78279	1-56396-389-2
No. 320 Basic Space Science (Lagos, Nigeria 1993)	94-79350	1-56396-328-0
No. 321 The First NREL Conference on Thermophotovoltaic Generation of Electricity (Copper Mountain, CO 1994)	94-72792	1-56396-353-1
No. 322 Atomic Processes in Plasmas Ninth APS Topical Conference (San Antonio, TX)	94-72923	1-56396-411-2
No. 323 Atomic Physics 14 Fourteenth International Conference on Atomic Physics (Boulder, CO 1994)	94-73219	1-56396-348-5
No. 324 Twelfth Symposium on Space Nuclear Power and Propulsion (Albuquerque, NM 1995)	94-73603	1-56396-427-9
No. 325 Conference on NASA Centers for Commercial Development of Space (Albuquerque, NM 1995)	94-73604	1-56396-431-7
No. 326 Accelerator Physics at the Superconducting Super Collider (Dallas, TX 1992-1993)	94-73609	1-56396-354-X
No. 327 Nuclei in the Cosmos III Third International Symposium on Nuclear Astrophysics (Assergi, Italy 1994)	95-75492	1-56396-436-8
No. 328 Spectral Line Shapes, Volume 8 12th ICSLS (Toronto, Canada 1994)	94-74309	1-56396-326-4
No. 329 Resonance Ionization Spectroscopy 1994 Seventh International Symposium (Bernkastel-Kues, Germany 1994)	95-75077	1-56396-437-6
No. 330 E.C.C.C. 1 Computational Chemistry F.E.C.S. Conference (Nancy, France 1994)	95-75843	1-56396-457-0
No. 331 Non-Neutral Plasma Physics II (Berkeley, CA 1994)	95-79630	1-56396-441-4

	Title	L.C. Number	ISBN
No. 332	X-Ray Lasers 1994 Fourth International Colloquium (Williamsburg, VA 1994)	95-76067	1-56396-375-2
No. 333	Beam Instrumentation Workshop (Vancouver, B. C., Canada 1994)	95-79635	1-56396-352-3
No. 334	Few-Body Problems in Physics (Williamsburg, VA 1994)	95-76481	1-56396-325-6
No. 335	Advanced Accelerator Concepts (Fontana, WI 1994)	95-78225	1-56396-476-7 (set) 1-56396-474-0 (Book) 1-56396-475-9 (CD-Rom)
No. 336	Dark Matter (College Park, MD 1994)	95-76538	1-56396-438-4
No. 337	Pulsed RF Sources for Linear Colliders (Montauk, NY 1994)	95-76814	1-56396-408-2
No. 338	Intersections Between Particle and Nuclear Physics 5th Conference (St. Petersburg, FL 1994)	95-77076	1-56396-335-3
No. 339	Polarization Phenomena in Nuclear Physics Eighth International Symposium (Bloomington, IN 1994)	95-77216	1-56396-482-1
No. 340	Strangeness in Hadronic Matter (Tucson, AZ 1995)	95-77477	1-56396-489-9
No. 341	Volatiles in the Earth and Solar System (Pasadena, CA 1994)	95-77911	1-56396-409-0
No. 342	CAM -94 Physics Meeting (Cacun, Mexico 1994)	95-77851	1-56396-491-0
No. 343	High Energy Spin Physics Eleventh International Symposium (Bloomington, IN 1994)	95-78431	1-56396-374-4
No. 344	Nonlinear Dynamics in Particle Accelerators: Theory and Experiments (Arcidosso, Italy 1994)	95-78135	1-56396-446-5
No. 345	International Conference on Plasma Physics ICPP 1994 (Foz do Iguaçu, Brazil 1994)	95-78438	1-56396-496-1
No. 346	International Conference on Accelerator-Driven Transmutation Technologies and Applications (Las Vegas, NV 1994)	95-78691	1-56396-505-4
No. 347	Atomic Collisions: A Symposium in Honor of Christopher Bottcher (1945-1993) (Oak Ridge, TN 1994)	95-78689	1-56396-322-1

Title	L.C. Number	ISBN
No. 348 Unveiling the Cosmic Infrared Background (College Park, MD, 1995)	95-83477	1-56396-508-9
No. 349 Workshop on the Tau/Charm Factory (Argonne, IL, 1995)	95-81467	1-56396-523-2
No. 350 International Symposium on Vector Boson Self-Interactions (Los Angeles, CA 1995)	95-79865	1-56396-520-8
No. 351 The Physics of Beams Andrew Sessler Symposium (Los Angeles, CA 1993)	95-80479	1-56396-376-0
No. 352 Physics Potential and Development of $\mu^+\mu^-$ Colliders: Second Workshop (Sausalito, CA 1994)	95-81413	1-56396-506-2
No. 353 13th NREL Photovoltaic Program Review (Lakewood, CO 1995)	95-80662	1-56396-510-0
No. 354 Organic Coatings (Paris, France, 1995)	96-83019	1-56396-535-6
No. 355 Eleventh Topical Conference on Radio Frequency Power in Plasmas (Palm Springs, CA 1995)	95-80867	1-56396-536-4
No. 356 The Future of Accelerator Physics (Austin, TX 1994)	96-83292	1-56396-541-0
No. 357 10th Topical Workshop on Proton-Antiproton Collider Physics (Batavia, IL 1995)	95-83078	1-56396-543-7
No. 358 The Second NREL Conference on Thermophotovoltaic Generation of Electricity	95-83335	1-56396-509-7
No. 359 Workshops and Particles and Fields and Phenomenology of Fundamental Interactions (Puebla, Mexico 1995)	96-85996	1-56396-548-8
No. 360 The Physics of Electronic and Atomic Collisions XIX International Conference (Whistler, Canada, 1995)	95-83671	1-56396-440-6
No. 361 Space Technology and Applications International Forum (Albuquerque, NM 1996)	95-83440	1-56396-568-2
No. 362 Two-Center Effects in Ion-Atom Collisions (Lincoln, NE 1994)	96-83379	1-56396-342-6
No. 363 Phenomena in Ionized Gases XXII ICPIG (Hoboken, NJ, 1995)	96-83294	1-56396-550-X
No. 364 Fast Elementary Processes in Chemical and Biological Systems (Villeneuve d'Ascq, France, 1995)	96-83624	1-56396-564-X

	Title	L.C. Number	ISBN
No. 365	Latin-American School of Physics XXX ELAF Group Theory and Its Applications (México City, México, 1995)	96-83489	1-56396-567-4
No. 366	High Velocity Neutron Stars and Gamma-Ray Bursts (La Jolla, CA 1995)	96-84067	1-56396-593-3
No. 367	Micro Bunches Workshop (Upton, NY, 1995)	96-83482	1-56396-555-0
No. 368	Acoustic Particle Velocity Sensors: Design, Performance and Applications (Mystic, CT, 1995)	96-83548	1-56396-549-6
No. 369	Laser Interaction and Related Plasma Phenomena (Osaka, Japan 1995)	96-85009	1-56396-445-7
No. 370	Shock Compression of Condensed Matter-1995 (Seattle, WA 1995)	96-84595	1-56396-566-6
No. 371	Sixth Quantum 1/f Noise and Other Low Frequency Fluctuations in Electronic Devices Symposium (St. Louis, MO, 1994)	96-84200	1-56396-410-4
No. 372	Beam Dynamics and Technology Issues for + - Colliders 9th Advanced ICFA Beam Dynamics Workshop (Montauk, NY, 1995)	96-84189	1-56396-554-2
No. 373	Stress-Induced Phenomena in Metallization (Palo Alto, CA 1995)	96-84949	1-56396-439-2
No. 374	High Energy Solar Physics (Greenbelt, MD 1995)	96-84513	1-56396-542-9
No. 375	Chaotic, Fractal, and Nonlinear Signal Processing (Mystic, CT 1995)	96-85356	1-56396-443-0
No. 376	Chaos and the Changing Nature of Science and Medicine: An Introduction (Mobile, AL 1995)	96-85220	1-56396-442-2
No. 377	Space Charge Dominated Beams and Applications of High Brightness Beams (Bloomington, IN 1995)	96-85165	1-56396-625-7
No. 378	Surfaces, Vacuum, and Their Applications (Cancun, Mexico 1994)	96-85594	1-56396-418-X
No. 379	Physical Origin of Homochirality in Life (Santa Monica, CA 1995)	96-86631	1-56396-507-0
No. 380	Production and Neutralization of Negative Ions and Beams / Production and Application of Light Negative Ions (Upton, NY 1995)	96-86435	1-56396-565-8
No. 381	Atomic Processes in Plasmas (San Francisco, CA 1996)	96-86304	1-56396-552-6

	Title	L.C. Number	ISBN
No. 382	Solar Wind Eight (Dana Point, CA 1995)	96-86447	1-56396-551-8
No. 383	Workshop on the Earth's Trapped Particle Environment (Taos, NM 1994)	96-86619	1-56396-540-2
No. 384	Gamma-Ray Bursts (Huntsville, AL 1995)	96-79458	1-56396-685-9
No. 385	Robotic Exploration Close to the Sun: Scientific Basis (Marlboro, MA 1996)	96-79560	1-56396-618-2
No. 386	Spectral Line Shapes, Volume 9 13th ICSLS (Firenze, Italy 1996)		1-56396-656-5
No. 387	Space Technology and Applications International Forum (Albuquerque, NM 1997)	96-80254	1-56396-679-4 (Case set) 1-56396-691-3 (Paper set)
No. 388	Resonance Ionization Spectroscopy 1996 Eighth International Symposium (State College, PA 1996)	96-80324	1-56396-611-5
No. 389	X-Ray and Inner-Shell Processes 17th International Conference (Hamburg, Germany 1996)	96-80388	1-56396-563-1
No. 390	Beam Instrumentation Proceedings of the Seventh Workshop (Argonne, IL 1996)	97-70568	1-56396-612-3
No. 391	Computational Accelerator Physics (Williamsburg, VA 1996)	97-70181	1-56396-671-9
No. 392	Applications of Accelerators in Research and Industry: Proceedings of the Fourteenth International Conference (Denton, TX 1996)	97-71846	1-56396-652-2
No. 393	Star Formation Near and Far Seventh Astrophysics Conference (College Park, MD 1996)	97-71978	1-56396-678-6
No. 394	NREL/SNL Photovoltaics Program Review Proceedings of the 14th Conference— A Joint Meeting (Lakewood, CO 1996)	97-72645	1-56396-687-5
No. 395	Nonlinear and Collective Phenomena in Beam Physics (Arcidosso, Italy 1996)	97-72970	1-56396-668-9
No. 396	New Modes of Particle Acceleration— Techniques and Sources (Santa Barbara, CA 1996)	97-72977	1-56396-728-6
No. 397	Future High Energy Colliders (Santa Barbara, CA 1997)	97-73333	1-56396-729-4

Title	L.C. Number	ISBN
No. 398 Advanced Accelerator Colliders Seventh Workshop (Lake Tahoe, CA 1996)	97-72788	1-56396-697-2 (set) 1-56396-727-8 (cloth) 1-56396-726-X (CD-Rom)
No. 399 The Changing Role of Physics Departments (College Park, MD 1996)	97-74866	1-56396-698-0
No. 400 High Energy Physics First Latin Symposium (Yucatan, México 1996)	97-73971	1-56396-686-7
No. 401 Thermophotovoltaic Generation of Electricity Third NREL Conference (Colorado Springs, CO 1997)	97-74374	1-56396-734-0
No. 402 Astrophysical Implications of the Laboratory Study of Presolar Materials (St. Louis, MO 1996)	97-74679	1-56396-664-6
No. 403 Radio Frequency Power in Plasmas 12th Topical Conference (Savannah, GA 1997)	97-74472	1-56396-709-X
No. 404 Future Generations Photovoltaic Technologies First NREL Conference (Denver, CO 1997)	97-74386	1-56396-704-9
No. 405 Beam Stability and Nonlinear Dynamics (Santa Barbara, CA 1996)	97-74676	1-56396-731-6
No. 406 Laser Interaction and Related Plasma Phenomena 13th International Conference (Monterey, CA 1997)	97-76763	1-56396-696-4
No. 407 Deep Inelastic Scattering and QCD 5th International Workshop (Chicago, IL 1997)	97-74677	1-56396-716-2
No. 408 The Ultraviolet Universe at Low and High Redshift (College Park, MD 1997)	97-76762	1-56396-708-1
No. 409 Dense 2-Pinches 4th International Conference (Vancouver, Canada 1997)	97-76959	1-56396-610-7
No. 410 Proceedings of the 4th Compton Symposium (Williamsburg, VA 1997)	97-77179	1-56396-659-X (set)
No. 411 Applied Non-Linear Dynamics Near the Millenium (San Diego, CA 1997)	97-77035	1-56396-736-7